Soft Methods in Probability, Statistics and Data Analysis

Advances in Soft Computing

Editor-in-chief
Prof. Janusz Kacprzyk
Systems Research Institute
Polish Academy of Sciences
ul. Newelska 6
01-447 Warsaw, Poland
E-mail: kacprzyk@ibspan.waw.pl
http://www.springer.de/cgi-bin/search-bock.pl?series=4240

Esko Turunen
Mathematics Behind Fuzzy Logic
1999. ISBN 3-7908-1221-8

Robert Fullér
Introduction to Neuro-Fuzzy Systems
2000. ISBN 3-7908-1256-0

Robert John and Ralph Birkenhead (Eds.)
Soft Computing Techniques and Applications
2000. ISBN 3-7908-1257-9

Mieczysław Kłopotek, Maciej Michalewicz
and Sławomir T. Wierzchoń (Eds.)
Intelligent Information Systems
2000. ISBN 3-7908-1309-5

Peter Sinčák, Ján Vaščák, Vladimír Kvasnička
and Radko Mesiar (Eds.)
The State of the Art in Computational Intelligence
2000. ISBN 3-7908-1322-2

Bernd Reusch and Karl-Heinz Temme (Eds.)
Computational Intelligence in Theory and Practice
2001. ISBN 3-7908-1357-5

Robert John and Ralph Birkenhead (Eds.)
Developments in Soft Computing
2001. ISBN 3-7908-1361-3

Mieczysław A. Kłopotek, Maciej Michalewicz
and Sławomir T. Wierzchoń (Eds.)
Intelligent Information Systems 2001
2001. ISBN 3-7908-1407-5

Antonio Di Nola and Giangiacomo Gerla (Eds.)
Lectures on Soft Computing and Fuzzy Logic
2001. ISBN 3-7908-1396-6

Tadeusz Trzaskalik and Jerzy Michnik (Eds.)
Multiple Objective and Goal Programming
2002. ISBN 3-7908-1409-1

James J. Buckley and Esfandiar Eslami
An Introduction to Fuzzy Logic and Fuzzy Sets
2002. ISBN 3-7908-1447-4

Ajith Abraham and Mario Köppen (Eds.)
Hybrid Information Systems
2002. ISBN 3-7908-1480-6

Lech Polkowski
Rough Sets
2002. ISBN 3-7908-1510-1

Mieczysław A. Kłopotek, Sławomir T. Wierzchoń,
and Maciej Michalewicz (Eds.)
Intelligent Information Systems 2002,
2002. ISBN 3-7908-1509-8

Przemysław Grzegorzewski
Olgierd Hryniewicz
María Ángeles Gil
Editors

Soft Methods in Probability, Statistics and Data Analysis

With 66 Figures
and 17 Tables

Springer-Verlag Berlin Heidelberg GmbH

Dr. Przemysław Grzegorzewski
Prof. Dr. Olgierd Hryniewicz
Polish Academy of Sciences
Systems Research Institute
ul. Newelska 6
01-447 Warsaw
Poland
pgrzeg@ibspan.waw.pl
hryniewi@ibspan.waw.pl

Professor María Ángeles Gil
Universidad de Oviedo
Facultad de Ciencias
Departamento de Estadística e I.O. y D.M.
C/Calvo Sotelo, s/n
33007 Oviedo
Spain
angeles@pinon.ccu.uniovi.es

ISSN 1615-3871
ISBN 978-3-7908-1526-9

Cataloging-in-Publication Data applied for
Die Deutsche Bibliothek – CIP-Einheitsaufnahme
Soft methods in probability, statistics and data analysis / ed.: Przemysław Grzegorzewski ...

  (Advances in soft computing)
  ISBN 978-3-7908-1526-9   ISBN 978-3-7908-1773-7 (eBook)
  DOI 10.1007/978-3-7908-1773-7

http://www.springer.de

© Springer-Verlag Berlin Heidelberg 2002
Originally published by Physica-Verlag Heidelberg New York in 2002
Softcover reprint of the hardcover 1st edition 2002

Softcover Design: Erich Kirchner, Heidelberg

SPIN 10889155   88/2202-5 4 3 2 1 0 – Printed on acid-free paper

Foreword

Data analysis is one of the most frequent activities of scientists, managers, engineers, and other professionals. Special scientific disciplines such as statistics and probability theory have been developed to serve this purpose. Probability theory provides mathematical models that are used for data modelling. On the other hand, statistics offers special procedures to solve practical problems. Thousands of successful applications of statistical methods confirm its applicability in many areas of science, engineering, medicine, business, etc. In all these applications, and their respective mathematical models, there exists one crucial assumption: randomness is the only source of uncertainty which has to be analyzed using statistical methods. For many years this assumption has not been questioned, neither by statisticians nor by practitioners. This situation has been changing during the last twenty or more years as data analysis has been extensively used in such areas as social science, economy, etc.

In traditional statistics we are dealing with objects measured in an objective way. Therefore, the results of statistical experiments are represented by precise numerical data. The paradigm of the existence of precise data has been changed together with the application of statistical methods for the analysis of data generated by humans. In many practical cases the attempts to be "as precise as possible" have been obviously unsuccessful. Statisticians have found that many statistical data they have had to deal with were inherently imprecise and vague. Many specialists have realized that even in the cases which traditionally have been treated as crisp there is a need to introduce new models that would be closer to reality. Practical problems of data analysis require various mathematical models and statistical tools. Specific methods should be used for imprecise linguistic data analysis, others for the description of subjective opinions, yet other methods for the analysis of missing data, etc. To describe and analyze all these problems new methods that are tailored for a particular type of imprecise data have to be used.

During the last thirty years a considerable number of papers have been published with the aim to extend the existing theory of probability and mathematical statistics. The common feature of those attempts is to "soften" the classical theory. Some "softening" approaches utilize concepts and techniques developed in such theories like fuzzy sets theory, rough sets, possibility theory, theory of belief functions and imprecise probabilities, etc.

First attempts to describe different types of uncertainty, vagueness and lack of precision were considered by the statistical community as an assault against probability and statistics. Even now, many traditional statisticians view newly emerged theories dealing with those uncertainties as antagonistic to classical probability and statistics. Fortunately, this is far from being true. Recent developments in the theories mentioned above have indicated that they are complementary to classical probability and statistics, and in many cases may be regarded as their generalizations.

The soft methods in probability and statistics have obviously many common features. Interesting mathematical models and methods have been proposed in the frameworks of different theories. Therefore, we have decided to establish a new conference oriented on people who are involved in the theoretical and applied research, to bring together experts representing different approaches used in soft probability and statistics. Most of the papers presented in this book have been presented at the First International Workshop on Soft Methods in Probability and Statistics, SMPS'2002, held in Warsaw in September 2002.

The papers presented in this book have been grouped in four parts. The volume starts with an introductory part with more general papers by distinguished scientists, Professors Didier Dubois, Henri Prade and Lotfi Zadeh, devoted to quantitative possibility theory and perception-based theory of probabilistic reasoning.

The second section is devoted to fundamental problems of "soft" probability, e.g. modeling uncertain and imprecise events, cardinality of fuzzy sets in random environment, intersection of probability theory and fuzzy sets, fuzzy measures in the framework of the probability theory, problems of independence in a fuzzy logic framework, analysis of random sets, problems of the theory of imprecise probabilities (lower and upper probabilities, lower and upper previsions), mutual relations between of the Dempster-Shafer theory and the theory of probability, etc.

In the next section papers have been collected which are devoted to fuzzy statistical problems considered from both frequentist and Bayesian point of view, papers on probabilistic fuzzy reasoning and different applications of fuzzy stochastic models.

Fuzzy stochastic models are not the only possible models for the analysis of data. Papers belonging to the last section present different approaches in data analysis. There are papers showing fuzzy counterparts of well known statistical problems such as time series analysis and conjoint analysis, papers dedicated to problems of fuzzy aggregation of data, identification of fuzzy measures from sample information, fusion of expert and learnt information, papers on rough sets approach to data mining and classification problems. Other papers describe solutions of practical problems using non-standard statistical methods.

The editors would like to express their thanks to all authors. A particular acknowledgment goes to Professor Janusz Kacprzyk, the Editor-in-Chief of this series, for benevolent promotion and support during publishing this volume. We also deeply thank Edyta Mrówka B.Sc. for her editorial support in preparing the final version of the volume.

Warsaw, September 2002

Przemysław Grzegorzewski
Olgierd Hryniewicz
Maria Angeles Gil

Contents

Introductory Papers

Quantitative Possibility Theory and its Probabilistic   3
Connections
     D. Dubois, H. Prade

Toward a Perception-Based Theory of Probabilistic   27
Reasoning with Imprecise Probabilities
     L. A. Zadeh

Soft Methods in Probability - Fundamentals

Independence and Conditioning in a Connectivistic Fuzzy   65
Logic Framework
     J. Aguilar-Martin

About Ideals on the Set of Fuzzy Measures   76
     A. G. Bronevich, A. N. Karkishchenko

Operators for Convolution of Fuzzy Measures   84
     A. G. Bronevich, A. E. Lepskiy

Probabilities of Fuzzy Events Based on Scalar Cardinalities   92
     J. Casasnovas, F. Roselló

On the Variation of the f-Inequality of a Random Variable   98
     I. Cascos-Fernández, M. López-Díaz, M. A. Gil-Álvarez

Probability of Intuitionistic Fuzzy Events   105
     P. Grzegorzewski, E. Mrówka

A New Calculus for Linguistic Prototypes in Data Analysis   116
     J. Lawry

Upper Probabilities and Selectors of Random Sets   126
     E. Miranda, I. Couso, P. Gil

On Multivalued Logic and Probability Theory   134
     B. Riečan

Reversing the Order of Integration in Iterated Expectations   140
of Compact Convex Random Sets
     L. J. Rodríguez-Muñiz, M. López-Díaz, M. A. Gil

Lower Previsions for Unbounded Random Variables   146
    M. C. M.Troffaes, G. de Cooman

A Hierarchical Uncertainty Model under Incomplete   156
Information
    L. V. Utkin

Imprecise Calculation with the Qualitative Information   164
about Probability Distributions
    L. U. Utkin

Approximation of Belief Functions by Minimizing   170
Euclidean Distances
    T. Weiler, U. Bodenhofer

Variants of Defining the Cardinalities of Fuzzy Sets   178
    M. Wygralak

Soft Methods in Statistics - Fuzzy Stochastic Models

Probabilistic Reasoning in Fuzzy Rule-Based Systems   189
    J. van den Berg, U. Kaymak, W. M. van den Bergh

Acceptance Sampling Plans by Variables for Vague Data   197
    P. Grzegorzewski

Possibilistic Approach to Bayes Statistical Decisions   207
    O. Hryniewicz

Fuzzy Comparison of Frequency Distributions   219
    M. Last, A. Kandel

Test of One-Sided Hypotheses on the Expected Value of a   228
Fuzzy Random Variable
    M. Montenegro, A. Colubi, M. R. Casals,
    M. A. Gil-Álvarez

Blackwell Sufficiency and Fuzzy Experiments   236
    A. Wünsche

Optimal Stopping in Fuzzy Stochastic Processes and its   244
Application to Option Pricing in Financial Engineering
    Y. Yoshida

Soft Methods in Data Analysis - Fuzzy, Rough and Other Approaches

Fuzzy Conjoint Analysis 255
M. Bhattacharyya

Grade Analysis of Repeated Multivariate Measurements 266
A. Ciok

A Contribution to Stochastic Modeling of Volcanic 274
Petroleum Reservoir
S. Darwis

Improving the Quality of Association Rule Mining by Means 281
of Rough Sets
D. Delic, H. J. Lenz, M. Neiling

Towards Fuzzy c-means Based Microaggregation 289
J. Domingo-Ferrer, V. Torra

Statistical Profiles of Words for Ontology Enrichment 295
A. Faatz, C. Seeberg, R. Steinmetz

Quest on New Applications for Dempster-Schafer Theory: 302
Risk Analysis in Project Profitability Calculus
M. A. Kłopotek, S. T. Wierzchoń

Decision Making Based on Informational Variables 310
J. L. Kulikowski

An Algorithm for Identifying Fuzzy Measures with Ordinal 321
Information
P. Miranda, M. Grabisch

Exploring an Alternative Method for Handling 329
Inconsistency in the Fusion of Expert and Learnt
Information
J. Rossiter

Approximate Reasoning Schemes: Classifiers for 338
Computing with Words
A. Skowron, M. S. Szczuka

Using Consensus Methods to User Classification in 346
Interactive Systems
J. Sobecki, N. T. Nguyen

Aggregation Methods to Evaluate Multiple Protected 355
Versions of the Same Confidential Data Set
A. Valls, V. Torra, J. Domingo-Ferrer

On Multivariate Fuzzy Time Series Analysis and Forecasting   363
   B. Wu, Y. Y. Hsu

Introductory Papers

Introductory Paper

Quantitative Possibility Theory and its Probabilistic Connections

Didier Dubois and Henri Prade

IRIT-CNRS, Universite Paul Sabatier, Toulouse, France
E-mail: e-mail: dubois@irit.fr

Abstract. Possibility theory is a representation framework general enough to model various kinds of information items: numbers, intervals, consonant random sets, special kind of probability families, as well as linguistic information, and uncertain formulae in logical settings. This paper focuses on quantitative possibility measures cast in the setting of imprecise probabilities. Recent results on possibility/probability transformations are recalled. The probabilistic interpretation of possibility measures sheds some light on defuzzification methods and suggests a common framework for fuzzy interval analysis and calculations with random parameters.

1 Introduction

Possibility theory refers to the study of maxitive and minitive set-functions, such that the possibility degree of a disjunction of events is the maximum or the minimum of the possibility degrees of events in the disjunction. In 1978, Zadeh published a paper proposing a theory of possibility based on fuzzy sets. The aim was to propose an approach to model flexible restrictions constructed from vague pieces of information, described by means of fuzzy sets. Possibility theory can also be viewed as a non-classical theory of uncertainty, different from probability theory, an idea which was actually first proposed and formalised by the economist Shackle (1961) already in the late forties. The comparison between possibility and probability theories is made easy by the parallelism of the constructs, which is not the case for fuzzy sets and probability. As a mathematical object, maxitive set functions have been already studied by (Shilkret, 1971). Possibility theory can also be viewed as a graded extension of modal logic where the dual notions of possibility and necessity already exist for a long time, in an all-or-nothing format. The notion "more possible than" was actually first modelled by David Lewis (1979), in the setting of modal logics of counterfactuals, by means of a complete preordering relation among events satisfying some prescribed properties. On finite sets, these relations are completely characterised by their restrictions to singletons. This notion was independently rediscovered in (Dubois, 1986) in the setting of decision theory, in an attempt to propose counterparts of comparative probability relations for possibility theory. Since then, maxitive set-functions or equivalent notions have emerged as a key tool in various domains, such as belief revision (Spohn, 1988), non monotonic reasoning (Benferhat et al. 1997), game theory (the so-called unanimity games, Shapley, 1971), imprecise probabilities (Walley, 1991), etc. Due to the ordinal nature of the basic axioms of possibility and necessity functions, there is no commitment to numerical possibility and necessity degrees. So there are basically two kind of possibility theories: quantitative and qualitative. In quantitative theories, degrees of possibility are numbers and one must indicate where they come from, in order to understand their meaning. Qualitative possibility theory inherits from Lewis proposals and is closely connected to non-monotonic reasoning as shown by Benferhat et al. (1997) for instance. All possibility theories agree on the maxitivity axiom for possibility functions:

$$\Pi (A \cup B) = \max (\Pi (A), \Pi (B)) \tag{1}$$

but they disagree on the conditioning operation.

This paper is devoted to a survey of results in quantitative possibility theory, which has close connections to probability theory and statistics, contrary to qualitative possibility theory. There are not so many extensive works on possibility theory. The first book on this topic (Dubois and Prade, 1988) emphasises the close links between possibility theory and fuzzy sets, and mainly deals with numerical possibility and necessity measures. It already points out their links with probability theory. Klir and Folger (1988) focus on the fact that possibility theory is a special case of belief function theory, with again a numerical flavour. A more recent detailed survey by the authors (Dubois and Prade, 1998) distinguishes between quantitative and qualitative sides of the theory. Basic mathematical aspects of possibility theory are studied at length by De Cooman (1997) More recently this author has investigated possibility theory as a special case of imprecise subjective probability (De Cooman, 2001). This paper relies on (Dubois Nguyen and Prade, 2000) and some recent findings in probability-possibility transformations.

2   Possibility as Plausibility

In the same way as probabilities can be interpreted in different ways (e.g., frequentist view vs. subjective view), possibility theory can support various interpretations. Basically there are four ideas each of which can be conveyed by the word 'possibility'. First is the idea of *feasibility*, such as ease of achievement, also referring to the solution to a problem, satisfying some constraints. At the linguistic level this meaning is at work in expressions such as "it is possible to solve this problem". Another notion of possibility is that of *plausibility*, referring to the propensity of events to occur. At the grammatical level this semantics is expressed by means of sentences such as "it is possible *that* the train arrives on time". Yet another view of possibility is logical and it refers to *consistency* with available information. Namely, stating that a proposition is possible means that it does not contradict this information. It is an all-or-nothing version of plausibility. The last semantics of possibility is deontic, whereby possible means *allowed*, permitted by the law. In this paper we shall focus on plausibility.

2.1 The Logical View

Possibility as logical consistency has been put forward by Yager (1980). Namely, let a piece of incomplete information be given as a set E such that "$x \in E$" is known for sure where x is some ill-known object. Let set E refer to the information possessed by an agent. It is incomplete insofar as E contains more than one element, i.e., the value of x can be any one (but only one) of the elements in E. Given such a piece of information, a set-function Π_E is built from E by the following procedure

$$\Pi_E(A) = \begin{cases} 1 & \text{if } A \cap E \neq \emptyset \quad (x \in A \text{ and } x \in E \text{ are consistent}) \\ 0 & \text{otherwise} \quad\quad (A \text{ and } E \text{ are mutually exclusive}) \end{cases}$$

Clearly, $\Pi_E(A) = 1$ means that given that $x \in E$, $x \in A$ is possible because the intersection between set A and set E is not empty, while $\Pi_E(A) = 0$ means that $x \in A$ is impossible knowing that $x \in E$. It is easy to verify that Π_E satisfies the maxitivity axiom (1).

2.2 Objective vs. Subjective Possibility

Hacking (1975) also pointed out that possibility can be understood either as an objective notion (referring to properties of the physical world) or as an epistemic one (referring to the state of knowledge of an agent). Feasibility and plausibility can be envisaged from both points of view. Possibility, as objective feasibility, means *physically easy to achieve*, as in the sentence "it is possible *for* Hans to eat six eggs for breakfast". Physical possibility has been advocated by Zadeh (1978) so as to justify the axiomatic rule of possibility measures (1). The degree of ease of some action that produces $A \cup B$ is given by the easiest of two actions, one that produces A and one that produces B. The idea of ease of attainment often comes along with the idea of *preference*: considering mutually exclusive alternatives, the most feasible(s) one(s) (in some sense) is/are usually preferred. Hence, subjective feasibility can be interpreted as the willingness of an agent to make a decision. Then, possibility refers to choice preference. As shown in Dubois Fargier and Prade (1996), the whole calculus of possibility theory can be interpreted in the light of preference modeling and optimization theory as much as in terms of partial belief.

The epistemic notion of plausibility has been put forward by Shackle (1961). He proposed a calculus of degrees of "potential surprise" that matches the possibilistic framework. Following Shackle, the potential surprise attached to an event by an agent reflects the level of contradiction between the occurrence of this event and the agent's knowledge. Plausibility here means lack of surprise and embodies the above mentioned logical view of consistency with available knowledge. Shackle claimed that decisions are guided at least as much by expectations that look possible as by those that look probable, because probabilities are often not available. This is the subjectivist side of possibility in relation to uncertainty modelling. Actually, this kind of possibility, like probability, has objectivist sides as well. Rather early works (Wang, 1983, Dubois and Prade, 1986) have developed a frequentist view of possibility, which bridges the gap between possibility theory and the mathematics of probability theory. Possibility degrees then offer a simple approach to imprecise (set-valued) statistics, in terms of upper bounds of frequency. However possibility degrees were introduced in terms of ease of attainment and flexible constraints by Zadeh and as an epistemic uncertainty notion by Shackle, with little reference to probability and/or statistics.

2.3 Necessity

Plausibility is dually related to certainty, in the sense that the certainty of an event reflects a lack of plausibility of its opposite. This is a striking difference with probability which is self-dual. The expression *It is not probable that "not A"* means that *It is probable that A*, while *It is not possible that "not A"* does not mean *It is possible that A*. It has a stronger meaning: *It is necessary that A*. Conversely, *it is possible that A* does not entail anything about the possibility nor the impossibility of *"not A"*. Here we are dealing with a dual pair possibility/ necessity.

A semantic analysis of *necessity* can be done paralleling the one of possibility. There exists physical necessity as well as epistemic necessity (the latter coincides with the notion of *certainty*). Objective necessity refers to laws of nature: if we drop an object it is necessary that it falls. Subjective necessity viewed as the dual of preference means *priority*. Objective necessity refers to lower bounds of frequency measurements, while subjective certainty means belief, acceptance, probable provability. Namely believing a proposition means accepting to reason using this proposition as if it were true, until some new information refutes the proposition. A proposition is certain if and only if it logically derives from the available knowledge. Hence necessity also conveys a logical meaning in terms of deduction, just as possibility is a matter of logical consistency (Yager, 1980). The certainty of the event "$x \in A$", knowing that "$x \in E$", is then evaluated by the following index, called necessity measure:

$$N_E(A) = \begin{cases} 1 & \text{if } E \subseteq A \\ 0 & \text{otherwise} \end{cases}$$

Clearly the information $x \in E$ logically entails $x \in A$ when E is contained in A, so that certainty applies to events that are logically entailed by the available evidence. It can be easily seen that $N_E(A) = 1 - \Pi_E(\text{not } A)$, i.e., A is necessarily true if and only if 'not A' is impossible.

3   The Possibilistic Representation of Incomplete Data

Consider an ill known parameter x. Let U denote the referential where x takes its values. The available information on the actual value of x is supposed to be modelled by a possibility distribution π_x . It is a mapping from U to a totally ordered plausibility scale, here $[0,1]$. A possibility distribution can be viewed as a representation of the more or less plausible values of an unknown quantity x. These values are assumed to be mutually exclusive, since x takes on only one value (its true value). Since one of the elements of U is the true value of x, $\pi_x(u^*) = 1$ for at least one value $u^* \in U$. This is the normalisation condition, that claims that at least one value is viewed as totally possible. If u and u' are such that $\pi_x(u) > \pi_x(u')$, u is considered to be a more plausible value than u'. When $\pi_x(u) = 0$, then x cannot take on value u. The possibility degree of an event A, understood as a subset of U is

$$\Pi(A) = \sup_{u \in A} \pi_x(u) \tag{2}$$

It is computed on the basis of the most plausible values of x in A, neglecting other realisations. A possibility distribution π_x is at least as informative (more specific) as another one π'_x if and only if $\pi_x \le \pi'_x$ (see, e.g., Yager, 1992). In particular, if $\forall u \in U, \pi_x(u^*) = 1, \pi_x$ contains no information at all (since it expresses that any value in u is possible for x).

Remark. The possibility measure defined in (2) satisfies a strong form of maxitivity (1) for the union of infinite families of sets. On infinite sets, axiom (1) alone does not imply the existence of a possibility distribution satisfying (2). For instance, consider the natural integers, and a set function assigning possibility 1 to infinite subsets of integers, possibility 0 to finite subsets. This function is a possibility measure in the sense of (1) but does not fulfil (2).

The possibilistic representation is capable of modelling several kinds of imprecise information within a unique setting.

3.1   Intervals

The simplest form of a possibility distribution on a numerical interval U is the characteristic function of a sub-interval I of U, i.e., $\pi_x(u) = 1$ if $x \in I$, 0 otherwise. This type of possibility distribution is naturally obtained from experts stating that "x lies between a and b". This way of expressing knowledge is more natural than giving a point-value u^* for x right away, because it allows for some imprecision. However this binary representation is not entirely satisfactory. If the interval is too narrow, the piece of information is not so reliable. When $\pi_x(u) = 0$ for some u, it means that $x = u$ is impossible. This is too strong, and one is then tempted to use wide uninformative intervals. Sometimes, even the widest, safest interval does not rule out some residual possibility that the value of x lies outside it.

3.2 Confidence Intervals

It is more satisfactory to describe imprecise information by means of several intervals with various levels of confidence. A possibility distribution π_x can then represent a finite family of nested confidence subsets $\{A_1, A_2, ..., A_m\}$ where $A_i \subset A_{i+1}$, $i = 1, m - 1$. Each confidence subset A_i is attached a positive confidence level λ_i. The set of possibility values $\{\pi(u) \mid u \in U\}$ is then finite. The links between the confidence levels λ_i's and the degrees of possibility are defined by postulating λ_i is the degree of necessity (i.e. certainty) of A_i which is defined as $N(A_i) = 1 - \Pi(A_i^c)$ where $\Pi(A_i^c)$ is the degree of possibility of the complement A_i^c of A_i (Dubois and Prade, 1988). This entails that $\lambda_1 \leq \ldots \leq \lambda_m$ due to the monotonicity of the necessity function N. The possibility distribution equivalent to the family $\{(A_1, \lambda_1), (A_2, \lambda_2), \ldots, (A_m, \lambda_m)\}$ is defined as the least informative possibility distribution π that obeys the constraints $\lambda_i = N(A_i)$, $i = 1, m$. It comes down to maximizing the degrees of possibility $\pi(u)$ for all u in U, subject to these constraints. The solution is unique and is

$$\forall u, \pi_x(u) = \begin{cases} 1 & \text{if } u \in A_1 \\ \min_{i: u \notin A_i} 1 - \lambda_i & \text{otherwise} \end{cases}$$

which also reads $\pi_x(u) = min_i \, max(1 - \lambda_i, A_i(u))$, where $A_i(\cdot)$ is the characteristic function of A_i. This solution is the least committed one with respect to the available data, since by allowing the greatest possibility degrees in agreement with the constraints, it defines the least restrictive possibility distribution. Conversely, the family $(A_1, \lambda_1), (A_2, \lambda_2), \ldots, (A_m, \lambda_m)\}$ of confidence intervals can be reconstructed from the possibility distribution π_x. Suppose that the set of possibility values $\pi_x(u)$ is $\{\alpha_1 = 1, \alpha_2 \geq \alpha_3 \geq \ldots \geq \alpha_m\}$ and let $\alpha_{m+1} = 0$. Then

$$A_i = \{u \mid \pi_x(u) \geq \alpha_i\}, \quad \lambda_i = 1 - \alpha_{i+1}, \quad \forall i = 1, \ldots, m.$$

In particular $\lambda_m = 1$ and A_m is the subset which for sure contains x; moreover, $A_m = U$ if no strict subset of U surely includes x. This analysis extends to an infinite nested set of confidence intervals $\{(A_\alpha, 1 - \alpha), \alpha > 0\}$, where $N(A_\alpha) = 1 - \alpha$, and $\pi_x(u) = inf_\alpha \, max(\alpha, A_\alpha(u))$.

3.3 Random Sets

Letting $p_i = \alpha_i - \alpha_{i+1}, \forall i = 1, \ldots, m$ note that

$$\forall u, \quad \pi_x(u) = \sum_{i: u \in A_i} p_i \tag{3}$$

The sum of weights $p_1, \ldots, p_m$ is 1. Hence the possibility distribution can be cast in the setting of random sets (Dubois and Prade, 1982), and more precisely the theory of evidence (Shafer, 1987). From a mathematical point of view, the information modelled by π_x can be viewed as a nested random set $\{(A_i, p_i), i = 1, m\}$, which allows for imprecision (reflected by the size of the A_i) and uncertainty (the p_i). And p_i is the probability that the source supplies exactly A_i as a faithful representation of the available knowledge of x (it is *not* the probability that x belongs to A_i). The random set view of possibility theory is developed in details in (Gebhardt and Kruse, 1993, 1994a,b). Namely given a bunch of imprecise (not necessarily nested) observations, equation (3) supplies an approximate representation of the data (see also Joslyn, 1997).

3.4 Imprecise Probability

The level of confidence λ_i can also be conveniently interpreted as a lower bound on the probability that the true value of x hits A_i. Then the possibility distribution encodes the family of probability measures $\mathbf{P} = \{P, P(A_i) \geq \lambda_i, i = 1, m\}$ because the possibility measure $\Pi(B)$ coincides with the upper probability $P^*(B) = \sup\{P(B), P \in \mathbf{P}\}$ while the necessity measure $N(B) = \inf\{P(B), P \in \mathbf{P}\}$ is the lower probability; see (Dubois and Prade, 1992, De Cooman and Aeyels, 1999) for details. These intervals can be interpreted in terms of fractiles of a probability distribution.

3.5 Likelihood Functions

Yet another interpretation of the possibility distribution π consists in viewing it as a likelihood function, that is, identifying $\pi(u)$ to the probability $P(u_m \mid u)$ that the source indicates the measured value u_m, when the actual value of x is u. Indeed suppose only $P(u_m \mid u)$ is known, $\forall u \in U$. The probability $P(u_m \mid u)$ is understood as the likelihood of $x = u$, in the sense that the greater $P(u_m \mid u)$ the more $x = u$ is plausible. In particular, note that $\forall A \subseteq U$

$$\min_{u \in A} P(u_m \mid u) \leq P(u_m \mid A) \leq \max_{u \in A} P(u_m \mid u)$$

Identifying $\pi(u)$ to $P(u_m|u)$, the upper bound of the probability $P(u_m \mid A)$ is a possibility measure (see Dubois *et al.*, 1997). Hence there is a strong similarity between maximum likelihood reasoning and possibilistic reasoning in the finite case. The degree of possibility that $x = u$ in the face the measurement u_m can be defined as $\pi_m(u) = \pi_x(u|u_m) = P(u_m \mid u)$. This setting is adapted to measurements, for instance.

In general $\sup_u P(u_m \mid u) \neq 1$ since in this approach, the normalisation with respect to u is not warranted. And in the continuous case $P(u_m \mid u)$ may be greater than 1. Yet, it is natural to assume that $\max_u P(u_m \mid u) = 1$. It corresponds to existing practice in statistics whereby the likelihood function is renormalized via a proportional rescaling (Edwards, 1972). It means there is at least one value $x = u$ that makes the observation $x_m = u_m$ completely possible; in other words u_m is a completely relevant observation for the set of assumptions U. Indeed, if for instance $\max_u P(u_m \mid u) = 0$, it would mean that it is impossible to observe $x_m = u_m$ for any value $x = u \in U$.

3.6 Linguistic Information

Human-originated information, even when it pertains to orders of magnitude, is often linguistic. In some contexts, only witnesses can supply information, and they use words. Zadeh (1965, 1979) proposed the theory of fuzzy sets as a tool for mathematical modelling of linguistic information, especially on numerical universes. A possibility distribution can be identified to the membership function of a fuzzy set modelling linguistic terms like "tall" (for heights) "hot" (for temperatures) "far" (for distances), etc. of possible values of a quantity x. So linguistic information taking the form of statements like "this man is tall", "the room temperature is hot", "the robot is far from the table" can be modelled by means of possibility distributions as well. However, the semantics of the possibility degrees is then described in terms of distance to fuzzy set prototypes, whose membership values is maximal. If the piece of information is of the form "x is F", where F is a fuzzy set, the closer a value u to a prototype u^* of F the more plausible it is, which is expressed by letting $\pi_x(u) = F(u)$, where $F(\cdot)$ is the membership function of F, such that $F(u^*) = 1$. Nguyen (1994) related fuzzy linguistic terms to random sets. De Cooman and Walley (1999) interpreted linguistic possibility distributions in terms of imprecise subjective possibilities.

4 Numerical Possibilistic Conditioning

Conditioning in possibility theory has been studied as a counterpart to probabilistic conditioning. However there is no longer a unique meaningful definition. And the main difference between numerical and qualitative possibility theories lies in the conditioning process. The notion of conditional possibility measure goes back to Hisdal (1978) who introduced the set function $\Pi(\cdot \mid A)$ through the equality

$$\forall B, B \cap A \neq \emptyset, \quad \Pi(A \cap B) = \min(\Pi(B \mid A), \Pi(A)) \qquad (4)$$

In order to overcome the existence of several solutions to this equation, the conditional possibility measure can be defined, as proposed by Dubois and Prade (1988), as the least specific solution to this equation, that is, when $\Pi(A) > 0$,

$$\Pi(B \mid A) = \begin{cases} 1 & \text{if} \quad \Pi(A \cap B) = \Pi(A) \\ \Pi(A \cap B) & \text{otherwise} \end{cases}$$

The only difference with conditional probability is that the renormalisation via division is changed into a simple move of the most plausible elements in A to 1. The conditioning equation agrees with a purely ordinal view of possibility theory and makes sense in a finite setting only. However, this form of conditioning applied to infinite numerical settings creates discontinuities, and the infinite maxitivity axiom is then not preserved by conditioning. Especially $\Pi(B \mid A) = \sup_{u \in B} \pi(u \mid A)$ may fail to hold for non-compact events B (De Cooman, 1997).

The use of the product instead of min in the conditioning equation equation (4) avoids discontinuity problems. In close agreement with probability theory, it leads to

$$\forall B, B \cap A \neq \emptyset, \quad \Pi(B \cap A) = \frac{\Pi(A \cap B)}{\Pi(A)}$$

provided that $\Pi(A) \neq 0$. Then $N(B \mid A) = 1 - \Pi(B^c \mid A)$. This is formally like Dempster rule of conditioning, specialised to possibility measures, i.e., consonant plausibility measures of Shafer (1976). See De Baets *et al.* (1999) for a complete mathematical study of possibilistic conditioning, leading to the unicity of the product-based notion, in the infinite setting. The possibilistic counterpart of Bayes theorem looks formally the same as in probability theory:

$$\Pi(B \mid A) \cdot \Pi(A) = \Pi(A \mid B) \cdot \Pi(B)$$

Considering the problem of testing hypothesis A against its complement, upon observing B. The expression of $\Pi(B \mid A)$ in terms of $\Pi(A \mid B)$, $\Pi(A \mid B^c)$, $\Pi(B)$, $\Pi(B^c)$ differs from the probabilistic form, due to the the maxitivity axiom . Namely

$$\Pi(B \mid A) = \frac{\Pi(A \mid B) \cdot \Pi(B)}{\max\left(\Pi(A \mid B) \cdot \Pi(B), \Pi(A \mid B^c) \cdot \Pi(B^c)\right)}$$

The uninformed case, with uniform possibility $\Pi(B) = \Pi(B^c) = 1$ leads to compute

$$\Pi(B \mid A) = \frac{\Pi(A \mid B)}{\max\left(\Pi(A \mid B), \Pi(A \mid B^c)\right)}$$

and compare it to $\Pi(B^c \mid A)$. It corresponds to some existing practice in statistics, called likelihood ratio tests, whereby the likelihood function is renormalized via a proportional rescaling (Edwards, 1972; Barnett, 1973). This approach has been recently successfully developed for use in practical applications by Lapointe and Bobee (2000).

Viewing possibility degrees as upper bounds of probabilities leads to Bayesian conditionalization of possibility measures. Notice that in probability theory, $P(B \mid A)$ is an increasing function of both $P(B \cap A)$ and $P(A^c \cup B)$

and this function is exactly $f(x, y) = \frac{x}{(x+1-y)}$. Then a natural counterpart of Bayesian conditioning in quantitative possibility theory is

$$\Pi(B \mid A) = \frac{\Pi(A \cap B)}{\Pi(A \cap B) + 1 - \Pi(A^c \cup B)} = \frac{\Pi(A \cap B)}{\Pi(A \cap B) + N(A \cap B^c)}$$

Then, the dual conditional necessity is such that

$$N(B \mid A) = 1 - \Pi(B^c \mid A) = \frac{N(A \cap B)}{N(A \cap B) + \Pi(A \cap B^c)}$$

This view of conditioning, that we shall call Bayesian possibilistic conditioning (Dubois and Prade, 1997; Walley, 1996), is in accordance with imprecise probabilities since $\Pi(B \mid A) = \sup \{P(B \mid A), P(A) > 0, P \leq \Pi\}$. Bayesian conditioning preserves consonance of possibility measures and the corresponding conditional possibility distribution has support A ($\pi(u \mid A) = 0$ if $u \notin A$) and, if $u \in A$:

$$\pi(u \mid A) = \max\left(\pi(u), \frac{\pi(u)}{\pi(u) + N(A)}\right)$$

which indicates that the result is less specific than π on A and coincides with the characteristic function of A if $N(A) = 0$. It contrasts with Dempster conditioning which always supplies more specific results than the above. See De Cooman (2001) for a detailed study of this form of conditioning.

5 Possibility-Probability Transformations

Possibility distributions may also come from the transformation of a probability distribution. Namely, given a unimodal probability density p on the real line, a (nested) set of confidence intervals of p can be encoded as a possibility distribution. The problem of transforming a possibility distribution into a probability distribution and conversely has received more and more attention in the past years (Klir, 1990; Dubois *et al.* 1993). This question is meaningful in the scope of uncertainty combination with heterogeneous sources (some supplying statistical data, other linguistic data, for instance).

5.1 Basic Principles

The starting point is that some consistency exists between possibilistic and probabilistic representations of uncertainty. Zadeh (1978) defined the degree of consistency between a possibility distribution π and a probability distribution p as follows: $p(\pi) = \sum_{i=1}^{n} \pi_i p_i$. It is the probability of the fuzzy event whose membership function is π. However Zadeh also described the consistency principle between possibility and probability in an informal way, whereby what is probable should be possible. Dubois and Prade (1980) have

13

translated this requirement via the inequality $\Pi(A) \geq P(A)$ that founds the interpretation of possibility measures as upper probability bounds. P and Π are then said to be consistent.

In the finite case, if we let $\pi(u_i) = \pi_i$ and $P(\{u_i\}) = p_i$ and assume $p_1 \geq \ldots \geq p_n \geq p_{n+1} = 0$, then P and Π are consistent if and only if $\pi_i \geq \sum_{j=1}^{n} p_j$, $\forall i = 1, \ldots, n$ (Delgado and Moral, 1987). More conditions for consistency could be added. For instance one may also require that the orderings induced by the probability and the possibility distributions on U be the same. However we cannot require the same condition for events since the possibility ordering is generally coarser than the probability ordering on finite sets.

Possibility and probability theories do not have the same descriptive power. It is clear that there are some states of information that probability can describe while possibility cannot (e.g., total randomness). The converse is true: a single probability distribution cannot express ignorance, as advocated at length by Dubois, Prade and Smets (1996). It can be advocated that the possibilistic representation of ignorance is weaker than the probabilistic representation, in the sense that the first is additive and supplies precise degrees of confidence and the other relies on an ordinal structure induced by the consonance assumption and only provides bounds on degrees of probability. It does not mean that precise probabilistic representations subsume possibilistic representations, since possibilistic representations can capture weaker states of information that probability distributions cannot model.

There are two basic approaches to possibility/probability transformations. They respect probability-possibility consistency. One, due to Klir (see Klir, 1990; Geer and Klir, 1992) is based on a principle of information invariance, the other (Dubois et al., 1993) is based on optimizing information content.

In Klir's view, the transformation should be based on three assumptions:

- A scaling assumption that forces each value π_i to be a function of $\frac{p_i}{p_1}$ (where $p_1 \geq p_2 \geq \ldots \geq p_n$, that can be ratio-scale, interval scale, Log-interval scale transformations, etc.
- An uncertainty invariance assumption according to which the entropy $H(p)$ should be numerically equal to the measure of information $E(\pi)$ contained in the transform π of p. $E(\pi)$ can be the logarithmic imprecision index of Higashi and Klir (1982), for instance.
- Transformations should satisfy the consistency condition $\pi(u) \geq p(u)$, $\forall u$, stating that what is probable must be possible.

The uncertainty invariance equation $E(\pi) = H(p)$, along with a scaling transformation assumption (e.g., $\pi(x) = \alpha p(x) + \beta, \forall x$), reduces the problem of computing π from p to that of solving an algebraic equation with one or two unknowns.

Klir's assumptions are debatable. First the scaling assumption leads to assume that $\pi(u)$ is a function of $p(u)$ only. This pointwiseness assumption may conflict with the probability/possibility consistency principle that requires $\Pi \geq P$ for all events. See Dubois and Prade (1980, pp. 258-259) for an example of such a violation. Then, the nice link between possibility and probability, casting possibility measures in the setting of upper and lower probabilities cannot be maintained.

The second and the most questionable prerequisite assumes that possibilistic and probabilistic information measures are commensurate. The basic idea is that the choice between possibility and probability is a mere matter of translation between languages "neither of which is weaker or stronger than the other" (quoting Klir and Parviz, 1992). It means that entropy and imprecision capture the same facet of uncertainty, albeit in different guises. The alternative approach recalled below does not make this assumption.

5.2 Transformations Based on Optimizing Information Content

If we accept that possibility distributions are weaker representations of uncertainty than probability distributions, the transformation problem can still be stated in a clear way. Namely going from possibility to probability leads to increase the informational content of the considered representation, while going the other way around means a loss of information. However the adopted transformations must be as little arbitrary as possible. Hence, the principles behind the two following transformations are different and are not always the converse of each other (Dubois et al. 1993):

- *From possibility to probability:*
  A generalised Laplacean indifference principle is adopted: the weights m_i bearing on the nested family of levels cuts of π are uniformly distributed on the elements of these cuts. This transformation, already proposed by Dubois and Prade (1982) comes down to selecting the gravity center of the set $P = \{P \mid \forall A, \ P(A) \leq \Pi(A)\}$ of probability distributions dominated by Π. This transformation also coincides with the so-called pignistic transformation of belief functions (Smets 1990) and the Shapley value in game theory, where a cooperative game can be viewed as a non additive set function (Shapley, 1971). The rationale behind this transformation is to minimize arbitrariness by preserving the symmetry properties of the representation. If we let $\pi(u_i) = \pi_i$ and $p(u_i) = p_i$ and assume $\pi_1 \geq \ldots \geq \pi_n \geq \pi_{n+1} = 0$, the transformation yields

$$p_i = \sum_{j=1}^{i} \frac{(\pi_j - \pi_{j-1})}{j} \tag{5}$$

- *From probability to possibility:*
  in this case, the rationale of the transformation is not the same according to whether the probability distribution we start with is subjective or objective.

In the case of a statistically induced probability distribution, the rationale is to preserve as much information as possible; hence we select as the result of the transformation of P, the most specific element of the set $F(P)$ of possibility measures dominating P. This most specific element is generally unique if P induces a linear ordering on U. Otherwise if there are equiprobable elements, unicity is preserved if equipossibility of the corresponding elements is enforced. In particular, uniform probability is then transformed into uniform possibility (Delgado and Moral, 1987).

In the finite case, if we let $\pi(u_i) = \pi_i$ and $p(u_i) = p_i$ and assume $p_1 > p_2 > \ldots > p_n > p_{n+1} = 0$ the transformation into a possibility distribution takes the following form

$$\pi_i = \sum_{j=i}^{n} p_j \tag{6}$$

where $\pi_1 = 1 > \ldots > \pi_n$. However if $p_1 = \ldots = p_n = \frac{1}{n}$ then selecting any linear ordering of elements and applying the above formula gives a most specific possibility distribution consistent with p. More generally if $E_1, \ldots, E_k$ is the well-ordered partition of U induced by p (elements in E_i have the same probability which is greater than the one of elements in E_{i+1}), then the most specific possibility distributions consistent with p are given by (6) applied to any linear ordering of U coherent with $E_1, \ldots, E_k$ (arbitrarily reordering elements within each E_i).

In the case of a subjective probability, the rationale is very different. It is assumed that a subjective probability supplied by an agent is only a trace of the agent's belief because it is forced to be additive by the rules of exchangeable bets. For instance the agent provides a uniform probability distribution whether (s)he knows nothing about the concerned phenomenon, or if (s)he knows the concerned phenomenon is purely random. More generally, it is assumed that the agent entertains beliefs that can be modelled by belief functions in the sense of the Transferable Belief Model (Smets & Kennes, 1994). In that framework, the agent uses a probability function induced by his or her beliefs, using the pignistic transformation (Smets, 1990) or Shapley value. Then the transformation from a subjective probability consists in reconstructing the underlying belief function. There are clearly several belief functions whose Shapley value is prescribed. Dubois et al. (2001) have proposed to consider the least informative of those, in the sense of a non-specificity index. They prove that the least informative belief function is based on a possibility distribution, previously suggested in (Dubois and Prade, 1983):

$$\pi_i = \sum_{j=1}^{n} \min(p_j, p_i) \tag{7}$$

Equation (7) gives results that are less specific than (6). It is the transformation converse to (5).

5.3 Confidence Intervals

Applied to continuous universes, the transformation of a unimodal objective probability density p with strictly monotonic sides into a possibility distribution is closely related to the notion of confidence interval. First it is a known fact that the confidence interval of length L, i. e., the interval with maximal probability is $I_L = [a_L, a_L + L]$ such that $p(a_L) = p(a_L + L)$. This interval has degree of confidence $P(I_L)$(often taken as 95%). The most specific possibility distribution consistent with p is π such that $\forall L > 0$, $\pi(a_L) = \pi(a_L + L) = 1 - P(I_L)$. Hence the α-cut of the optimal (most specific) π is the $(1 - \alpha)$- confidence interval of p (Dubois et al., 1993). These confidence intervals are nested around the mode of p, viewed as the "most frequent value". Going from objective probability to possibility means adopting a representation of uncertainty in terms of confidence intervals.

More recently Mauris et al. (2001) have found more results along this line (see also Dubois et al., 2002) for symmetric densities. Noticeably, each side of the optimal possibilistic transform is convex and there is no derivative for the mode of π. Hence given a probability density on a bounded interval $[a, b]$, the triangular fuzzy number whose core is the mode of π and the support is $[a, b]$ is an upper approximation of p regardless of its shape. In the case of a uniform distribution on $[a, b]$, any triangular fuzzy number with support $[a, b]$ provides a most specific upper approximation. These results justify the use of triangular fuzzy numbers as fuzzy counterparts to uniform probability distributions. This setting is adapted to sensor measurements. Well-known inequalities of probability theory, such as Chebyshev and Camp-Meidel ones, can also be viewed as possibilistic approximations of probability functions, since they provide families of (loose) intervals. However they provide shapes of fuzzy numbers that can act as generic possibilistic approximations of probability functions, regardless of their shapes.

6 Quantitative Possibility and Choquet Integrals

Possibility and necessity measures are very special cases of Choquet capacities and can encode families of probabilities. The integral of function F from a set U, equiped with a Choquet capacity , to the reals can be evaluated with respect to a fuzzy measure M using a Choquet integral, defined as follows:

$$E_M(F) = \int_0^1 M\left(F_\alpha\right) d\alpha$$

See Denneberg (1994) for a mathematical introduction. When $M = P$, a probability measure, it reduces to a Lebesgue integral. When $M = \Pi$, a possibility measure, or $M = N$, a necessity measure, it reads

$$E_M(F) = \int_0^1 N\left(F_\alpha\right) d\alpha = \int_0^1 \inf\left\{F\left(u\right) : \pi\left(F\left(u\right)\right) \geq \alpha\right\} d\alpha$$

$$E_\Pi(F) = \int_0^1 \Pi\left(F_\alpha\right) d\alpha = \int_0^1 \sup\left\{F\left(u\right) : \pi\left(F\left(u\right)\right) \geq \alpha\right\} d\alpha$$

If F is the membership function of a fuzzy set, the latter equation is a definition of the possibility of fuzzy events different from Zadeh's (1978). But the maxitivity of $E_\Pi(F)$ w.r.t. to F is not preserved.

In the finite setting, $U = \{1, 2, \ldots, n\}$, these expressions read (Dubois and Prade, 1985; Grabisch *et al.*, 1995):

$$E_N\left(F\right) = \sum_{i=1}^n F\left(u_{\sigma(i)}\right) \left(\max_{k \leq i} \pi\left(u_{\sigma(k)}\right)\right) - \max_{k < i} \pi\left(u_{\sigma(k)}\right)$$

$$= \sum_{i=1}^n \left(F\left(u_{\sigma(i)}\right) - F\left(u_{\sigma(i-1)}\right)\right) \min_{k < i}\left(1 - \pi\left(u_{\sigma(k)}\right)\right)$$

$$= \sum_{i=1}^n \left(\pi_i - \pi_{i+1}\right) \min\left\{F\left(u\right) : u \in A_i\right\}$$

$$E_\Pi\left(F\right) = \sum_{i=1}^n F\left(u_{\sigma(i)}\right) \left(\max_{k \geq i} \pi\left(u_{\sigma(k)}\right)\right) - \max_{k > i} \pi\left(u_{\sigma(k)}\right)$$

$$= \sum_{i=1}^n \left(F\left(u_{\sigma(i)}\right) - F\left(u_{\sigma(i-1)}\right)\right) \max_{k \geq i}\left(\pi\left(u_{\sigma(k)}\right)\right)$$

$$= \sum_{i=1}^n \left(\pi_i - \pi_{i+1}\right) \max\left\{F\left(u\right) : u \in A_i\right\}$$

where, without loss of generality, $\pi_1 = 1 \geq \ldots \geq \pi_n \geq \pi_{n+1} = 0$, and A_i is the π_i-cut of F, and $F\left(u_{\sigma(1)}\right) \leq \ldots \leq F\left(u_{\sigma(n)}\right)$. The interval $[E_N(F), E_\Pi(F)]$ is the range of the expectations of F with respect to probability measures P such that $\Pi \geq P$. To any function F from U to the reals, associate the functions F^+ and F^- defined by :

$$F^-\left(i\right) = \min\left\{F\left(u\right) : u \in A_i\right\}$$
$$F^+\left(i\right) = \max\left\{F\left(u\right) : u \in A_i\right\}$$

Since the A_i are nested, F^- is decreasing and F^+ is decreasing in the wide sense and $E_N(F) = E_N(F^-)$, $E_\Pi(F) = E_\Pi(F^+)$. Possibilistic expectations of F are also regular expectations of F^- or F^+ w.r.t. a probability function defined by $p_i = (\pi_i - \pi_{i+1})$. Moreover, F^- and p are comonotonic, and so are F^- and G^- (resp. F^+ and G^+) for any F and G. It is well-known that Choquet integrals are additive for comonotonic functions. Possibilistic integrals are thus additive for larger subsets of functions. For instance, if F and G are such that $(F+G)^- = F^- + G^-$ then $E_N(F+G) = E_N(F) + E_N(G)$. More on possibilistic Choquet integrals can be found in (De Cooman, 2001).

7 Fuzzy Intervals in the Probabilistic Setting

A fuzzy interval is a fuzzy set M of real numbers whose cuts are nested intervals, usuallly closed ones (see Dubois, Kerre *et al.*, 2000) for an extensive survey. Relating fuzzy intervals to probability is based on the fact that fuzzy sets can be viewed as consonant plausibility functions in the sense of Shafer (1976), or imprecise probabilities. there are actually three probabilistic views of a fuzzy interval.

(i) In the *imprecise probability view*, M encodes a set of probability measures. The upper and lower distribution functions are limits of distribution functions in this set (Dubois and Prade, 1987). Let $[m^*, m_*]$ be the core of M. The upper distribution function F^* is:

$$\forall a, \quad F^*(a) = \Pi_M((-\infty, a])$$
$$= \sup\{M(x) : x \leq a\} = \begin{cases} M(a) & \text{if} \quad a \leq m_* \\ 1 & \text{otherwise.} \end{cases}$$

Similarly, the lower distribution function F_* such that:

$$\forall a, \quad F_*(a) = N_M((-\infty, a]) = 1 - \Pi_M([a, \infty))$$
$$= \inf\{1 - M(x) : x > a\} = \begin{cases} 0 & \text{if} \quad a < m^* \\ 1 - \lim_{x \to a} M(x) & \text{otherwise.} \end{cases}$$

The upper distribution function F^* matches the increasing part of the membership function of M. The lower distribution function F_* reflects the decreasing part of the membership function of M of wich it is the fuzzy complement.

(ii) In the *random set view*, M encodes the one point coverage function of a random interval, defined by a probability measure on the unit interval (for instance the uniformly distributed one) and a family of nested intervals, via a multivalued mapping from $(0, 1]$ to r, following Dempster (1967). For instance a fuzzy interval M can be obtained by carrying the Lebesgue measure from the unit interval to the real line via the multiple-valued mapping that assigns to each level $\alpha \in (0, 1]$ the cut M_α. (see Dubois and Prade, 1987, Heilpern, 1992, for instance). The fuzzy interval M is then viewed as a random interval I_M and $M(a) = P(a \in I_M)$ the probability that a belongs to a realization of I_M. This view presupposes that a probability measure P could have been obtained from the available statistics if outcomes had been precisely observed.

(iii) M stands for a *pair of PDFs view*, whereby M is defined by two random variables x^- and x^+ with distributions F^* and F_* yielding the random interval $[x^-, x^+]$ with possibly independent end-points (See Heilpern, 1992, 1997; Gil, 1992). Namely let $M_\lambda = [m_\lambda^*, m_{*\lambda}]$ be the cuts of M. One may have x^- and x^+ depend on a single parameter λ such that $[x^-(\lambda), x^+(\lambda)] = [m_\lambda^*, m_{*\lambda}]$, or the generated intervals may appear in an asymmetric way: e.g., $[x^-(\lambda), x^+(\lambda)] = [m_\lambda^*, m_{*1-\lambda}]$. Or one may have x^- and x^+ depending on two independent parameters λ and λ' such that $[x^-(\lambda), x^+(\lambda')] = [m_\lambda^*, m_{*\lambda}]$. Note that the intersection of all such generated intervals is not empty so as to ensure the normalization of M.

The first view is more general than the second one, and can be understood as the specification of an imprecise probability function; and the third view is also more general, since the second view corresponds to a random process that delivers nested intervals as a whole (hence a clear dependence between endpoints) while the third view is one where end-points of the interval are separately generated, possibly in an independent way. A thorough development of each point of view is a matter of further research.

7.1 The Mean Interval and Defuzzification

The simplest non-fuzzy substitute of the fuzzy interval is its core (or its mode when it is a fuzzy number). In Dubois and Prade (1980), what is called mean value of a fuzzy number is actually its modal value. Under the random set interpretation of a fuzzy interval, upper and lower mean values of M in the sense of Dempster (1967), can be defined, i.e., $E^*(M)$ and $E_*(M)$, respectively, such that (Dubois and Prade, 1987; Heilpern, 1992):

$$E^*(M) = \int_0^1 (\sup M_\alpha) \, d\alpha;$$

$$E_*(M) = \int_0^1 (\inf M_\alpha)\, d\alpha.$$

Note that these expressions are Choquet integrals with respect to the possibility and the necessity measures induced by M. The mean *mean interval* of a fuzzy interval M is defined as $E(M) = [E_*(M), E^*(M)]$. It is also the interval containing the mean values of all random variables compatible with M (i.e., $P \in P_M$). That the mean value of a fuzzy interval is an interval seems to be intuitively satisfactory. Particularly the mean interval of a (regular) interval $[a, b]$ is this interval itself. The same mean interval obtains in the three probabilistic settings for fuzzy intervals.

The upper and lower mean values (Equations (10.8) or (10.9)) are additive with respect to the fuzzy addition, since they satisfy, for u.s.c. fuzzy intervals (Dubois and Prade, 1987; Heilpern, 1992):

$$E^*(M \oplus N) = E^*(M) + E^*(N);$$
$$E_*(M \oplus N) = E_*(M) + E_*(N).$$

This property is a consequence of the additivity of Choquet integral for the sum of comonotonic functions .

Finding a scalar representative value of a fuzzy interval is often called defuzzification in the literature of fuzzy control (See Yager and Filev, 1993, 1994, and Van Leekwijk and Kerre, 1999 for extensive overviews). Various proposals exist:

-) the *mean of maxima* (MOM), which is the middle point in the core of the fuzzy interval M,
-) the *center of gravity*. This is the center of gravity of the support of M, weighted by the membership grade.
-) the *center of area (median)*: This is the point of the support of M that equally divides the area under the membership function.

The MOM sounds natural as a representative of a fuzzy interval M in the scope of possibility theory where values of highest possibility are considered as default plausible values. This is in the particular case when the maximum is unique. However it makes sense especially if a revision process is in order, that may affect the MOM, by deleting the most plausible values upon arrival of new information. In a numerical perspective, the MOM clearly does not exploit all the information contained in M since it neglects the membership function. Yager and Filev (1993) present a general methodology for extracting characteristic values from fuzzy intervals. They show that all methods come down to a possibility-probability transformation followed by the extraction of characteristic value such as a mean value. Note that the MOM, the center of gravity and the center of area come down to renormalizing the fuzzy interval as a probability distribution and computing its mode, expected value or its median, respectively. These approaches are debatable, because the renormalization technique is itself arbitrary since the obtained probability may not belong to P_M, the set of probabilities dominated by the possibility measure attached to M (Dubois and Prade, 1980).

In view of the quantitative possibility setting, it seems that the most natural defuzzication proposal is the *middle point of the mean interval* (Yager, 1981)

$$
\begin{aligned}
E^{\#}(M) &= \int_0^1 \frac{(\inf M_\alpha + \sup M_\alpha)}{2} d\alpha \\
&= \frac{E_*(M) + E^*(M)}{2}
\end{aligned}
$$

Only the mean interval accounts for the specific possibilistic nature of the fuzzy interval. The choice of the middle point expresses a neutral attitude of the user and extends the MOM to an average mean of cuts. Other choices are possible, for instance using a weighted average of $E^*(M)$ and $E_*(M)$.

$E(M)$ has a natural interpretation in terms of simulation of a fuzzy variable, an idea originally due to Kaufmann (1980) and Yager (1982). Chanas and Nowakowski (1988) investigate this problem in greater detail. Namely, consider the two step random generator which selects a cut at random (by choosing $\alpha \in (0,1]$), and a number in the cut. The corresponding random quantity is $x(\alpha, \lambda) = \lambda \cdot \inf M_\alpha + (1 - \lambda) \cdot \sup M_\alpha$. The mean value of this random variable is $E^{\#}(M)$. It corresponds to the mean value of the Shapley value $p_M \in P_M$ obtained by considering cuts as uniformly distributed probabilities (Dubois, Prade and Sandri, 1993) :

$$
p_M(x) = \int_0^1 \frac{M_\lambda(x)}{(\sup M_\lambda - \inf M_\lambda)} d\lambda
$$

p_M is in fact, the center of gravity of P_M. The mean value $E^{\#}(M)$ is linear in the sense of fuzzy addition and scalar multiplication (Fortemps and Roubens, 1996).

7.2 Calculations with Possibilistic Variables

The extension principle of Zadeh (1975) can be explained in a random set view. Consider a two place function f. If a joint possibility relating two variables x_1 and x_2 is separable, i.e., $\pi = \min(\pi_1, \pi_2)$, then the possibility distribution π^f of $f(\pi_1, \pi_2)$ is

$$
\pi^f(\{v\}) = \begin{cases} \sup\{\min(\pi_1(u_1), \pi_2(u_2)) : f(u_1, u_2) = v\} & \text{if } f^{-1}(\{v\}) \neq \emptyset \\ 0 & \text{otherwise} \end{cases}
$$

In the setting of random sets, the above notion of joint possibility distributions relies on a dependence assumption between confidence levels. It presupposes that if the first random process delivers a cut $(A_1)_\lambda$ of π_1 then the other one delivers $(A_2)_\lambda$ for the same value of λ. Two nested random sets with associated one-point coverage functions π_1 and π_2 then produce a nested random set of one-point coverage function $\min(\pi_1, \pi_2)$. In other words, it comes down to working with confidence intervals with the same levels of confidence.

On the contrary, the assumption that cuts $(A_1)_\lambda$ and $(A_2)_v$ may be jointly observed will not lead to a nested random set, hence is not equivalent to a joint possibility distribution (Dubois and Prade, 1991). Assuming independence between the choice of confidence levels leads to compute π^f by means of a counterpart of Dempster rule of combination of belief functions, where the intersection of focal sets is changed into interval computations on cut intervals. The result can be approximated by a possibility distribution obtained by a sup-product extension principle (Dubois and Prade, 1990) This setting encompasses both fuzzy interval and random variable computation.

8 Conclusion

Quantative possibility theory seems to be a promising framework for probabilistic reasoning under incomplete information. This is because some families of probability measures can be encoded by possibility distributions. The simplicity of possibility distributions make them attractive for practical applications of imprecise probabilities (De Cooman, 2001). Of course the probabilistic view is only one among other interpretive settings for possibility measures. Besides, cognitive studies for the empirical evaluation of possibility theory have recently appeared (Raufaste, Da Silva Neves, 1998). Their experiments suggest "that human experts might behave in a way that is closer to possibilistic predictions than probabilistic ones". The cognitive validation of possibility theory is clearly an important issue for a better understanding of when possibility theory is most appropriate.

References

1. Barnett V. (1973). Comparative Statistical Inference, J. Wiley, New York.
2. Benferhat S., Dubois D. and Prade H. (1997a). Nonmonotonic reasoning, conditional objects and possibility theory, Artificial Intelligence, **92**, 259-276.
3. Chanas S. and Nowakowski M. (1988). Single value simulation of fuzzy variable, Fuzzy Sets and Systems, **25**, 43-57.
4. De Baets B., Tsiporkova E. and Mesiar R. (1999). Conditioning in possibility with strict order norms, Fuzzy Sets and Systems, **106**, 221-229.
5. De Cooman G. (1997). Possibility theory - Part I: Measure- and integral-theoretics groundwork; Part II: Conditional possibility; Part III: Possibilistic independence, Int. J. of General Systems, **25**(4), 291-371.

6. De Cooman G. (2001) Integration and conditioning in numerical possibility theory. Annals of Math. and AI, **32**, 87-123.
7. De Cooman G. and Aeyels D. (1996). On the coherence of supremum preserving upper previsions, Proc. of the 6th Inter. Conf. Information Processing and Management of Uncertainty in Knowledge-Based Systems (IPMU'96), Granada, 1405-1410.
8. De Cooman G., Aeyels D. (1999). Supremum-preserving upper probabilities. Information Sciences, **118**, 173 -212.
9. Delgado M. and Moral S. (1987). On the concept of possibility-probability consistency, Fuzzy Sets and Systems, **21**, 311-318.
10. Dempster A. P. (1967). Upper and lower probabilities induced by a multivalued mapping, Ann. Math. Stat., **38**, 325-339.
11. Denneberg D. (1994). Nonadditive Measure and Integral, Kluwer Academic, Dordrecht, The Netherlands.
12. Dubois D. (1986). Belief structures, possibility theory and decomposable confidence measures on finite sets, Computers and Artificial Intelligence (Bratislava), **5**, 403-416.
13. Dubois D., Fargier H; and Prade H. (1996b). Possibility theory in constraint satisfaction problems: Handling priority, preference and uncertainty, Applied Intelligence, **6**, 287-309.
14. Dubois D., Kerre E., Mesiar R., Prade H. Fuzzy interval analysis. In: Fundamentals of Fuzzy Sets, Dubois,D. Prade,H., Eds: Kluwer , Boston, Mass , The Handbooks of Fuzzy Sets Series , 483-581 , 2000.
15. Dubois D., Moral S. and Prade H. (1997). A semantics for possibility theory based on likelihoods, J. of Mathematical Analysis and Applications, **205**, 359-380.
16. Dubois D. and Prade H. (1980). Fuzzy Sets and Systems: Theory and Applications, Academic Press, New York.
17. Dubois and Prade H. (1982) On several representations of an uncertain body of evidence," in Fuzzy Information and Decision Processes, M.M. Gupta, and E. Sanchez, Eds., North-Holland, Amsterdam, 1982, pp. 167-181.
18. Dubois D. and Prade H. (1983) Unfair coins and necessity measures: towards a possibilistic interpretation of histograms. Fuzzy Sets and Systems, **10**, 15-20.
19. Dubois D. and Prade H. (1985). Evidence measures based on fuzzy information, Automatica, **21**, 547-562.
20. Dubois D. and Prade H. (1986). Fuzzy sets and statistical data, Europ. J. Operations Research, **25**, 345-356.
21. Dubois D. and Prade H. (1987). The mean value of a fuzzy number, Fuzzy Sets and Systems, **24**, 279-300.
22. Dubois D. and Prade H. (1988). Possibility Theory, Plenum Press, New York.
23. Dubois D. and Prade H. (1990) Consonant approximations of belief functions. Int. J. Approximate Reasoning, 4, 419-449.
24. Dubois D. and Prade H. (1991). Random sets and fuzzy interval analysis, Fuzzy Sets and Systems, **42**, 87-101.
25. Dubois D. and Prade H. (1992). When upper probabilities are possibility measures, Fuzzy Sets and Systems, **49**, 65-74.
26. Dubois D. and Prade H. (1997) Bayesian conditioning in possibility theory, Fuzzy Sets and Systems, **92**, 223-240.

27. Dubois D. and Prade H. (1998). Possibility theory: Qualitative and quantitative aspects, Handbook of Defeasible Reasoning and Uncertainty Management Systems - Vol. 1 (Gabbay D.M. and Smets P., eds.), Kluwer Academic Publ., Dordrecht, 169-226.

28. Dubois D., Nguyen H. T., Prade H. (2000) Possibility theory, probability and fuzzy sets: misunderstandings, bridges and gaps. In: Fundamentals of Fuzzy Sets, (Dubois, D. Prade,H., Eds.), Kluwer , Boston, Mass., The Handbooks of Fuzzy Sets Series , 343-438.

29. Dubois D., Foulloy L., Mauris G., Prade H. (2002), Possibility/probability transformations, triangular fuzzy sets, and probabilistic inequalities. Proc. IPMU conference, Annecy, France.

30. Dubois D., Prade H. and Sandri S. (1993). On possibility/probability transformations. In: Fuzzy Logic. State of the Art, (R. Lowen, M. Roubens, eds.), Kluwer Acad. Publ., Dordrecht, 103-112.

31. Dubois D., Prade H. and Smets P. (1996). Representing partial ignorance, IEEE Trans. on Systems, Man and Cybernetics, **26**, 361-377.

32. Dubois D., Prade H. and Smets P. (2001). New semantics for quantitative possibility theory. Proc. ESQARU 2001, Toulouse, LNAI 2143, Springer-Verlag, p. 410-421, 19-21.

33. Edwards W. F. (1972). Likelihood, Cambridge University Press, Cambridge, U.K.

34. Fortemps P. and Roubens M. (1996). Ranking and defuzzification methods based on area compensation, Fuzzy Sets and Systems, **82**, 319-330.

35. Gebhardt J. and Kruse R. (1993). The context model - an intergating view of vagueness and uncertainty. Int. J. Approximate Reasoning, **9**, 283-314.

36. Gebhardt J. and Kruse R. (1994a) A new approach to semantic aspects of possibilistic reasoning. Symbolic and Quantitative Approaches to Reasoning and Uncertainty (M. Clarke et al. Eds.), Lecture Notes in Computer Sciences Vol. 747, Springer Verlag, 151-160.

37. Gebhardt J. and Kruse R. (1994b) On an information compression view of possibility theory. Proc 3rd IEEE Int. Conference on Fuzzy Systems. Orlando,Fl., 1285-1288.

38. Geer J.F. and Klir G.J. (1992). A mathematical analysis of information-preserving transformations between probabilistic and possibilistic formulations of uncertainty, Int. J. of General Systems, **20**, 143-176.

39. Gil M. A. (1992). A note on the connection between fuzzy numbers and random intervals, Statistics and Probability Lett., **13**, 311-319.

40. Gonzalez A. (1990). A study of the ranking function approach through mean values, Fuzzy Sets and Systems, **35**, 29-43.

41. Grabisch M., Murofushi T. and Sugeno M. (1992). Fuzzy measure of fuzzy events defined by fuzzy integrals, Fuzzy Sets and Systems, **50**, 293-313.

42. Hacking I. (1975). All kinds of possibility, Philosophical Review, **84**, 321-347.

43. Heilpern S. (1992). The expected value of a fuzzy number, Fuzzy Sets and Systems, **47**, 81-87.

44. Heilpern S. (1997). Representation and application of fuzzy numbers, Fuzzy Sets and Systems, **91**, 259-268.

45. Higashi and Klir G. (1982). Measures of uncertainty and information based on possibility distributions, Int. J. General Systems, **8**, 43-58.

46. Hisdal E. (1978). Conditional possibilities independence and noninteraction, Fuzzy Sets and Systems, **1**, 283-297.

47. Joslyn C. (1997). Measurement of possibilistic histograms from interval data, Int. J. of General Systems, **26**(1-2), 9-33.
48. Kaufmann A (1980). La simulation des ensembles flous, CNRS Round Table on Fuzzy Sets, Lyon, France (unpublished proceedings).
49. Klir G.J. (1990). A principle of uncertainty and information invariance, Int. J. of General Systems, **17**, 249-275.
50. Klir G.J. and Folger T. (1988). Fuzzy Sets, Uncertainty and Information, Prentice Hall, Englewood Cliffs, NJ.
51. Klir G.J. and Parviz B. (1992). Probability-possibility transformations: A comparison, Int. J. of General Systems, **21**, 291-310.
52. Lapointe S. Bobee B. (2000). Revision of possibility distributions: A Bayesian inference pattern, Fuzzy Sets and Systems, **116**, 119-140
53. Mauris G., Lasserre V., Foulloy L. (2001). A fuzzy appproach for the expression of uncertainty in measurement. Int. J. Measurement, **29**, 165-177.
54. Lewis D. L. (1979). Counterfactuals and comparative possibility, Ifs (Harper W. L., Stalnaker R. and Pearce G., eds.), D. Reidel, Dordrecht, 57-86.
55. Raufaste E. and Da Silva Neves R. (1998). Empirical evaluation of possibility theory in human radiological diagnosis, Proc. of the 13th Europ. Conf. on Artificial Intelligence (ECAI'98) (Prade H., ed.), John Wiley & Sons, 124-128.
56. Saade J. J. and Schwarzlander H. (1992). Ordering fuzzy sets over the real line: An approach based on decision making under uncertainty, Fuzzy Sets and Systems, **50**, 237-246.
57. Shackle G. L.S. (1961). Decision, Order and Time in Human Affairs, (2nd edition), Cambridge University Press, UK.
58. Shafer G. (1976). A Mathematical Theory of Evidence, Princeton University Press, Princeton.
59. Shafer G. (1987). Belief functions and possibility measures, Analysis of Fuzzy Information - Vol. I: Mathematics and Logic (Bezdek J.C., Ed.), CRC Press, Boca Raton, FL, 51-84.
60. Shapley S. (1971). Cores of convex games, Int. J. of Game Theory, **1**, 12-26.
61. Shilkret N. (1971). Maxitive measure and integration, Indag. Math., **33**, 109-116.
62. Smets P. (1990). Constructing the pignistic probability function in a context of uncertainty, Uncertainty in Artificial Intelligence 5 (Henrion M. et al., Eds.), North-Holland, Amsterdam, 29-39.
63. Smets P. and Kennes R. (1994). The transferable belief model, Artificial Intelligence, **66**, 191-234.
64. Spohn W. (1988). Ordinal conditional functions: A dynamic theory of epistemic states, Causation in Decision, Belief Change and Statistics (Harper W. and Skyrms B., Eds.), 105-134.
65. Van Leekwijk W. and Kerre E. (1999). Defuzzification: criteria and classification. Fuzzy Sets and Systems., **118**, 159-178.
66. Walley P. (1991). Statistical Reasoning with Imprecise Probabilities, Chapman and Hall.
67. Walley P. (1996). Measures of uncertainty in expert systems, Artificial Intelligence, **83**, 1-58.
68. Walley P. and de Cooman G.(1999) A behavioural model for linguistic uncertainty, Information Sciences, **134**, 1-37.

69. Wang P.Z. (1983). From the fuzzy statistics to the falling random subsets, Advances in Fuzzy Sets, Possibility Theory and Applications (Wang P.P., Eds.), Plenum Press, New York, 81-96.
70. Yager R.R. (1980). A foundation for a theory of possibility, J. Cybernetics, **10**, 177-204.
71. Yager R. R. (1981). A procedure for ordering fuzzy subsets of the unit interval, Information Sciences, **24**, 143-161.
72. Yager R.R. (1992). On the specificity of a possibility distribution, Fuzzy Sets and Systems, **50**, 279-292.
73. Yager R. R. and Filev D. (1993). On the issue of defuzzification and selection based on a fuzzy set, Fuzzy Sets and Systems, **55**, 255-271.
74. Zadeh L.A. (1965). Fuzzy sets, Information and Control, **8**, 338-353.
75. Zadeh L. A. (1975). The concept of a linguistic variable and its application to approximate reasoning, Information Sciences, Part I: **8**, 199-249; Part II: **8**, 301-357; Part III: **9**, 43-80.
76. Zadeh L. A. (1978). Fuzzy sets as a basis for a theory of possibility, Fuzzy Sets and Systems, **1**, 3-28.
77. Zadeh L.A. (1979a).A theory of approximate reasoning, Machine Intelligence, Vol. 9 (Hayes J. E., Michie D. and Mikulich L. I., eds.), John Wiley & Sons, New York, 149-194.

Toward a Perception-Based Theory of Probabilistic Reasoning with Imprecise Probabilities*

Lotfi A. Zadeh

Computer Science Division and the Electronics Research Laboratory,
Department of EECS,
University of California, Berkeley, CA 94720-1776, USA

Abstract. The perception-based theory of probabilistic reasoning which is out-lined in this paper is not in the traditional spirit. Its principal aim is to lay the groundwork for a radical enlargement of the role of natural languages in probability theory and its applications, especially in the realm of decision analysis. To this end, probability theory is generalized by adding to the theory the capability to operate on perception-based information, e.g., "Usually Robert returns from work at about 6 p.m", or "It is very unlikely that there will be a significant increase in the price of oil in the near future". A key idea on which perception-based theory is based is that the meaning of a proposition, p, which describes a perception, may be expressed as a generalized constraint of the form X isr R, where X is the constrained variable, R is the constraining relation and isr is a copula in which r is a discrete variable whose value defines the way in which R constrains X. In the theory, generalized constraints serve to define imprecise probabilities, utilities and other constructs, and generalized constraint propagation is employed as a mechanism for reasoning with imprecise probabilities as well as for computation with perception-based in-formation.

Key words: Perception-based information; Fuzzy set theory; Fuzzy logic; Gener-alized constraints; Constraint languages

1 Introduction

Interest in probability theory has grown markedly during the past decade. Underlying this growth is the ballistic ascent in the importance of informa-tion technology. A related cause is the concerted drive toward automation of decision-making in a wide variety of fields ranging from assessment of creditworthiness, biometric authentication, and fraud detection to stock mar-ket forecasting, and management of uncertainty in knowledge-based systems. Probabilistic reasoning plays a key role in these and related applications.

A side effect of the growth of interest in probability theory is the widen-ing realization that most real-world probabilities are far from being precisely

---

* Reprinted from Journal of Statistical Planning and Inference 105 (2002), 233–264, with kind permission from Elsevier Science

known or measurable numbers. Actually, reasoning with imprecise probabilities has a long history (Walley, 1991) but the issue is of much greater importance today than it was in the past, largely because the vast increase in the computational power of information processing systems makes it practicable to compute with imprecise probabilities to perform computations which are far more complex and less amenable to precise analysis than computations involving precise probabilities.

Transition from precise probabilities to imprecise probabilities in probability theory is a form of generalization and as such it enhances the ability of probability theory to deal with real-world problems. The question is: Is this mode of generalization sufficient? Is there a need for additional modes of generalization? In what follows, I argue that the answers to these questions are, respectively, No and Yes. In essence, my thesis is that what is needed is a move from imprecise probabilities to perception-based probability theory-a theory in which perceptions and their descriptions in a natural language play a pivotal role.

The perception-based theory of probabilistic reasoning which is outlined in the following is not in the traditional spirit. Its principal aim is to lay the groundwork for a radical enlargement in the role of natural languages in probability theory and its applications, especially in the realm of decision analysis.

For convenience, let PT denote standard probability theory of the kind found in textbooks and taught in courses. What is not in dispute is that standard probability theory provides a vast array of concepts and techniques which are highly effective in dealing with a wide variety of problems in which the available information is lacking in certainty. But alongside such problems we see many very simple problems for which PT offers no solutions. Here are a few typical examples:

1. What is the probability that my tax return will be audited?
2. What is the probability that my car may be stolen?
3. How long does it take to get from the hotel to the airport by taxi?
4. Usually Robert returns from work at about 6 p.m. What is the probability that he is home at 6:30 p.m.?
5. A box contains about 20 balls of various sizes. A few are small and several are large. What is the probability that a ball drawn at random is neither large nor small?

Another class of simple problems which PT cannot handle relates to commonsense reasoning (Kuipers, 1994; Fikes and Nilsson, 1971; Smithson, 1989; Shen and Leitch, 1992; Novak et al., 1992; Krause and Clark, 1993) exemplified by

6. Most young men are healthy; Robert is young. What can be said about Robert's health?

7. Most young men are healthy; it is likely that Robert is young. What can be said about Robert's health?

8. Slimness is attractive; Cindy is slim. What can be said about Cindy's attractiveness?

Questions of this kind are routinely faced and answered by humans. The answers, however, are not numbers; they are linguistic descriptions of fuzzy perceptions of probabilities, e.g., not very high, quite unlikely, about 0.8, etc. Such answers cannot be arrived at through the use of standard probability theory. This assertion may appear to be in contradiction with the existence of a voluminous literature on imprecise probabilities (Walley, 1991). In may view, this is not the case.

What are the sources of difficulty in using PT? In Problems 1 and 2, the difficulty is rooted in the basic property of conditional probabilities, namely, given $P(X)$, all that can be said about $P(X|Y)$ is that its value is between 0 and 1, assuming that Y is not contained in X or its complement. Thus, if I start with the knowledge that 1% of tax returns are audited, it tells me nothing about the probability that my tax return will be audited. The same holds true when I add more detailed information about myself, e.g., my profession, income, age, place of residence, etc. The Internal Revenue Service may be able to tell me what fraction of returns in a particular category are audited, but all that can be said about the probability that my return will be audited is that it is between 0 and 1. The tax-return-audit example raises some non-trivial issues which are analyzed in depth in a paper by Nguyen et al. (1999).

A closely related problem which does not involve probabilities is the following.

Consider a function, $y = f(x)$, defined on an interval, say $[0; 10]$, which takes values in the interval $[0; 1]$. Suppose that I am given the average value, a, of f over $[0; 10]$, and am asked: What is the value of f at $x = 3$? Clearly, all I can say is that the value is between 0 and 1.

Next, assume that I am given the average value of f over the interval $[2; 4]$, and am asked the same question. Again, all I can say is that the value is between 0 and 1. As the length of the interval decreases, the answer remains the same so long as the interval contains the point $x = 3$ and its length is not zero. As in the previous example, additional information does not improve my ability to estimate $f(3)$. The reason why this conclusion appears to be somewhat counterintuitive is that usually there is a tacit assumption that f is a smooth function. In this case, in the limit the average value will converge to $f(3)$. Note that the answer depends on the way in which smoothness is defined.

In Problem 3, the difficulty is that we are dealing with a time series drawn from a nonstationary process. When I pose the question to a hotel clerk, he/she may tell me that it would take approximately 20-25 min. In giving this answer, the clerk may take into consideration that it is raining lightly and that as a result it would take a little longer than usual to get to the airport. PT does not have the capability to operate on the perception-based information that "it is raining lightly" and factor-in its effect on the time of travel to the airport.

In problems 4-8, the difficulty is more fundamental. Specifically, the problem is that PT-as stated above-has no capability to operate on perceptions described in a natural language, e.g., "usually Robert returns from work at about 6 p.m.", or "the box contains several large balls " or "most young men are healthy". This is a basic shortcoming that will be discussed in greater detail at a later point.

What we see is that standard probability theory has many strengths and many limitations. The limitations of standard probability theory fall into several categories. To see the min a broad perspective, what has to be considered is that a basic concept which is immanent in human cognition is that of partiality. Thus, we accept the reality of partial certainty, partial truth, partial precision, partial possibility, partial knowledge, partial understanding, partial belief, partial solution and partial capability, whatever it may be. Viewed through the prism of partiality, probability theory is, in essence, a theory of partial certainty and randombehavior. What it does not address-at least not explicitly-is partial truth, partial precision and partial possibility-facets which are distinct from partial certainty and fall within the province of fuzzy logic (FL) (Zadeh, 1978; Dubois and Prade, 1988; Novak, 1991; Klir and Folger, 1988; Reghis and Roventa, 1998; Klir and Yuan, 1995; Grabisch et al., 1995). This observation explains why PT and FL are, for the most part, complementary rather than competitive (Zadeh, 1995; Krause and Clark, 1993; Thomas, 1995).

A simple example will illustrate the point. Suppose that Robert is three-quarters German and one-quarter French. If he were characterized as German, the characterization would be imprecise but not uncertain. Equivalently, if Robert stated that he is German, his statement would be partially true; more specifically, its truth value would be 0.75. Again, 0.75 has no relation to probability.

Within probability theory, the basic concepts on which PT rests do not reflect the reality of partiality because probability theory is based on two-valued Aristotelian logic. Thus, in PT, a process is random or not random; a time series is stationary or not stationary; an event happens or does not happen; events A and B are either independent or not independent; and so on. The denial of partiality of truth and possibility has the effect of seriously restricting the ability of probability theory to deal with those problems in which truth and possibility are matters of degree.

A case in point is the concept of an event. A recent Associated Press article carried the headline, "Balding on Top Tied to Heart Problems; Risk of disease is 36 percent higher, a study finds". Now it is evident that both balding on top, and heart problems, are matters of degree or, more concretely, are fuzzy

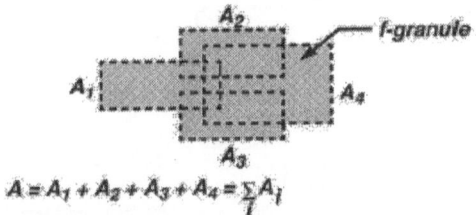

$$A = A_1 + A_2 + A_3 + A_4 = \sum_i A_i$$

f-granularity is a reflection of the bounded ability of sensory organs and, ultimately, the brain, to resolve detail and store information

Fig. 1. f-Granularity (fuzzy granularity).

events, as defined in Zadeh (1968), Kruse and Meyer (1987) and Wang and Klir (1992). Such events are the norm rather than exception in real-world settings. And yet, in PT the basic concept of conditional probability of an event B given an event A is not defined when A and B are fuzzy events.

Another basic, and perhaps more serious, limitation is rooted in the fact that, in general, our assessment of probabilities is based on information which is a mixture of measurements and perceptions (Vallee, 1995; Barsalou, 1999). Reflecting the bounded human ability to resolve detail and store information, perceptions are intrinsically imprecise. More specifically, perceptions are f-granular (Zadeh, 1979, 1997), that is: (a) perceptions are fuzzy in the sense that perceived values of variables are not sharply defined and (b) perceptions are granular in the sense that perceived values of variables are grouped into granules, with a granule being a clump of points drawn together by indistinguishability, similarity, proximity or functionality (Fig. 1). For example, the fuzzy granules of the variable Age might be young, middle-aged and old (Fig. 2). Similarly, the fuzzy granules of the variable Probability might be likely, not likely, very unlikely, very likely, etc.

Perceptions are described by propositions expressed in a natural language. For example

- Dana is young,
- it is a warm day,
- it is likely to rain in the evening,
- the economy is improving,
- a box contains several large balls, most of which are black.

An important class of perceptions relates to mathematical constructs such as functions, relations and counts. For example, a function such as shown in Fig. 3 may be described in words by a collection of linguistic rules (Zadeh, 1973, 1975, 1996).

In particular, a probability distribution, e.g., discrete-valued probability distribution of Carol's age, P^*, may be described in words as

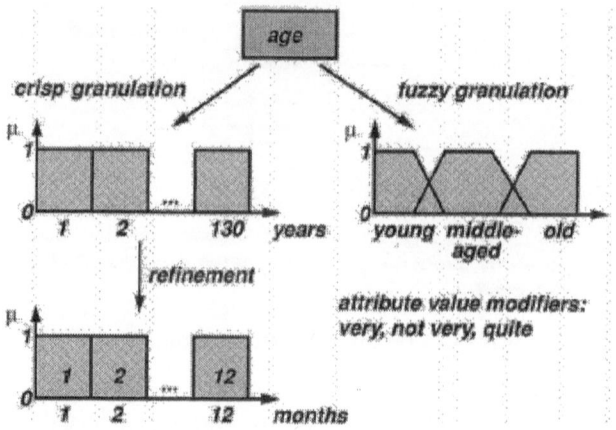

Fig. 2. Crisp and fuzzy granulation of *Age*.

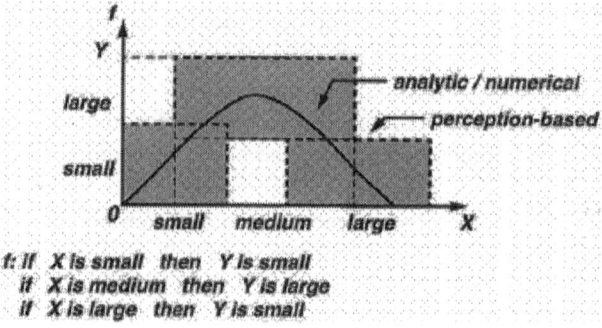

f: if X is small then Y is small
if X is medium then Y is large
if X is large then Y is small

Fig. 3. Coarse description of a function by a collection of linguistic rules. Linguistic representation is perception-based.

$$\text{Prob}\{ \text{ Carol is } young \} \text{ is } low,$$
$$\text{Prob}\{ \text{ Carol is } middle\text{-}aged \} \text{ is } high,$$
$$\text{Prob}\{ \text{ Carol is } old \} \text{ is } low$$

or as a linguistic rule-set

$$\text{if } Age \text{ is young then } P^* \text{ is } low,$$
$$\text{if Age is } middle\text{-}aged \text{ then } P^* \text{ is } high,$$
$$\text{if Age is } old \text{ then } P^* \text{ is } low.$$

For the latter representation, using the concept of a fuzzy graph (Zadeh, 1996, 1997), which will be discussed later, the probability distribution of

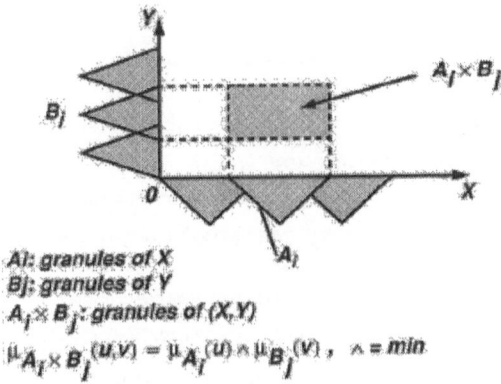

Fig. 4. Cartesian granulation. Granulation of X and Y induces granulation of (X, Y).

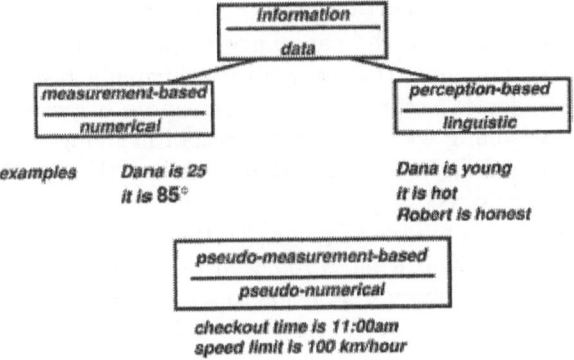

Fig. 5. CStructure of information: measurement-based, perception-based and pseudo-measurement based information.

Carol's age may be represented as a fuzzy graph and written as

$$P^* = young \times low + middle - aged \times high + old \times low$$

which, as shown in Fig. 4, should be interpreted as a disjunction of cartesian products of linguistic values of *Age* and *Probability* (Zadeh, 1997; Pedrycz and Gomide, 1998).

An important observation is in order. If I were asked to estimate Carol's age, it would be unrealistic to expect that I would come up with a numerical probability distribution. But I would be able to describe my perception of the probability distribution of Carol's age in a natural language in which *Age* and *Probability* are represented-as described above-as linguistic, that is, granular variables (Zadeh, 1973, 1975, 1996, 1997).

34

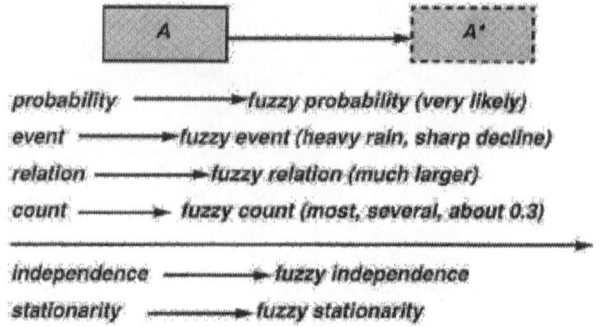

Fig. 6. f-Generalization (fuzzification). Fuzzification is a mode of generalization from crisp concepts to fuzzy concepts.

Information which is conveyed by propositions drawn from a natural language will be said to be perception-based (Fig. 5).

In my view, the most important shortcoming of standard probability theory is that it does not have the capability to process perception-based information. It does not have this capability principally because there is no mechanism in PT for (a) representing the meaning of perceptions and (b) computing and reasoning with representations of meaning.

To add this capability to standard probability theory, three stages of generalization are required.

The first stage is referred to as f-generalization (Zadeh, 1997). In this mode of generalization, a point or a set is replaced by a fuzzy set. f-generalization of standard probability theory, PT, leads to a generalized probability theory which will be denoted as PT+. In relation to PT, PT+ has the capability to deal with

1. fuzzy numbers, quantifiers and probabilities, e.g., about 0.7, most, not very likely,
2. fuzzy events, e.g., warm day,
3. fuzzy relations, e.g., much larger than,
4. fuzzy truths and fuzzy possibilities, e.g., very true, quite possible.

In addition, PT+ has the potential-as yet largely unrealized-to fuzzify such basic concepts as independence, stationarity and causality. A move in this direction would be a significant paradigmshift in probability theory.

The second stage is referred to as f.g-generalization (fuzzy granulation) (Zadeh, 1997). In this mode of generalization, a point or a set is replaced by a granulated fuzzy set (Fig. 6). For example, a function, f, is replaced by its fuzzy graph, f^* (Fig. 7). f.g-generalization of PT leads to a generalized probability theory denoted as PT++.

PT++ adds to PT+ further capabilities which derive from the use of granulation. They are, mainly

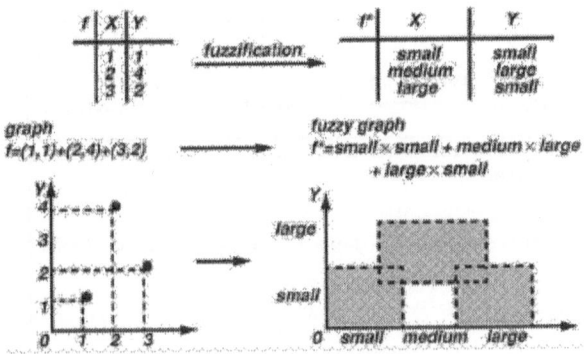

Fig. 7. Fuzzy graph of a function. A fuzzy graph is a generalization of the concept of a graph of a function.

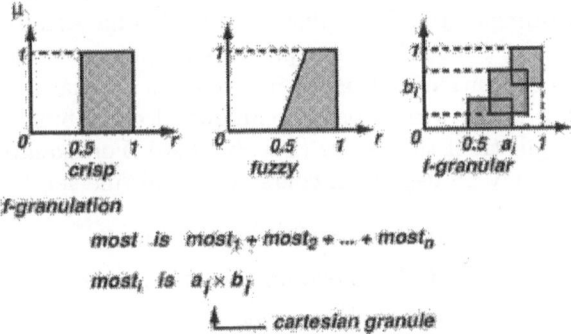

Fig. 8. Representation of most. Crisp, fuzzy and f-granular.

1. linguistic (granular) variables,
2. linguistic (granular) functions and relations,
3. fuzzy rule-sets and fuzzy graphs,
4. granular goals and constraints,
5. granular probability distributions.

As a simple example, representation of the membership function of the fuzzy quantifier *most* (Zadeh, 1983) in PT, PT+ and PT++ is shown in Fig. 8.

The third stage is referred to a p-generalization (perceptualization). In this mode of generalization, what is added to PT++ is the capability to process perception-based information through the use of the computational theory of perceptions (CTP) (Zadeh, 1999, 2000). p-generalization of PT leads to what will be referred to as perception-based probability theory (PT$_P$).

The capability of PT$_P$ to process perception-based information has an important implication. Specifically, it opens the door to a major enlargement

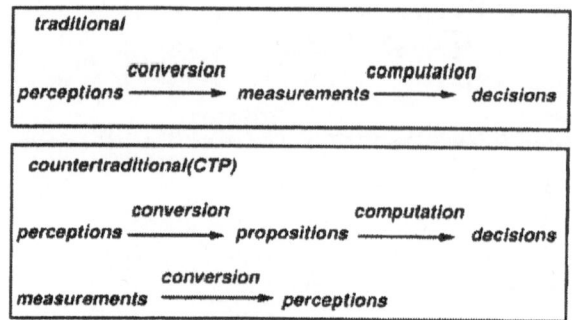

Fig. 9. Countertraditional conversion of measurements into perceptions. Traditionally, perceptions are converted into measurements.

of the role of natural languages in probability theory. As a simple illustration, instead of describing a probability distribution, P, analytically or numerically, as we normally do, P could be interpreted as a perception and described as a collection of propositions expressed in a natural language. A special case of such description is the widely used technique of describing a function via a collection of linguistic if-then rules (Zadeh, 1996). For example, the function shown in Fig. 7 may be described coarsely by the rule-set

$$f : \text{if } X \text{ is } small \text{ then } Y \text{ is } smal,$$
$$\text{if } X \text{ is } medium \text{ then } Y \text{ is } large,$$
$$\text{if } X \text{ is } large \text{ then } Y \text{ is } small,$$

with the understanding that the coarseness of granulation is a matter of choice.

In probability theory, as in other fields of science, it is a long-standing tradition to deal with perceptions by converting them into measurements. PT_p does not put this tradition aside. Rather, it adds to PT a countertraditional capability to convert measurements into perceptions, or to deal with perceptions directly, when conversion of perceptions into measurements is infeasible, unrealistic or counterproductive (Fig. 9).

There are three important points that are in need of clarification. First, when we allude to an enlarged role for natural languages in probability theory, what we have in mind is not a commonly used natural language but a subset which will be referred to as a precisiated natural language (PNL). In essence, PNL is a descriptive language which is intended to serve as a basis for representing the meaning of perceptions in a way that lends itself to computation. As will be seen later, PNL is a subset of a natural language which is equipped with constraint-centered semantics and is translatable into what is referred to as the generalized constraint language (GCL). At this point, it will suffice to observe that the descriptive power of PNL is much higher than that of the subset of a natural language which is translatable into predicate logic.

The second point is that in moving from measurements to perceptions, we move in the direction of lesser precision. The underlying rationale for this move is that precision carries a cost and that, in general, in any given situation there is a tolerance for imprecision that can be exploited to achieve tractability, robustness, lower cost and better rapport with reality.

The third point is that perceptions are more general than measurements and PT_p is more general that PT. Reflecting its greater generality, PT_p has a more complex mathematical structure than PT and is computationally more intensive. Thus, to exploit the capabilities of PT, it is necessary to have the capability to perform large volumes of computation at a low level of precision.

Perception-based probability theory goes far beyond standard probability theory both in spirit and in content. Full development of PT_p will be a long and tortuous process. In this perspective, my paper should be viewed as a sign pointing in a direction that departs from the deep-seated tradition of according more respect to numbers than to words.

Basically, perception-based probability theory may be regarded as the sum of standard probability theory and the computational theory of perceptions. The principal components of the computational theory of perceptions are (a) meaning representation and (b) reasoning. These components of CTP are discussed in the following sections.

2 The basics of perception-based probability theory; the concept of a generalized constraint

As was stated already, perception-based probability theory may be viewed as a p-generalization of standard probability theory. In the main, this generalization adds to PT the capability to operate on perception-based information through the use of the computational theory of perceptions. What follows is an informal precis of some of the basic concepts which underlie this theory.

To be able to compute and reason with perceptions, it is necessary to have a means of representing their meaning in a form that lends itself to computation. In CTP, this is done through the use of what is called constraint-centered semantics of natural languages (CSNL) (Zadeh, 1999).

A concept which plays a key role in CSNL is that of a generalized constraint (Zadeh, 1986). Introduction of this concept is motivated by the fact that conventional crisp constraints of the form $X \in C$, where X is a variable and C is a set, are insufficient to represent the meaning of perceptions.

A generalized constraint is, in effect, a family of constraints. An unconditional constraint on a variable X is represented as

$$X \text{ isr } R, \tag{1}$$

38

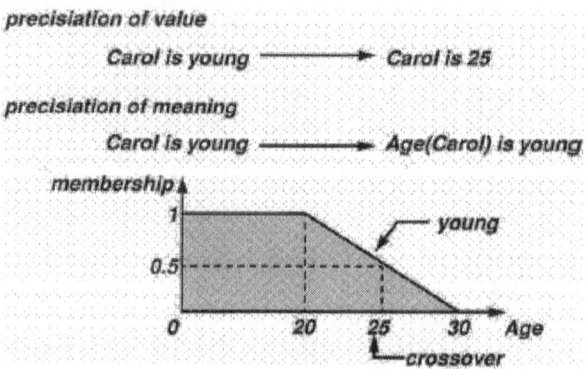

Fig. 10. Membership function of young (context-dependent). Two modes of pre-
cisiation.

where R is the constraining relation and is$r$, pronounced as ezar, is a variable
copula in which the discrete-valued variable r defines the way in which R
constrains X.

The principal constraints are the following:

$r :=$ equality constraint; $X = R$

$r : blank$ possibilistic constraint; X is R; R is the possibility
distribution of X (Zadeh; 1978; Dubois and Prade; 1988)

$r : v$ veristic constraint; X is$v$ R; R is the verity distribution of
X (Zadeh; 1999)

$r : p$ probabilistic constraint; X is$p$ R; R is the probability
distribution of X

$r : pv$ probability-value constraint; X ispv R; X is the probability
of a fuzzy event (Zadeh, 1968) and R is its value

$r : rs$ random set constraint; X isrs R; R is the fuzzy-set-valued
probability distribution of X

$r : fg$ fuzzy graph constraint; X isfg R; X is a function and R is
its fuzzy graph

$r : u$ usuallity constraint; X is$u$ R; means: usually (X is R).

As an illustration, the constraint

$$\text{Carol is } young$$

in which *young* is a fuzzy set with a membership function such as shown
in Fig. 10, is a possibilistic constraint on the variable $X : Age(Carol)$. This
constraint defines the possibility distribution of X through the relation

$$\text{Poss}\{X = u\} = \mu_{young}(u),$$

where u is a numerical value of *Age*; μ_{young} is the membership function of
young; and $\text{Poss}\{X = u\}$ is the possibility that Carol's age is u.

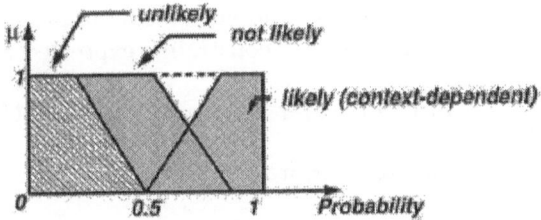

Fig. 11. Membership function of *likely* (context-dependent).

Fig. 11. Membership function of *likely* (context-dependent).

The veristic constraint

$$X \text{ isv } R \tag{2}$$

means that the verity (truth value) of the proposition $\{X = u\}$ is equal to the value of the verity distribution R at u. For example, in the proposition "Alan is half German, quarter French and quarter Italian", the verity of the proposition "Alan is German" is 0.5. It should be noted that the numbers 0.5 and 0.25 are not probabilities.

The probabilistic constraint

$$X \text{ is } p \ N(m; \sigma^2) \tag{3}$$

means that X is a normally distributed random variable with mean m and variance σ^2.

The proposition

$$p : \text{it is likely that Carol is young} \tag{4}$$

may be expressed as the probability-value constraint

$$\text{Prob}\{Age(\text{Carol}) \text{is } young\} \text{is } likely. \tag{5}$$

In this expression, the constrained variable is X :Prob$\{Age(\text{Carol})$ is $young\}$ and the constraint

$$X \text{ is } likely \tag{6}$$

is a possibilistic constraint in which *likely* is a fuzzy probability whose membership function is shown in Fig. 11.

In the random-set constraint, X is a fuzzy-set-valued random variable. Assuming that the values of X are fuzzy sets $\{Ai; i = 1, \ldots, n\}$ with respective probabilities $p1, \ldots, p_n$, the random-set constraint on X is expressed symbolically as

$$X \text{ isrs}(p_1 \backslash A_1 + \ldots + p_n \backslash A_n). \tag{7}$$

It should be noted that a random-set constraint may be viewed as a combination of (a) a probabilistic constraint, expressed as

$$X \text{ is } p(p_1 \backslash u_1 + \ldots + p_n \backslash u_n); \quad u_i \in U \tag{8}$$

and a possibilistic constraint expressed as

$$(X, Y) \text{ is } R, \tag{9}$$

where R is a fuzzy relation defined on $U \times V$, with membership function $\mu_R : U \times V \to [0; 1]$.

If A_i is a section of R, defined as in Zadeh (1997) by

$$\mu_{A_i}(v) = \mu_R(u_i, v), \tag{10}$$

then the constraint on Y is a random-set constraint expressed as

$$Y \text{ isrs } (p_1 \backslash A_1 + \ldots + p_n \backslash A_n). \tag{11}$$

Another point that should be noted is that the concept of a random-set constraint is closely related to the Dempster-Shafer theory of evidence (Dempster, 1967; Shafer, 1976) in which the focal sets are allowed to be fuzzy sets (Zadeh, 1979).

In the fuzzy-graph constraint

$$X \text{ isfg } R,$$

the constrained variable, X , is a function, f, and R is a fuzzy graph (Zadeh, 1997) which plays the role of a possibility distribution of X . More specifically, if $f : U \times V \to [0; 1]$ and $A_i;\ i = 1, \ldots, m$ and $B_j,\ j = 1, \ldots, n$, are, respectively, fuzzy granules in U and V (Fig. 12), then the fuzzy graph of f is the disfunction of cartesian products (granules) $U_i \times V_j$, expressed as

$$f^* = \sum_{i=1, j=1}^{m,n} U_i \times V_j,$$

with the understanding that the symbol $\sum$ should be interpreted as the union rather than as an arithmetic sum, and U_i and V_j take values in the sets $\{A_1, \ldots, A_m\}$ and $\{B_1, \ldots, B_n\}$, respectively.

A fuzzy graph of f may be viewed as an approximate representation of f. Usually, the granules A_i and B_j play the role of values of linguistic variables. Thus, in the case of the function shown in Fig. 7, its fuzzy graph may be expressed as

$$f^* = small \times small + medium \times large + large \times small. \tag{12}$$

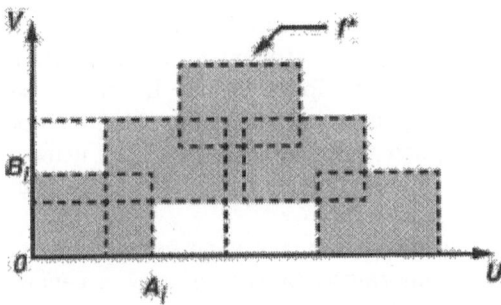

Fig. 12. Fuzzy-graph constraint. f^* is a fuzzy graph which is an approximate representation of f.

Equivalently, if f is written as $Y = f(X)$, then f^* may be expressed as the rule-set

$$f^* : \text{if } X \text{ is } small \text{ then } Y \text{ is } small,$$
$$\text{if } X \text{ is } medium \text{ then } Y \text{ is } large,$$
$$\text{if } X \text{ is } large \text{ then } Y \text{ is } small,$$

This rule-set may be interpreted as a description-in a natural language-of a perception of f.

The usuallity constraint is a special case of the probability-value constraint. Thus,

$$X \text{ isu } A, \tag{13}$$

should be interpreted as an abbreviation of

$$usually \ (X \text{ is } A), \tag{14}$$

which in turn may be interpreted as

$$\text{Prob}\{X \text{ is } A\} \text{ is } usually,$$

with *usually* playing the role of a fuzzy probability which is close to 1. In this sense, A is a *usual* value of X . More generally, A is a *usual* value of X if the fuzzy probability of the fuzzy event $\{X \text{ is } A\}$ is close to one and A has high specificity, that is, has a tight possibility distribution, with tightness being a context-dependent characteristic of a fuzzy set. It is important to note that, unlike the concept of the expected value, the usual value of a random variable is not uniquely determined by its probability distribution. What this means is that the usual value depends on the calibration of the context-dependent natural language predicates "close to one" and "high specificity".

The difference between the concepts of the expected and usual values goes to the heart of the difference between precise and imprecise probability theories. The expected value is precisely defined and unique. The usual value is context-dependent and hence is not unique. However, its definition is precise if the natural language predicates which occur in its definition are defined precisely by their membership functions. In this sense, the concept of the usual value has a flexibility that the expected value does not have. Furthermore, it may be argued that the concept of the usual value is closer to our intuitive perception of "expected value" than the concept of the expected value as it is defined in PT.

In the foregoing discussion, we have focused our attention on unconditional generalized constraints. More generally, a generalized constraint may be conditional, in which case it is expressed in a generic form as an if-then rule

$$\text{if } X \text{ isr } R \text{ then } Y \text{ iss } S, \tag{15}$$

or, equivalently, as

$$Y \text{ iss } S \text{ if } X \text{ isr } R. \tag{16}$$

Furthermore, a generalized constraint may be exception-qualified, in which case it is expressed as

$$X \text{ isr } R \text{ unless } Y \text{ iss } S. \tag{17}$$

A generalized rule-set is a collection of generalized if-then rules which collectively serve as an approximate representation of a function or a relation. Equivalently, a generalized rule-set may be viewed as a description of a perception of a function or a relation.

As an illustration, consider a function, $f : (U \times V) \to [0; 1]$, expressed as $Y = f(X)$, where U and V are the domains of X and Y, respectively. Assume that U and V are granulated, with the granules of U and V denoted, respectively, as A_i; $i = 1, \ldots, m$, and B_j, $j = 1, \ldots, n$. Then, a generic form of a generalized rule set may be expressed as

$$f^* : \{\text{if } X \text{ isr } U_i \text{ then } Y \text{ iss } V_j\} \quad i = 1, \ldots, m; \quad j = 1, \ldots, n, \tag{18}$$

where U_i and V_j take values in the sets $\{A_1, \ldots, A_m\}$ and $\{B_1, \ldots, B_n\}$, respectively. In this expression, f^* represents a fuzzy graph of f.

A concept which plays a key role in the computational theory of perceptions is that of the Generalized Constraint Language, GCL (Zadeh, 1999). Informally, GCL is a meaning-representation language in which the principal semantic elements are generalized constraints. The use of generalized constraints as its semantic elements makes a GCL a far more expressive language than conventional meaning-representation languages based on predicate logic.

3 Meaning-representation: constraint-centered semantics of natural languages

In perception-based probability theory, perceptions-and, in particular, perceptions of likelihood, dependency, count and variations in time and space-are described by propositions drawn from a natural language. To mechanize reasoning with perceptions, it is necessary to have a method of representing the meaning of propositions in a way that lends itself to computation. In the computational theory of perceptions, a system that is used for this purpose is called the constraint-centered semantics of natural language (CSNL) (Zadeh, 1999).

Meaning-representation is a central part of every logical system. Why, then, is it necessary to introduce a system that is significantly different from the many meaning representation methods that are in use? The reason has to do with the intrinsic imprecision of perceptions and, more particularly, with their f-granularity. It is this characteristic of perceptions that puts them well beyond the expressive power of conventional meaning-representation methods, most of which are based on predicate logic.

To illustrate, consider the following simple perceptions:

- Ann is much younger than Mary.
- A box contains black and white balls of various sizes. Most are large. Most of the large balls are black.
- Usually it is rather cold in San Francisco during the summer.
- It is very unlikely that there will be a significant increase in the price of oil in the near future.

Conventional meaning-representation methods do not have the capability to represent the meaning of such perceptions in a form that lends itself to computation.

A key idea which differentiates CSNL from conventional methods is that the meaning of a proposition, p, drawn from a natural language, is represented as a generalized constraint, with the understanding that the constrained variable and the constraining relation are, in general, implicit rather than explicit in p. For example, in the proposition

$$p : \text{likely that Kate is } young,$$

the constraint is possibilistic; the constrained variable is the probability that Kate is young; and the constraining relation is *likely*.

The principal ideas and assumptions which underlie CSNL may be summarized as follows:

1. Perceptions are described by propositions drawn from a natural language.
2. A proposition, p, may be viewed as an answer to a question. In general, the question is implicit and not unique. For example, the proposition "Carol is young" may be viewed as an answer to the question: "How old is Carol", or as the answer to "Who is young?"
3. A proposition is a carrier of information.
4. The meaning of a proposition, p, is represented as a generalized constraint which defines the information conveyed by p.

44

5. Meaning-representation is viewed as translation from a language into the GCL.

In CSNL, translation of a proposition, p, into GCL is equated to explicitation of the generalized constraint which represents the meaning of p. In symbols

$$p\frac{\text{translation}}{\text{explicitation}}X \text{ isr } R. \tag{19}$$

The right-hand member of this relation is referred to as a canonical form of p, written as $CF(p)$. Thus, the canonical form of p places in evidence (a) the constrained variable which, in general, is implicit in p; (b) the constraining relation, R; and (c) the copula variable r which defines the way in which R constrains X.

The canonical form of a question, q, may be expressed as

$$CF(q) : X \text{ isr } ?R \tag{20}$$

and read as "What is the generalized value of X ?"

Similarly, the canonical form of p, viewed as an answer to q, is expresed as

$$CF(p) : X \text{ isr } R \tag{21}$$

and reads "The generalized value of X isr R". As a simple illustration, if the question is "How old is Carol?", its canonical form is

$$CF(q) : Age(Carol) \text{ is } ?R. \tag{22}$$

Correspondingly, the canonical form of

$$p : \text{Carol is young} \tag{23}$$

is

$$CF(p) : Age(Carol) \text{ is } young. \tag{24}$$

If the answer to the question is

$$p : \text{it is likely that Carol is young} \tag{25}$$

then

$$CF(p) : \text{Prob}\{Age(Carol) \text{ is } young\} \text{ is } likely \tag{26}$$

More explicitly, if $Age(Carol)$ is a random variable with probability density g, then the probability measure (Zadeh, 1968) of the fuzzy event "Carol is young" may be expressed as

$$\int_0^{120} \mu_{young}(u)g(u)du, \tag{27}$$

where μ_{young} is the membership function of *young*. Thus, in this interpretation the constrained variable is the probability density g, and, as will be seen later, the membership function of the constraining relation is given by

$$\mu_R(g) = \mu_{likely}\left(\int_0^{120} \mu_{young}(u)g(u)du\right). \qquad (28)$$

A concept which plays an important role in CSNL is that of cardinality, that is, the count of elements in a fuzzy set (Zadeh, 1983; Ralescu, 1995; Haffek, 1998). Basically, there are two ways in which cardinality can be defined: (a) crisp cardinality and (b) fuzzy cardinality (Zadeh, 1983; Ralescu et al., 1995; Ralescu, 1995). In the case of (a), the count of elements in a fuzzy set is a crisp number; in the case of (b) it is a fuzzy number. For our purposes, it will suffice to restrict our attention to the case where a fuzzy set is defined on a finite set and is associated with a crisp count of its elements.

More specifically, consider a fuzzy set A defined on a finite set $U = \{u_1, \ldots, u_n\}$ through its membership function $\mu_A : U \to [0;1]$. The sigma-count of A is defined as

$$\sum Count(A) = \sum_{i=1}^{n} \mu_A(u_i). \qquad (29)$$

If A and B are fuzzy sets defined on U, then the relative sigma-count, $\sum Count(A = B)$, is defined as

$$\sum Count(A/B) = \frac{\sum_{i=1}^{n} \mu_A(u_i) \wedge \mu_B(u_i)}{\sum_{i=1}^{n} \mu_B(u_i)}, \qquad (30)$$

where $\wedge = min$, and summations are arithmetic.

As a simple illustration, consider the perception

$$p : \text{most Swedes are tall.}$$

In this case, the canonical form of p may be expressed as

$$CF(p) : \sum Count(tall . Swedes \: / \: Swedes) \text{ is } \frac{1}{n}\sum_{i=1}^{n} \mu_{tall . Swede}(u_i), \quad (31)$$

where u_i is the height of the ith Swede and $\mu_{tall . Swede}(u_i)$ is the grade of membership of the ith Swede in the fuzzy set of tall Swedes.

In a general setting, how can a given proposition, p, be expressed in its canonical form? A framework for translation of propositions drawn from a natural language into GCL is partially provided by the conceptual structure of test-score semantics (Zadeh, 1981). In this semantics, X and R are defined by procedures which act on an explanatory database, ED, with ED playing the role of a collection of possible worlds in possible world semantics (Cresswell, 1973). As a very simple illustration, consider the proposition (Zadeh, 1999)

$$p : \text{Carol lives in a small city near San Francisco}$$

and assume that the explanatory database consists of three relations:

$$ED = POPULATION[Name; Residence] \tag{32}$$
$$+ SMALL[City; \mu] + NEAR[City1; City2; \mu].$$

In this case,

$$X = Residence(Carol) =_{Residence} POPULATION[Name = Carol], \tag{33}$$
$$R = SMALL[City;] \cap_{City1} NEAR[City2 = SanFrancisco]. \tag{34}$$

In R, the first constituent is the fuzzy set of small cities; the second constituent is the fuzzy set of cities which are near San Francisco; and $\cap$ denotes the intersection of these sets. Left subscripts denote projections, as defined in Zadeh (1981).

There are many issues relating to meaning-representation of perception-based information which go beyond the scope of the present paper. The brief outline presented in this section is sufficient for our purposes. In the following section, our attention will be focused on the basic problem of reasoning based on generalized constraint propagation. The method which will be outlined contains as a special case a basic idea suggested in an early paper of Good (1962). A related idea was employed in Zadeh (1955).

4 Reasoning based on propagation of generalized constraints

One of the basic problems in probability theory is that of computation of the probability of a given event from a body of knowledge which consists of information about the relevant functions, relations, counts, dependencies and probabilities of related events.

As was alluded to earlier, in many cases the available information is a mixture of measurements and perceptions. Standard probability theory provides a vast array of tools for dealing with measurement-based information. But what is not provided is a machinery for dealing with information which is perception-based. This limitation of PT is exemplified by the following elementary problems-problems in which information is perception-based.

1. X is a normally distributed random variable with small mean and small variance. Y is much larger than X. What is the probability that Y is neither small nor large?

2. Most Swedes are tall.

Most Swedes are blond.

What is the probability that a Swede picked at random is tall and blond?

3. Consider a perception-valued times series

$$T = \{t_1, t_2, t_3, \ldots\},$$

in which the t_i's are perceptions of, say temperature, e.g., warm, very warm, cold,.... For simplicity, assume that the t_i's are independent and identically distributed. Furthermore, assume that the t_i's range over a finite set of linguistic values, $A_1; A_2, \ldots, A_n$, with respective probabilities $P_1, \ldots, P_n$. What is the average value of T?

To be able to compute with perceptions, it is necessary, as was stressed already, to have a mechanism for representing their meaning in a form that lends itself to computation. In the computational theory of perceptions, this purpose is served by the constraint-centered semantics of natural languages. Through the use of CSNL, propositions drawn from a natural language are translated into the GCL.

The second stage of computation involves generalized constraint propagation from premises to conclusions. Restricted versions of constraint propagation are considered in Zadeh (1979), Bowen et al. (1992), Dubois et al. (1993), Katai et al. (1992) and Yager (1989). The main steps in generalized constraint propagation are summarized in the following. As a preliminary, a simple example is analyzed.

Assume that the premises consist of two perceptions:

p1 : most Swedes are tall,

p2 : most Swedes are blond.

and the question, q, is: What fraction of Swedes are tall and blond? This fraction, then, will be the linguistic value of the probability that a Swede picked at random is tall and blond.

To answer the question, we first convert $p_1; p_2$ and q into their canonical forms:

$$CF(p_1) : \sum Count(tall . Swedes/Swedes) \text{ is } most, \tag{35}$$

$$CF(p_2) : \sum Count(blond . Swedes/Swedes) \text{ is } most, \tag{36}$$

$$CF(q) : \sum Count(tall \cap blond . Swedes/Swedes) \text{ is } ?Q, \tag{37}$$

where Q is the desired fraction.

Next, we employ the identity (Zadeh, 1983)

$$\sum Count(A \cap B) + \sum Count(A \cup B) = \sum Count(A) + \sum Count(B), \tag{38}$$

in which A and B are arbitrary fuzzy sets. Fromth is identity, we can readily deduce that

$$\sum Count(A) + \sum Count(B) - 1 \leq \sum Count(A \cap B) \qquad (39)$$
$$\leq \min(\sum Count(A), \sum Count(B)),$$

with the understanding that the lower bound is constrained to lie in the interval $[0; 1]$. It should be noted that the identity in question is a generalization of the basic identity for probability measures

$$P(A \cap B) + P(A \cup B) = P(A) + P(B). \qquad (40)$$

Using the information conveyed by canonical forms, we obtain the bounds

$$2most - 1 \leq \sum Count(tall \cap blond . Swedes/Swedes) \leq most, \qquad (41)$$

which may be expressed equivalently as

$$\sum Count(tall \cap blond . Swedes/Swedes) \text{ is } \leq most \cap \geq (2most - 1). \quad (42)$$

Now

$$\leq most = [0; 1] \qquad (43)$$

and

$$\geq (2most - 1) = 2most - 1, \qquad (44)$$

in virtue of monotonicity of $most$ (Zadeh, 1999).

Consequently,

$$\sum Count(tall \cap blond . Swedes/Swedes) \text{ is } 2most - 1 \qquad (45)$$

and hence the answer to the question is

$$a : (2most - 1)SwedesT \text{ are } tall \text{ and } blond. \qquad (46)$$

In a more general setting, the principal elements of the reasoning process are the following.

1. Question (query), q. The canonical form of q is assumed to be

$$X \text{ isr } ?Q. \qquad (47)$$

2. Premises. The collection of premises expressed in a natural language constitutes the initial data set (IDS).

3. Additional premises which are needed to arrive at an answer to q. These premises constitute the external data set (EDS). Addition of EDS to IDS results in what is referred to as the augmented data set (IDS+).

Example. Assume that the initial data set consists of the propositions

$p1$: Carol lives near Berkeley,

$p2$: Pat lives near Palo Alto.

Suppose that the question is: How far is Carol from Pat? The external data set in this case consists of the proposition

distance between Berkeley and Palo Alto is approximately 45 miles. (48)

4. Through the use of CSNL, propositions in IDS+ are translated into the GCL. The resulting collection of generalized constraints is referred to as the augmented initial constraint set ICS+.
5. With the generalized constraints in ICS+ serving as antecedent constraints, the rules which govern generalized constraint propagation in CTP are applied to ICS+, with the goal of deducing a set of generalized constraints, referred to as the terminal constraint set, which collectively provide the information which is needed to compute q.

The rules governing generalized constraint propagation in the computational theory of perceptions coincide with the rules of inference in fuzzy logic (Zadeh, 1999, 2000). In general, the chains of inference in CTP are short because of the intrinsic imprecision of perceptions. The shortness of chains of inference greatly simplifies what would otherwise be a complex problem, namely, the problem of selection of rules which should be applied in succession to arrive at the terminal constraint set. This basic problem plays a central role in theorem proving in the context of standard logical systems (Fikes and Nilsson, 1971).

6. The generalized constraints in the terminal constraint set are re-translated into a natural language, leading to the terminal data set. This set serves as the answer to the posed question. The process of re-translation is referred to as linguistic approximation (Pedrycz and Gomide, 1998). Retranslation will not be addressed in this paper.

The basic rules which govern generalized constraint propagation are of the general form

$$
\begin{array}{c}
p_1 \\
p_2 \\
\vdots \\
\dfrac{p_k}{p_{k+1}}
\end{array}
\tag{49}
$$

where $p_1, \ldots, p_k$ are the premises and p_{k+1} is the conclusion. Generally, $k = 1$ or 2.

In a generic form, the basic constraint-propagation rules in CTP are expressed as follows (Zadeh, 1999):

1. *Conffunctive rule* 1:

$$\frac{\begin{array}{c} X \text{ isr } R \\ X \text{ iss } S \end{array}}{X \text{ ist } T.} \tag{50}$$

The different symbols $r; s, t$ in constraint copulas signify that the constraints need not be of the same type.

2. *Conffunctive rule* 2:

$$\frac{\begin{array}{c} X \text{ isr } R \\ Y \text{ iss } S \end{array}}{(X,Y) \text{ ist } T.} \tag{51}$$

3. *Disffunctive rule* 1:

$$\frac{\begin{array}{c} X \text{ isr } R \\ \text{or } X \text{ iss } S \end{array}}{X \text{ ist } T.} \tag{52}$$

4. *Disffunctive rule* 2:

$$\frac{\begin{array}{c} X \text{ isr } R \\ \text{or } Y \text{ iss } S \end{array}}{(X,Y) \text{ ist } T.} \tag{53}$$

5. *Proffective rule:*

$$\frac{(X,Y) \text{ isr } R}{(Y) \text{ iss } S.} \tag{54}$$

6. *Surffective rule:*

$$\frac{X \text{ isr } R}{(X,Y) \text{ iss } S.} \tag{55}$$

7. *Inversive rule:*

$$\frac{f(X) \text{ isr } R}{X \text{ iss } S,} \tag{56}$$

where $f(X)$ is a function of X.

From these basic rules the following frequently used rules may be derived:

8. *Compositional rule:*

$$\frac{\begin{array}{c} X \text{ isr } R \\ \text{or } (X,Y) \text{ iss } S \end{array}}{Y \text{ ist } T.} \tag{57}$$

9. *Generalized extension principle*:

$$\frac{f(X) \text{ isr } R}{g(X) \text{ iss } S,} \tag{58}$$

where f and g are given functions. The generalized extension principle is the principal rule of inference in fuzzy logic.

The generic rules lead to specialized rules for various types of constraints. In particular, for possibilistic constraints we have, for example (Pedrycz and Gomide, 1998)

Conffunctive rule 1:

$$\frac{\begin{array}{c} X \text{ is } R \\ X \text{ is } S \end{array}}{X \text{ is } R \cap S,} \tag{59}$$

where R and S are fuzzy sets and $R \cap S$ is their intersection.

Compositional rule:

$$\frac{\begin{array}{c} X \quad \text{is } R \\ (X, Y) \text{ is } S \end{array}}{Y \text{ is } R \bullet S,} \tag{60}$$

where $R \bullet S$ is the composition of R and S. If conjunction and disjunction are identified with min and max, respectively, then

$$\mu_{R \bullet S}(v) = \max_{u}(\min(\mu_R(u); \mu_S(u; v))),$$

where μ_R and μ_S are the membership functions of R and S.

Generalized extension principle (Fig. 13):

$$\frac{f(X) \text{ isr } R}{g(X) \text{ is } g\left(f^{-1}(R)\right),} \tag{61}$$

where

$$\mu_{g(f^{--1}(R))}(v) = \max_{u|v=g(u)} \mu_R\left(f(u).\right) \tag{62}$$

Compositional rule for probabilistic constraints (Bayes' rule):

$$\frac{\begin{array}{c} X \quad \text{isp } R \\ X|Y \text{ isp } S \end{array}}{Y \text{ isp } R \bullet S,} \tag{63}$$

where $Y|X$ denotes Y conditioned on X, and $R \bullet S$ is the composition of the probability distributions R and S.

Compositional rule for probabilistic and possibilistic constraints (random-set constraint):

$$\frac{\begin{array}{c} X \quad \text{isp } R \\ (X, Y) \text{ is } S \end{array}}{Y \text{ isrs } T,} \tag{64}$$

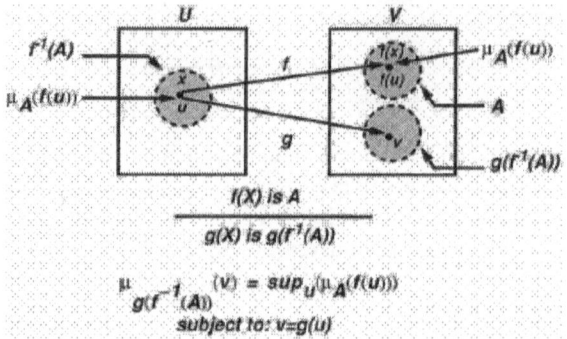

Fig. 13. Generalized extension principle. Constaint on $f(X)$ induces on $g(X)$.

where T is a random set. As was stated at an earlier point, if X takes values in a finite set $\{u_1, \ldots, u_n\}$ with respective probabilities $p_1, \ldots, p_n$, then the constraint X isp R may be expressed compactly as

$$X \text{ isp } \left(\sum_{i=1}^{n} p_i \backslash u_i \right). \tag{65}$$

When X takes a value u_i, the possibilistic constraint $(X;Y)$ is S induces a constraint on Y which is given by

$$Y \text{ is } S_i, \tag{66}$$

where S_i is a fuzzy set defined by

$$S_i = S(u_i, Y). \tag{67}$$

From this it follows that when X takes the values $u_1, \ldots, u_n$ with respective probabilities $p_1, \ldots, p_n$, the fuzzy-set-valued probability distribution of Y may be expressed as

$$Y \text{ isp } \left(\sum_{i=1}^{n} p_i \backslash S_i \right). \tag{68}$$

This fuzzy-set-valued probability distribution defines the random set T in the random-set constraint

$$Y \text{ isrs } T. \tag{69}$$

Conffunctive rule for random set constraints: For the special case in which R and S in the generic conjunctive rule are random fuzzy sets as defined above, the rule assumes a more specific form:

$$\frac{\begin{array}{c} X \text{ isr}s \sum_{i=1}^{m} p_i \backslash R_i \\ X \text{ isr}s \sum_{i=1}^{n} q_j \backslash S_j \end{array}}{X \text{ isr}s \sum_{i=1,j=1}^{m,n} p_i q_j \backslash (R_i \cap S_j)} \tag{70}$$

In this rule, R_i and S_i are assumed to be fuzzy sets. When R_i and S_i are crisp sets, the rule reduces to the Dempster rule of combination of evidence (Dempster, 1967; Shafer, 1976). An extension of Dempster's rule to fuzzy sets was described in a paper dealing with fuzzy information granularity (Zadeh, 1979). It should be noted that in (4.37) the right-hand member is not normalized, as it is in the Dempster-Shafer theory (Strat, 1992).

The few simple examples discussed above demonstrate that there are many ways in which generic rules can be specialized, with each specialization leading to a distinct theory in its own right. For example, possibilistic constraints lead to possibility theory (Zadeh, 1978; Dubois and Prade, 1988); probabilistic constraints lead to probability theory; and random-set constraints lead to the Dempster-Shafer theory of evidence. In combination, these and other specialized rules of generalized constraint propagation provide the machinery that is needed for a mechanization of reasoning processes in the logic of perceptions and, more particularly, in a perception-based theory of probabilistic reasoning with imprecise probabilities.

As an illustration, let us consider a simple problem that was stated earlier- a typical problem which arises in situations in which the decision-relevant information is perception-based. Given the perception: Usually Robert returns from work at about 6 p.m.; the question is: What is the probability that he is home at 6:30 p.m.? An applicable constraint-propagation rule in this case is the generalized extension principle. More specifically, let g denote the probability density of the time at which Robert returns from work. The initial data set is the proposition

p : usually Robert returns from work at about 6 p.m.

This proposition may be expressed as the usuallity constraint

$$X \text{ is}u\ 6^*, \tag{71}$$

where 6^* is an abbreviation for "about 6 p.m.", and X is the time at which Robert returns from work. Equivalently, the constraint in question may be expressed as

$$p : \text{Prob}\{X \text{ is } 6^*\} \text{ is } usually. \tag{72}$$

54

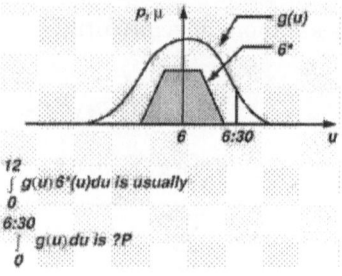

Fig. 14. Example of an electronically included eps figure

Using the definition of the probability measure of a fuzzy event (Zadeh, 1968), the constraint on g may be expressed as

$$\int_0^{12} g(u)\,\mu_{6^*}(u)\,du \text{ is } usually,\tag{73}$$

where $\mu_{6^*}(u)$ is the membership function of 6* (Fig. 14).

Let $P(g)$ denote the probability that Robert is at home at 6:30 p.m. This probability would be a number if g were known. In our case, information about g is conveyed by the given usuallity constraint. This constraint defines the possibility distribution of g as a functional:

$$\mu(g) = \mu_{usually}\left(\int_0^{12} g(u)\,\mu_{6^*}(u)\,du\right).\tag{74}$$

In terms of g, the probability that Robert is home at 6:30 p.m. may be written as a functional:

$$P(g) = \int_0^{6:30} g(u)\,du.\tag{75}$$

The generalized extension principle reduces computation of the possibility distribution of P to the solution of the variational problem

$$\mu_p(v) = \max_g \left(\mu_{usually}\left(\int_0^{12} g(u)\,\mu_{6^*}(u)\,du\right)\right)\tag{76}$$

subject to

$$v = \int_0^{6:30} g(u)\,du.$$

The reduction of inference to solution of constrained variational problems is a basic feature of fuzzy logic (Zadeh, 1979).

Solution of variational problems of form (4.43) may be simplified by a discretization of g. Thus, if u is assumed to take values in a finite set $U = \{u_1, \ldots, u_n\}$, and the respective probabilities are $p_1, \ldots, p_n$, then the variational problem(4.43) reduces to the nonlinear program

$$\mu_p(v) = \max_p \left(\mu_{usually} \left(\sum_{i=1}^{n} p_i \mu_{6^*}(u) \right) \right) \tag{77}$$

subject to

$$v = \sum_{j=1}^{m} P_j,$$

$$0 \leq P_j \leq 1,$$

$$\sum_{i=1}^{n} P_i = 1,$$

where $p = (p_1, \ldots, p_n)$, and m is such that $u_m = 6:30$.

In general, probabilities serve as a basis for making a rational decision. As an illustration, assume that I want to call Robert at home at 6:30 p.m. and have to decide on whether I should call him person-to-person or station-to-station. Assume that we have solved the variational problem(4.43) and have in hand the value of P defined by its membership function $\mu_P(v)$. Furthermore, assume that the costs of person-to-person and station-to-station calls are a and b, respectively.

Then the expected cost of a person-to-person call is

$$A = aP,$$

while that of a station-to-station call is

$$B = b,$$

where A is a fuzzy number defined by (Kaufmann and Gupta, 1985)

$$\mu_A(v) = a\mu_P(v).$$

More generally, if X is a random variable taking values in the set of numbers $U = \{a_1, \ldots, a_n\}$ with respective imprecise (fuzzy) probabilities $P_1, \ldots, P_n$, then the expected value of X is the fuzzy number (Zadeh, 1975; Kruse and Meyer, 1987)

$$E(X) = \sum_{i=1}^{n} a_i P_i. \tag{78}$$

The membership function of $E(X)$ may be computed through the use of fuzzy arithmetic (Kaufmann and Gupta, 1985; Mares, 1994). More specifically, if the membership functions of P_i are μ_i, then the membership function of $E(X)$ is given by the solution of the variational problem

$$\mu_{E(X)}(v) = max_{u_1,\dots,u}(\mu_{P_1}(u_1) \wedge \dots \wedge \mu_{P_n}(u_n)) \tag{79}$$

subject to the constraints

$$0 \leq u_i \leq 1,$$
$$\sum_{i=1}^{n} u_i = 1,$$
$$v = \sum_{i=1}^{n} a_i u_i.$$

Returning to our discussion of the Robert example, if we employ a generalized version of the principle of maximization of expected utility to decide on how to place the call, then the problem reduces to that of ranking the fuzzy numbers A and B. The problem of ranking of fuzzy numbers has received considerable attention in the literature (see Pedrycz and Gomide, 1998), and a number of ranking algorithms have been described.

Our discussion of the Robert example is aimed at highlighting some of the principal facets of the perception-based approach to reasoning with imprecise probabilities. The key point is that reasoning with perception-based information may be reduced to solution of variational problems. In general, the problems are computationally intensive, even for simple examples, but well within the capabilities of desktop computers. Eventually, novel methods of computation involving neural computing, evolutionary computing, molecular computing or quantum computing may turn out to be effective in computing with imprecise probabilities in the context of perception-based information.

As a further illustration of reasoning with perception-based information, it is instructive to consider a perception-based version of a basic problem in probability theory.

Let X and Y be random variables in U and V, respectively. Let f be a mapping from U to V. The basic problem is: Given the probability distribution of $X, P(X)$, what is the probability distribution of Y?

In the perception-based version of this problem it is assumed that what we know are perceptions of f and $P(X)$, denoted as f^* and $P^*(X)$, respectively. More specifically, we assume that X and f are granular (linguistic) variables and f^* is described by a collection of granular (linguistic) if-then rules:

$$f^* : \{\text{if } X \text{ is } A_i \text{ then } Y \text{ is } B_i\}, \quad i = 1, \dots, m, \tag{80}$$

where A_i and B_i are granules of X and Y, respectively (Fig. 12). Equivalently, f^* may be expressed as a fuzzy graph

$$f^* = \sum_{i=1}^{m} A_i \times B_i, \tag{81}$$

where $A_i \times B_i$ is a cartesian granule in $U \times V$. Furthermore, we assume that the perception of $P(X)$ is described as

$$P^*(X) \text{ is } \sum_{j=1}^{n} p_j \backslash C_j, \tag{82}$$

where the C_j are granules of U, and

$$p_j = \text{Prob}\{X \text{ is } C_j\}. \tag{83}$$

Now, let $f^*(C_j)$ denote the image of C_j . Then, application of the extension principle yields

$$f^*(C_j) = \sum_{i=1}^{m} m_{ij} \wedge B_i, \tag{84}$$

where the matching coeffcient, m_{ij} , is given by

$$m_{ij} = \sup(A_i \cap C_j), \tag{85}$$

with the understanding that

$$\sup(A_i \cap C_j) = \sup_{u}(\mu_{A_i}(u) \wedge \mu_{C_j}(u)), \tag{86}$$

where $u \in U$ and μ_{A_i} and μ_{C_j} are the membership functions of A_i and C_j, respectively.

In terms of $f^*(C_j)$, the probability distribution of Y may be expressed as

$$P^*(Y) \text{ is } \sum_{j=1}^{n} p_j \backslash f^*(C_j) \tag{87}$$

or, more explicitly, as

$$P^*(Y) \text{ is } \sum_{j=1}^{n} p_j \backslash \left(\sum_{i} m_{ij} \wedge B_i \right). \tag{88}$$

What these examples show is that computation with perception-based functions and probability distribution is both more general and more complex than computation with their measurement-based counterparts.

5 Concluding remarks

The perception-based theory of probabilistic reasoning which is outlined in this paper may be viewed as an attempt to add to probability theory a significant capability-a capability to operate on information which is perception-based. It is this capability that makes it possible for humans to perform a wide variety of physical and mental tasks without any measurements and any computations.

Perceptions are intrinsically imprecise, reelecting a fundamental limitation on the cognitive ability of humans to resolve detail and store information. Imprecision of perceptions places them well beyond the scope of existing meaning-representation and deductive systems. In this paper, a recently developed computational theory of perceptions is used for this purpose. Applicability of this theory depends in an essential way on the ability of modern computers to perform complex computations at a low cost and high reliability.

Natural languages may be viewed as systems for describing perceptions. Thus, to be able to operate on perceptions, it is necessary to have a means of representing the meaning of propositions drawn from a natural language in a form that lends itself to computation. In this paper, the so-called constraint-centered semantics of natural languages serves this purpose.

A conclusion which emerges from these observations is that to enable probability theory to deal with perceptions, it is necessary to add to it concepts and techniques drawn from semantics of natural languages. Without these concepts and techniques, there are many situations in which probability theory cannot answer questions that arise when everyday decisions have to be made on the basis of perception-based information. Examples of such questions are given in this paper.

A related point is that, in perception-based theory of probabilistic reasoning, imprecision probabilities. In particular, imprecision can occur on the level of events, counts and relations. More basically, it can occur on the level of definition of such basic concepts as random variable, causality, independence and stationarity. The concept of precisiated natural language may suggest a way of generalizing these and related concepts in a way that would enhance their expressiveness and operationality.

The confluence of probability theory and the computational theory of perceptions opens the door to a radical enlargement of the role of natural languages in probability theory. The theory outlined in this paper is merely a first step in this direction. Many further steps will have to be taken to develop the theory more fully. This will happen because it is becoming increasingly clear that real-world applications of probability theory require the capability to process perception-based information as a basis for rational decisions in an environment of imprecision, uncertainty and partial truth.

References

1. Barsalou, L.W., 1999. Perceptual symbol systems. Behav. Brain Sci. 22, 577-660.
2. Bowen, J., Lai, R., Bahler, D., 1992. Fuzzy semantics and fuzzy constraint networks. Proceedings of the First IEEE Conference on Fuzzy Systems, San Francisco, pp. 1009-1016.
3. Cresswell, M.J., 1973. Logic and Languages. Methuen, London.
4. Dempster, A.P., 1967. Upper and lower probabilities induced by a multivalued mapping. Ann. Math. Statist. 38, 325-339.
5. Dubois, D., Prade, H., 1988. Possibility Theory. Plenum Press, New York.
6. Dubois, D., Fargier, H., Prade, H., 1993. The calculus of fuzzy restrictions as a basis for flexible constraint satisfaction. Proceedings of the Second IEEE International Conference on Fuzzy Systems, San Francisco, pp. 1131-1136.
7. Fikes, R.E., Nilsson, N.J., 1971. STRIPS: a new approach to the application of theorem proving to problem solving. Artif. Intell. 2, 189-208.
8. Good, I.J., 1962. Subjective probability as the measure of a non-measurable set. In: Nagel, E., Suppes, P., Tarski, A. (Eds.), Logic, Methodology and Philosophy of Science. Stanford University Press, Stanford, pp. 319-329.
9. Grabisch, M., Nguyen, H.T., Walker, E.A., 1995. Fundamentals of Uncertainty Calculi with Applications to Fuzzy Inference. Kluwer, Dordrecht.
10. Haffek, P., 1998. Metamathematics of Fuzzy Logic. Kluwer, Dordrecht.
11. Katai, O., Matsubara, S., Masuichi, H., Ida, M., et al., 1992. Synergetic computation for constraint satisfaction problems involving continuous and fuzzy variables by using occam. In: Noguchi, S., Umeo, H. (Eds.), Transputer/Occam, Proceedings of the Fourth Transputer/Occam International Conference. IOS Press, Amsterdam, pp. 146-160.
12. Kaufmann, A., Gupta, M.M., 1985. Introduction to Fuzzy Arithmetic: Theory and Applications. Von Nostrand, New York.
13. Klir, G., Folger, T.A., 1988. Fuzzy Sets, Uncertainty, and Information. Prentice-Hall, Englewood CliFs, Nff.
14. Klir, G., Yuan, B., 1995. Fuzzy Sets and Fuzzy Logic. Prentice-Hall, Englewood Cliffs, Nff.
15. Krause, P., Clark, D., 1993. Representing Uncertain Knowledge. Kluwer, Dordrecht.
16. Kruse, R., Meyer, D., 1987. Statistics with Vague Data. Kluwer, Dordrecht.
17. Kuipers, B.J., 1994. Qualitative Reasoning. MIT Press, Cambridge.
18. Mares, M., 1994. Computation over Fuzzy Quantities. CRC Press, Boca Raton.
19. Nguyen, H.T., Kreinovich, V., Wu, B., 1999. Fuzzy/probability-fractal/smooth. Internat. J. Uncertainty Fuzziness Knowledge-based Systems 7, 363-370.
20. Novak, V., 1991. Fuzzy logic, fuzzy sets, and natural languages. Int. J. Gen. Systems 20, 83-97.
21. Novak, V., Ramik, M., Cerny, M., Nekola, J. (Eds.), 1992. Fuzzy Approach to Reasoning and Decision-Making. Kluwer, Boston.
22. Pedrycz, W., Gomide, F., 1998. Introduction to Fuzzy Sets. MIT Press, Cambridge.
23. Ralescu, A.L., Bouchon-Meunier, B., Ralescu, D.A., 1995. Combining fuzzy quantifiers. Proceedings of the International Conference of CFSA/IFIS/SOFT'95 on Fuzzy Theory and Applications, Taipei, Taiwan.

24. Ralescu, D.A., 1995. Cardinality, quantifiers and the aggregation of fuzzy criteria. Fuzzy Sets and Systems 69, 355-365.
25. Reghis, M., Roventa, E., 1998. Classical and Fuzzy Concepts in Mathematical Logic and Applications. CRC Press, Boca Raton.
26. Shafer, G., 1976. A Mathematical Theory of Evidence. Princeton University Press, Princeton.
27. Shen, Q., Leitch, R., 1992. Combining qualitative simulation and fuzzy sets. In: Faltings, B., Struss, P. (Eds.), Recent Advances in Qualitative Physics. MIT Press, Cambridge.
28. Smithson, M., 1989. Ignorance and Uncertainty. Springer, Berlin.
29. Strat, T.M., 1992. Representative applications of fuzzy measure theory: decision analysis using belief functions. In: Wang, Z., Klir, G. J. (Eds.), Fuzzy Measure Theory. Plenum Press, New York, pp. 285-310.
30. Thomas, S.F., 1995. Fuzziness and Probability. ACG Press, Wichita, KS.
31. Vallee, R., 1995. Cognition et Systeme. l'Interdisciplinaire Systeme(s), Paris.
32. Walley, P., 1991. Statistical Reasoning with Imprecise Probabilities. Chapman & Hall, London.
33. Wang, Z., Klir, G., 1992. Fuzzy Measure Theory. Plenum Press, New York.
34. Yager, R.R., 1989. Some extensions of constraint propagation of label sets. Internat. J. Approx. Reasoning 3, 417-435.
35. Zadeh, L.A., 1955. General filters for separation of signals and noise. Proceedings of the Symposium on Information Networks, Polytechnic Institute of Brooklyn, New York, pp. 31-49.
36. Zadeh, L.A., 1968. Probability measures of fuzzy events. J. Math. Anal. Appl. 23, 421-427.
37. Zadeh, L.A., 1973. Outline of a new approach to the analysis of complex systems and decision processes. IEEE Trans. Systems Man Cybernet. SMC-3, 28-44.
38. Zadeh, L.A., 1975. The concept of a linguistic variable and its application to approximate reasoning. Part I: Inf. Sci. 8, 199-249; Part II: Inf. Sci. 8, 301-357; Part III: Inf. Sci. 9, 43-80.
39. Zadeh, L.A., 1978. Fuzzy sets as a basis for a theory of possibility. Fuzzy Sets and Systems 1, 3-28.
40. Zadeh, L.A., 1979. Fuzzy sets and information granularity. In: Gupta, M., Ragade, R., Yager, R. (Eds.), Advances in Fuzzy Set Theory and Applications. North-Holland, Amsterdam, pp. 3-18.
41. Zadeh, L.A., 1981. Test-score semantics for natural languages and meaning representation via PRUF. In: Rieger, B. (Ed.), Empirical Semantics. Brockmeyer, Bochum, W. Germany, pp. 281-349.
42. Zadeh, L.A., 1983. A computational approach to fuzzy quantifiers in natural languages. Comput. Math. 9, 149-184.
43. Zadeh, L.A., 1986. Outline of a computational approach to meaning and knowledge representation based on a concept of a generalized assignment statement. In: Thoma, M., Wyner, A. (Eds.), Proceedings of the International Seminar on Artificial Intelligence and Man-Machine Systems. Springer, Heidelberg, pp. 198-211.
44. Zadeh, L.A., 1995. Probability theory and fuzzy logic are complementary rather than competitive. Technometrics 37, 271-276.
45. Zadeh, L.A., 1996. Fuzzy logic and the calculi of fuzzy rules and fuzzy graphs: a precis. Multivalued Logic 1, 1-38.

46. Zadeh, L.A., 1997. Toward a theory of fuzzy information granulation and its centrality in human reasoning and fuzzy logic. Fuzzy Sets and Systems 90, 111-127.
47. Zadeh, L.A., 1999. From computing with numbers to computing with words-from manipulation of measurements to manipulation of perceptions. IEEE Trans. Circuits Systems 45, 105-119.
48. Zadeh, L.A., 2000. Outline of a computational theory of perceptions based on computing with words. In:
49. Sinha, N.K., Gupta, M.M. (Eds.), Soft Computing and Intelligent Systems: Theory and Applications. Academic Press, London, pp. 3-22.

Soft Methods in Probability:

Fundamentals

Independence and Conditioning in a Connectivistic Fuzzy Logic Framework

Joseph Aguilar-Martin

LEA-SICA. ESAII, Universitat Politecnica de Catalunya ETSEIT, Rambla S
Nebridi,10 08222 TERRASSA, Catalonia, SPAIN
LAAS-CNRS, LEA-SICA, 7, av Colonel Roche, 31077 TOULOUSE Cedex,
FRANCE

Abstract. Probability is one of the most common examples of valuation function.
The relationship between subsets is expressed by their joint probability and gives
rise to the concept of statistical dependence. In fuzzy logic this relationship is
based in connection properties. The aim of this work is to analyse how valuations
about subsets may be made. In a first part some theorems are recalled in order to
situate the essential role of connectives as t-norms, then the equivalence between
Fuzzy Logic's and probabilistic spaces is proven when Frank's t-norms are used.
Dependence and independence between subsets are analysed under the light of the
previous results, and gives rise to a new concept of dependence degree, as well as
its possible use in uncertain reasoning schemes.

Keywords*: Probability, Fuzzy Logic, Connectivity.*

1 Introduction

The aim of this work is to extend the concept of statistical or probabilistic
dependence to valuations of subsets by means of fuzzy measures. Zadeh's Fuzzy
Logic is an example of space using the connectivistic properties of *maximum* and
minimum. Subsequent approaches use as logical connectives t-norms for the
conjunction, and consequently t-conorms for the disjunction, in order to preserve
compatibility with Boolean algebra.

   Most of mathematical concepts are well known in the different areas of study,
nevertheless we shall recall them for the sake of establishing the relations between
contrasted mathematical models: this has been mainly done by labelling
DEFINITIONS, and THEOREMS. To clarify the deductive links between
propositions stated in this paper, a theorem has been labelled RECIPROCAL, and
two straightforward consequences called COROLLARIES have been stated
without proofs.

2 Algebra and Probability

Let **E** be a lattice, its minimal and its maximal elements are denoted respectively *0* and *1*. As usual given x,y∈ **E** x∧y and x∨y mean respectively the minimum and the maximum in the pair {x,y}. A lattice is complemented if ∀ x, ∃ x' such that x∧x'=*0* and x∨x'=*1*.

A Boolean algebra is a distributive complemented lattice. The isomorphism between any algebra of subsets and a Boolean algebra is given by the Stone theorem, and authorises to replace [*1,0*] by [$\Omega,\varnothing$] , conjunction ∧ by intersection ∩ and disjunction ∨ by union ∪, complementation x' by set complementation **C**x

Definition 1. Let A be the Boolean algebra of the subsets of Ω, a *probability* p(x) is a positive real mapping such that:

1) $p(\varnothing)=0$ and $p(\Omega)=1$ (*1st. "Kolmogorov axiom*)
2) for any given pair x,y∈ A , $p(x\wedge y) + p(x\vee y) = p(x) + p(y)$

Let us examine some consequences of this definition:

- For any x∈ A , $0 \leq p(x) \leq 1$
- $p(x')=1-p(x)$,
- ∀ x,y∈ A ; $x\wedge y = \varnothing$, then $p(x\vee y) = p(x) + p(y)$. (*2nd "Kolmogorov axiom*)

The 3rd. "Kolmogorov axiom" adds the following limit property for any infinite collection of disjoint elements of A {x_i} such that :

$$\forall\ x_i,x_j\in A\ ,\ x_i\wedge x_j = \varnothing \quad \text{then} \quad p(\textstyle\bigvee_i x_i)= \sum_i p(x_i)$$

Finally the conditional probability of y given x is defined as $p(y/x) = p(x\wedge y)/p(x)$, and a subsequent notion is the *probabilistic independence* of a pair x,y∈ A, if $p(y/x) = p(y)$, or equivalently $p(x/y) = p(x)$.

3 Predicate Algebra and Semantic Systems

Definition 2. A predicate algebra L is a formal language where:

1) {xi} are variables or constants
2) $\pi(x_1, x_2, ... x_n)$ is a n-argument predicate
3) Finite set of connectives $\Gamma= \{c_j\}$
4) Given two predicates π_1 and π_2, $\pi_3 = \pi_1\ c_j\ \pi_2$ is a well formed formula (wwf) belonging to *L*.

A *semantic system* $S = \{L\ ,\ \mathbf{v}\}$ where *L* is a predicate algebra and **v** a valuation function **v**, $\mathbf{P}(\Omega) \rightarrow [0,1]$ called also "generalized truth function".

A *connectivistic space* consists on:

1) a subset Γ^* of Γ
2) a semantic system $K = \{L^*, v\}$ where L^* is a restriction of L to the wwf's involving only connectives in Γ^*,
3) a subset of elementary predicates A, such that $\forall \alpha \in A$ $v(\alpha)$ is known,
4) a set of connective functions $\{c_\kappa\}$ such that

$$\forall \pi_1, \pi_2 \in K, \forall \gamma \in \Gamma^*, \exists \kappa \text{ such that } v(\pi_1 \gamma \pi_2) = c_\kappa(v(\pi_1), v(\pi_2)).$$

The boolean semantic system is the a particular connectivistic space $\Gamma^* = \{N, \&, +\}$, where the valuation function is a dichotomy $P(U) \rightarrow \{0,1\}$, and N stands for negation (1-x), & for conjunction (min) and + for disjunction (max).

Given a finite family $\alpha = \{\alpha_i\}$ of elementary predicates, $A(\alpha)$ stands for its Lindenbaum algebra, i.e. the minimum set of predicates containing α and closed under operations in Γ^*, if v is a boolean function $A(\alpha)$ is its *boolean closure*.

4 Fuzzy Logic Connectivistic Spaces

The aim of this work is to analyse how valuations about subsets may be generalised on Ω including probability and fuzzy measures. It is clear from the above considerations that:

1) *Probability spaces* are described as a whole by global probability distributions. For two given subsets a and b, the knowledge of marginal valuations $v(a)$ and $v(b)$ does not suffice for the knowledge of joint valuations as $v(a\&b)$. A global probability measure must be given, and the relative position (intersection) of a and b, must be known.

2) *Connectivistic spaces* are semantic systems described separately for elementary predicates α, and by means of connective functions any compound predicate can be valuated, therefore any subset of the universe Ω can be valuated as a function of the marginal valuations of its components.

Not all functions v from $A(\alpha)$ to [0,1] are good candidates as connectives. It is necessary that some restrictions hold, in order to maintain the compatibility with Boolean spaces, they have been developed in the Fuzzy Logic framework.

Definition 3. f(x,y) is a *junctor* if:

J1 : f is commutative $f(x,y)=f(y,x)$
J2 : f is non-decreasing in both arguments
J3 : Ground condition: $f(x,0)=0$
J4 : Margin condition: $f(x,1)=x$

f(x,y) is *Archimedean* if $f(x,x)<x$

An associative junctor is called t-*norm* (triangular norm), i.e.

$$J5 : \mathbf{f}(x,\mathbf{f}(y,z))=\mathbf{f}(\mathbf{f}(x,y),z)$$

A 2-*copula* is a function satisfying at least J2, J3, and J4.

Theorem 1. *Schweizer-Sklar 1* [SCHWEIZER & SKLAR 1983]

Archimedean strictly increasing t-norms can be expressed by means of a bijective function $\mathbf{g}(.)$ called additive generator, as follows:

$$t(x,y)=\mathbf{g}(\,\mathbf{g}^{-1}(x)+\mathbf{g}^{-1}(y))$$

From now on we only will be concerned by continuous strictly increasing archimedean t-norms (*csiat-norms*) except for some limit situations.

Definition 4. $c(x_1,x_2, ...x_n)$ is a **n**-*copula* if:

  C1 : c is a n-increasing function
  C2 : c is grounded i.e. $c(-,-,-,...-,0,-,-,-,)=0$
  C3 : margins of c are identities, i.e. $c(1,1,1,1,..,1,x,1,1,1)=x$

The following theorem relies n-copulæ and probability spaces:

Theorem 2. *Frechet-Schweizer-Sklar* [SCHWEIZER & SKLAR 1983]

Given a n-dimensional probability space $\{\Omega,\mathbf{A},\mathbf{p}\}$, such that all marginal distributions $F_1,F_2, ...,F_n$ are continuous, there exist a unique n-copula c such that

$$F(x_1,x_2,...,x_n) = c(F_1(x_1), F_2(x_2),... F_n(x_n))$$

By associativity a n-norm t^{*n} is obtained by n-1 iterations of t. We shall analyse here the conditions for a t-norm to generate a n-copula.

Theorem 3. *Schweizer & Sklar 2* [SCHWEIZER & SKLAR 1983]

Let t be a csiat-norm, and $\mathbf{g}$ its additive generator, the n-place function t^{*n} is a n-copula if, and only if, for all m≤n the following *N-monotonicity condition* C4 holds:

$$C4 : \; \forall \; m \leq N \; ; \; (-1)^m.(d^m\mathbf{g}/dx^m) > 0$$

A Fuzzy Logic space defined by a t-norm can be related to a probabilistic space, only if the iterated n-norm is a n-copula. The exact nature of this correspondence will be analysed here.

5 Frank T-Norms and Copula

Let us introduce here an important family of t-norms called the Frank's Family:

Definition 5. The following function t_φ is called *Frank t-norm*, φ is its parameter;

$$-\infty \leq \varphi \leq \infty, \qquad t_\varphi(x,y) = \frac{1}{\varphi} \text{Log} \left[1 + \frac{(e^{\varphi x} - 1)(e^{\varphi y} - 1)}{(e^{\varphi} - 1)} \right].$$

The limit values are:

$$t_{-\infty}(x,y) = \min(x,y)$$
$$t_0(x,y) = x.y$$
$$t_{+\infty}(x,y) = \max(x+y-1,0) = luk(x,y)$$

It shall be noticed that for $|\varphi| < \infty$ all Frank's t-norms are csiat-norms, and therefore hold marginal bijectivity. The Frank theorem gives a characteristic property of Frank's t-norms essential for their interpretation as connectives.

The s-*norm* **s** is the dual of a t-norm **t** with respect to $N(x)=1-x$, canonical negation, : $s(x,y)=N(t(N(x),N(y)))= 1 - t(1-x, 1-y)$.

Theorem 4. *Frank* [FRANK 1979] and [KLEMENT 1981]

Given a csiat-norm **t**, the following additive property A1 holds if, and only if, it belongs to the Frank family:

$$A1 : t(x,y)+h(x,y)=x+y$$

Corollary 1. If a given csiat-norm **t** is a 2-copula, then **t** is a Frank's t-norm

Corollary 2. If t^{*r} is the r-th iterate of a t-norm **t**, then, t^{*r} is a r-copula , if and only if **t** is a Frank's t-norm and C4 holds for $n \leq r$.

6 Probability and Fuzzy Logic Connectivistic Spaces

A probability space can be constructed from a Semantic system, (Lindenbaum algebra) $A(\alpha)$, and a measurable function **p**. This construction assumes the knowledge of the function **p** on $P(\Omega)$.

Fuzzy Logic as introduced by Zadeh [ZADEH, 1965] leads to a space where conjunction is represented by the function **min**. Later other authors [SUGENO 1974], [YAGER 1985] extended this concept to more general functions, junctors as defined formerly, finally the concept of De Morgan triads, i.e. a t-norm, a negation and its dual co-norm, has been studied in [TRILLAS 1979] and [ESTEVA & PIERA 1982]. Connectivistic spaces based on csiat-norms are Fuzzy Logic based on De Morgan triads. An immediate consequence of last Theorems and Corollaries can be stated as follows:

Theorem 5. *Necessary Imbedding Condition,*
Let a connectivistic space K be characterised by a csiat-norm $t(x,y)$, the necessary condition for the existence of a probability space E such that $E{\subseteq}K$, is that t belongs to the Frank's family of t-norms

Proof. Let K be a connectivistic space such that there exist a probability space $E{\subseteq}K$, , and let us choose arbitrarily two propositions A and B of K, their valuations being a and b, and the joint valuation $t(a,b)$. If the probability measure in E is such that $p(A)=a$, $p(B)=B$ and $P(A\&B)=t(a,b)$, then there exist a 2-copula $c(.,.)$ such that : $P(A\&B)=c(p(A),p(B))=t(a,b)$.
Obviously $t(.,.)$ and $c(.,.)$ must be the same function because A and B have been chosen arbitrarily, therefore $t(.,.)$ must be a 2-copula, and by COROLARY 1 it must be a Frank's t-norm.

Reciprocal. *Sufficient Imbedding Condition*
Let $G*^M$ be the set of iterates $t*^M$ of the csiat-norm t, up to degree M-1, and let K_M be the restriction of K to the propositions involving at most M-1 times the same connective $\gamma{\in}\Gamma^*$, the sufficient condition for the existence of a probability space E such that $E{\in}K_M$, is that t belongs to the Frank's family of t-norms, and that the N-monotonicity condition C4 holds for N≤M.

Proof. Let $A_M = \{A_i\}_{i=1,...M}$ be a finite set of M propositions of K_M and let their valuations be a_i. The valuation of any compound proposition obtained by the conjunction of m<M propositions of A_M is such that :

$$C_m=A_1\&A_2\& ...\&A_i\&...\&A_m, \text{ then } c_m= t*^m(a_1,a_2, ...,a_i,...,a_m).$$

The valuations a_i can always be considered as marginal probabilities but c_m is a joint probability only if $t*^m$ is a m-copula, this is only possible if the N-monotonicity condition C4 holds.

7 Dependence and Independence

Independent pair of events
Let us consider, in a probabilistic space, a pair of independent events, then the following relation holds :

$$p(a\&b) = p(a).p(b) \text{ and } p(a+b)= p(a)+p(b)-p(a).p(b)$$

The t-norm *product* and its dual *probabilistic sum* are respectively used for the conjunction and the disjunction of events. This collection is a particular case of Sugeno's space where $p(a \lor b) = p(a) + p(b) - \lambda \cdot p(a) \cdot p(b)$, for the value $\lambda = 1$.

Causal collection of events

Let us consider, now, a collection C_X of events such that there exist a strong implication between any pair, i.e.:

$$\forall \{a,b\} \in C_X \times C_X \text{ either } a \subseteq b \text{ or } b \subseteq a, \text{ then the following relations hold:}$$
$$p(a\&b) = \min[p(a),p(b)] \text{ and } p(a+b) = \max[p(a),p(b)]$$

The t-norm **min** and its dual **max** are respectively used for the conjunction and the disjunction of events. Fuzzy valuated spaces as Zadeh introduced initially are collections of that type.

Lukasiewicz collection of events

Let us consider, now, a first collection C_1 of disjoint events, the following relations hold in C_1 :

$$\forall \{a,b\} \in C_1 \times C_1 \quad p(a\&b) = 0 \text{ and } p(a+b) = p(a) + p(b)$$

and a second collection C_2 of the pairs of complements of the events of C_1, i.e. $\forall a \in C_2, \exists a' \in C_1$ such that $a + a' = \Omega$, and $a \& a' = \varnothing$

Then the following relations hold in C_2 :

$$\forall \{a,b\} \in C_2 \quad p(a\&b) = p(a) + p(b) - 1 \text{ and } p(a+b) = 1$$

Let $C_z = C_1 \cup C_2$, and the above relations be condensed into:

$$\forall \{a,b\} \in C_z \times C_z \; p(a\&b) = \max[p(a) + p(b) - 1, 0],$$
$$p(a+b) = \min[\, [p(a) + p(b), 1]$$

The t-norm use here is known as *Lukasiewicz function* and its dual is the *bounded sum*, they are respectively used for the conjunction and the disjunction of events.

The three above collections of events, or propositions, are connectivistic spaces corresponding to the three more used t-norms: *product, minimum,* and *Lukasiewicz.* They can be looked as Fuzzy Logic called respectively, *probabilistic, Zadeh's* or *strict,* and *Lukasiewicz's.* It can be noticed [TRILLAS, ALSINA & VALVERDE, 1982], that only the second is a distributive lattice.

Frank's collection of events

Let us consider a Frank's t-norm t_φ and its co-norm $s_\varphi(x,y) = 1 - t_\varphi((1-x),(1-y))$
And a collection C_φ of events such that:

$$\forall \{a,b\} \in C_\varphi \times C_\varphi \quad p(a\&b) = t_\varphi[p(a),p(b)] \text{ and } p(a+b) = s_\varphi[p(a),p(b)]$$

According to the definition of Frank's t-norms the collections above mentioned correspond respectively for:

$$\varphi = -\infty \qquad \text{to causal collection } \mathbf{C_m}$$
$$\varphi = 0 \qquad \text{to independent collection } \mathbf{C_p}$$
$$\varphi = +\infty \qquad \text{to Lukasiewicz's collection } \mathbf{C_z}$$

Definition 6. Let us consider in a probability space two propositions a and b such that $\alpha = p(a)$, $\beta = p(b)$ and $\eta = p(a\&b)$. The *dependence degree* δ between two propositions is the value δ of φ such that $\eta = t_\delta(\alpha, \beta)$

For a given pair α, β the function $\mathbf{h}_{\alpha, \beta}(\varphi) = t_\varphi(\alpha, \beta)$ is monotonous and its inverse exists, therefore $\delta = \mathbf{h}_{\alpha, \beta}^{-1}(\eta)$

This definition of the dependence degree can be applied to any two events in the framework of classical probability, it is nevertheless clear that this degree is a non-probabilistic one, although it imbeds for $\varphi = 0$ the probabilistic independence. It graduates dependence between its two extreme situations: complete inclusion that can be interpreted as full causality, and complete exclusion, corresponding to complete contradiction.

8 Conditioning

Another consequence of the imbedding of fuzzy logic connectivistic spaces in probabilistic spaces is the possibility of interpreting, through a connectivistic point of view, the conditional valuations.

The classical definition of conditional probability measure is as follows:

$$p_y(x) = p(x/y) = p(x\&y) / p(y).$$

It can be easily noticed that it includes the definition of independence. It is clear that $p(x\&y) = p_y(x).p(y)$, let us assume that it is possible to find at least one proposition z, independent of y, such that:

$$p(z) = p_y(x), \text{ and therefore } p(x\&y) = p(z).p(y) .$$

Because of the independence $p(z\&y) = p(z).p(y)$, so $p(x\&y) = p(z\&y)$.

The world of events independent of y, will be called, *independent world anchored to* y, and z is a *representative* of x in that world. This notion is easily extended to the φ-dependence (dependence of degree φ) introduced above.

By these considerations, a new notion of conditional valuation that includes conditional probability can be introduced:

Definition 7. The φ-*conditional valuation* of proposition *a* with respect to proposition *b* is the valuation of a representative of a in the φ-world anchored to b.

Let us fix the value φ, the φ-world anchored to *b* contains all the collections of propositions x_φ such that:
$$v(x_\varphi \& b) = t_\varphi[v(x_\varphi), v(b)].$$

A proposition a_φ is a representative of a in the φ-world anchored to b if $v(a_\varphi \& b) = v(a\&b)$, its valuation is called φ-conditional valuation of *a* with respect to *b*.

This conditional valuation is obtained by solving with respect to $X = v_\varphi(a/b)$ the equation:

$$t_\varphi[\, X \, , v(b)] = v(a\&b) \tag{*}$$

If we only consider Frank's t-norms so that |φ|<∞, we can write:

$$v_\varphi(a/b) = t_\varphi^{-1}[v(a\&b)/v(b)] \tag{**}$$

A particular look to the infinite limit values of φ reflects the traditional difficulties encountered in [NGUYEN 1978] and others for the conditioning in classical fuzzy set theory, for the generalisation of Bayes-type rules [SHAFER & SHENOY 1988]. In that cases (*) holds but it is no longer possible to write (**), therefore an indirect technique could be proposed so as to establish $v_\varphi(a/b)$ for a φ finite and to analyse $\lim_{\varphi\to\infty}[v_\varphi(a/b)]$.

9 Connectivistic Reasoning under Uncertainty

A causal rule relating a proposition *a* to another proposition *b* taken as its uncertain consequence can be stated as : if *a* then *b* with value v(a/b).
Reasoning using the knowledge given by this rule consists in combining it with some real world observation that is interpreted as v(a) and gives some information about *b* or v(b) about *a*.

In [SUPPES 1966] , the extensions based on the interpretation of v(a/b)as the prob(a→b), have been proposed, and in [LOPEZ DE MANTARAS 1992] the possibilistic approach has been described.

We shall here use the conditional valuations in the *modus ponens* and *modus tollens* mechanisms as follows:

modus ponens	modus tollens
$v(a/b) \geq x$	$v(a/b) \geq x$
$v(a) \geq y$	$v(b) \leq z$
$v(b) = t_\varphi[\,v_\varphi(a/b),v(b)] \geq t_\varphi[\,x,\underline{y}]$	$v(b) = s_\varphi[\,(1-v_\varphi(a/b)),v(b)] \leq t_\varphi\,[(1-x),z]$

This mechanism allows adjusting the uncertain reasoning to known situations by means of the parameter φ.

10 Conclusion

The purpose of this paper was to analyse the formal links between two approaches of valuation theory, the first based in probability measures defined for any subset of the universe, and the second related to Fuzzy Logic and defined as connectivistic spaces. The main result is the necessary and partially sufficient conditions for imbedding a connectivistic space in a probabilistic world. When this imbedding is effective many interesting conclusions can be deduced, such as a new look to dependence and conditioning. Important consequences are: the definition of a general degree of dependence, and a new scope on conditioning in non-probabilistic environments, as uncertain reasoning in expert systems.

This approach was first presented in [AGUILAR-MARTIN 1985] and later developed in [AGUILAR-MARTIN 1988], finally a first draft of this paper was presented at the Prague Conference in 1990, published in [AGUILAR-MARTIN 1992]. It is more restrictive than the approach of from the approach of [M. E. HOFFMAN, M.E., MANEVITZ & E. K. WONG] as it is based uniquely in the properties of t-norms, and more specifically Frank t-norms, in the case where the description of events can be embedded in particular probability spaces according to the strict definition of conditional measures.

Many questions remain to be developed, particularly concerning the subsets of propositions using alternatively several logical connectives, as well as an accurate analysis of continuous spaces introducing densities.

References

[1] J. AGUILAR-MARTIN (1987): Una Interpretació Probabilistica de les Lògiques Difuses. Proceedings of the 6th Congrès Català de Lógica (April 1987, Barcelona), pp.49-52.
[2] J. AGUILAR-MARTIN (1988): Probabilistic and fuzzy relational semantic systems in propositional approximate reasoning. Proceedings of the IEEE, 8th I.S.M.V.L (May 1988, Palma de Mallorca), pp.205-209

[3] J. AGUILAR-MARTIN (1992): Probability, Fuzzy Logic and Connectivistic Spaces. Transactions of the Eleventh Prague Congerence (August 1990, Prague), ACADEMIA ed.

[4] J. AGUILAR-MARTIN & N. PIERA (1985) : Les connectifs mixtes: de nouveaux opérateurs d'association des variables dans la classification automatique avec apprentissage. in: Data analysis and Informatics IV pp253-265, E.Diday ed. Elsevier Science Publ. B.V. North-Holland

[5] F. ESTeVA & N. PIErA (1982) : Sobre àlgebres de De Morgan de difusos, Proceedings of the 1er Congrès Català de Lògica Matemàtica, Barcelona.

[6] M.J. FRANK (1979) : On the simultaneous Associativity of F(x,y) and x+y-F(x,y), Aequationes Math. n° 19.

[7] E.P. KLEMENT (1981) : Characterisation of Fuzzy measures constructed by means of triangular norms, Journal of Mathematical Analysis and Applications.

[8] M. E. HOFFMAN, M.E., MANEVITZ, E. K. WONG:, , L.M. and Wong, E.K. : Fuzzy Independence and Extended Conditional Probability, Information Sciences 90 (1996) 137-156.

[9] M. E. HOFFMAN, E. K. WONG: A Ridge-Following Algorithm for Finding the Skeleton of a Fuzzy Image. Information Sciences 105(1-4): 227-238 (1998)

[10] R. LOPEZ DE MANTARAS (1992) : A review of default and approximate reasoning and their relation, in , Aspects of the Scientific Cooperation among France, Italy and Spain on Information Technology,M. Furnari & al. eds, Bibliopolis - CNR, Napoli.

[11] H.T. NGUYEN (1978) : On Conditional Possibility Distributions, Fuzzy Sets and Systems, vol 1

[12] B. Schweizer & A. Sklar (1983) : Probabilistic Metric Spaces, book, North-Holland

[13] M. SuGENO (1974) : Subjective Evaluation of Fuzzy Objects, Proceedings IFAC Symposium on Stochastic Control, Budapest.

[14] G. SHAFER and P.P. SHENOY (1988) Bayesian and Belief Function Propagation, Working paper n° 192, School of Business, The University of Kansas, Lawrence.

[15] P. SUPPES (1966) Probabilistic inference and the concept of total evidence, in, Aspects of Inductive Logic, North Holland.

[16] E. TRILLAS (1979) : Sobre funciones de negación en la teoría de conjuntos difusos, Stochastica, Vol. III n° 1 pp 47-60

[17] E. TRILLAS , C. ALSINA & L. VALVERDE (1982) : Do we need max, min and 1-j in Fuzzy Set Theory, in Fuzzy Set and Possibility Theory, Pergamon Press, N.Y. , pp 275-297

[18] R.R. YAGER (1985) : On the relationships of methods of aggregation evidence in expert systems. Cybernetics and Systems 16, pp 1-21.

[19] L. ZADEH (1965) : Fuzzy Sets, Information and Control, Vol. 8 pp 338-353

About Ideals on the Set of Fuzzy Measures

Andrew G. Bronevich and Alexander N. Karkishchenko

Taganrog State University of Radio-Engineering Nekrasovskij bystreet, 44
Taganrog, RUSSIA

Abstract. This work is devoted to investigation of algebraic structures on the set of fuzzy measures that are closed relatively convex sum and multiplication of fuzzy measures. It is shown that the basic convex families of fuzzy measures considered in [1] own such properties.

1 Introduction

In this paper we consider classification and basic properties of fuzzy measure families, investigated in [1]. This classification is produced due to conception of ideal. Ideals are such convex families of fuzzy measures, which are closed relatively an operation of fuzzy measure's multiplication. It is shown that the basic fuzzy measure families that can be interpreted as lower probabilities own such a property.

The paper has the following structure. In section 2 we introduce the basic convex families of fuzzy measures that are analyzed in the article. Section 3 represents the result obtained in [1]. It gives the algebraic description of the convex family of all fuzzy measures by primitive measures. Section 4 bears subsidiary function. It gives us the well-known results about set-theoretical operations on filters, which are used below. And finally, sections 5,6 are directly devoted to the ideal description - its definition, ways of construction, including some algebraic operations on ideals.

This investigation can be considered as a development of the theory, where upper and lower estimates of probabilities are used for description imprecise and uncertain information [2–5].

2 The main notions and theoretical constructions

The set function g on a finite algebra $\Im = 2^X$ of a finite space X is called a *fuzzy measure* [6,7] if
1) $g(\emptyset) = 0$, $g(X) = 1$ (norming);
2) $g(A) \geq 0$ for $A \in \Im$ (non-negativeness);
3) $g(A) \leq g(B)$ if $A \subseteq B$ (monotonicity).

Lemma 1. *Let g be a fuzzy measure on the algebra $\Im$ of the space X. Then the set function $q(A) = \neg g(A) = 1 - g(\bar{A})$ is also a fuzzy measure on $\Im$.*

The fuzzy measure q from lemma 1 is called usually as a *dual* one to the generating measure g. It is clear that the duality relation among fuzzy measures is symmetric, i.e. the fuzzy measure g from the lemma 1 is dual to $\neg g$.

Consider the basic classes of fuzzy measures, which can be interpreted as lower estimations of probabilities [1].

The fuzzy measure g is called a *lower probability* if there exists a probability measure P on $\Im$ such as $g(A) \leq P(A)$ for all $A \in \Im$. Introduce the following notation: $g \leq q$ if $g(A) \leq q(A)$ for all $A \in \Im$.

Let g be a lower probability. This measure is called an *accurate lower probability* if for an arbitrary $A \in \Im$ it can be found a probability measure P, $g \leq P$, with the following property $P(A) = g(A)$.

The fuzzy measure g is called *superadditive* if it satisfies the following inequality: $g(A) + g(B) \leq g(A \cap B) + g(A \cup B)$ for all $A, B \in \Im$.

The fuzzy measure Bel is called a *belief measure* [3], if we can construct the non-negative set function m on $\Im$ and g can be represented by

$$Bel(A) = \sum_{B \in \Im | B \subseteq A} m(B). \tag{1}$$

3 Representation of fuzzy measure by the convex sum of primitive fuzzy measures

We establish some simple facts about the structure of the set of fuzzy measures on a finite algebra $\Im$ of a type 2^X (the space X has a finite number of elements). Let g_1 and g_2 be two fuzzy measures on $\Im$. Then their convex combination $g = \alpha_1 g_1 + \alpha_2 g_2$, $\alpha_1 + \alpha_2 = 1$, $\alpha_1, \alpha_2 \geq 0$, is also a fuzzy measure. Indeed, $g(\emptyset) = 0$, $g(X) = 1$ and if $A \subseteq B$, then $g(A) = \alpha_1 g_1(A) + \alpha_2 g_2(A) \leq \alpha_1 g_1(B) + \alpha_2 g_2(B) = g(B)$. Thus, the set of all fuzzy measures is convex.

Introduce the following definition. The fuzzy measure is called a *primitive measure* if its values belong to the set $\{0, 1\}$ for an arbitrary event. For more detailed description of the set of fuzzy measures we need some definitions of the theory of partially ordered sets (or posets). We shall consider the algebra $\Im$ as the poset with respect to the ordinary set-theoretical inclusion. Filter is such a subset $\mathbf{f}$ of $\Im$ that if $A \in \mathbf{f}$ and $A \subseteq B$ then $B \in \mathbf{f}$. We take by definition that neither filter contains $\emptyset$. However, each filter contains a set of minimal elements $\{A_1, A_2, ..., A_k\}$, i.e. such mutually incomparable elements that $\mathbf{f} = \{A \in \Im | \exists A_i \subseteq A\}$. Clearly minimal elements generate a filter $\mathbf{f}$: this fact is noted by $\mathbf{f} = \langle A_1, A_2, ..., A_k \rangle$. The filter $\mathbf{f}$ is principal if it is generated by one minimal element, i.e. $\mathbf{f} = \langle A \rangle$.

Lemma 2. *Let* $\mathbf{f}$ *be an arbitrary filter in* $\Im$ *and* $\eta_{\mathbf{f}}$ *be a characteristic function of the filter* $\mathbf{f}$, *i.e.*

$$\eta_{\mathbf{f}}(A) = \begin{cases} 1, & A \in \mathbf{f}, \\ 0, & A \notin \mathbf{f}. \end{cases}$$

78

Then $\eta_{\mathbf{f}}$ is a primitive measure. Conversely each primitive measure is associated with a certain filter $\mathbf{f}$.

Notice that among primitive measures we can extract primitive necessity measures [8] being associated with principal filters $\mathbf{f} = \langle A \rangle$ in $\Im$. Sometimes these measures are called Dirac measures. In particular, when $\langle A \rangle = \langle \{x_i\} \rangle$, $x_i \in X$, Dirac measure is a probability measure concentrated in the point x_i. The following theorems describe important characteristics of the convex set of fuzzy measures.

Theorem 1. *Any fuzzy measure g can be represented as a convex combination of primitive measures.*

Theorem 2. *Neither primitive measure can be represented by a convex sum of other primitive measures.*

One can easily prove that representation of fuzzy measure by the convex combination of primitive measures is not unique. This can be shown by examples. We also point out that plausibility (belief) measures can be represented in a form of a convex linear combination of primitive possibility (necessity) measures. Actually, consider the primitive necessity measure

$$\eta_{\langle A \rangle}(B) = \begin{cases} 1, A \subseteq B, \\ 0, A \not\subseteq B, \end{cases} A, B \in \Im.$$

Then, using expression (1), we obtain equality

$$Bel(B) = \sum_{A \in \Im} m(A)\eta_{\langle A \rangle}(B). \tag{2}$$

By analogy the formula for the plausibility measure can be derived. It is based on the convex sum of primitive possibility measures. One can prove [3] that the representation (2) is determined uniquely. It is natural to put a question - what kind of primitive measures are lower probabilities and what fuzzy measures can be represented by the convex sum of such measures. The following results give us the answer.

Lemma 3. *Let $\eta_{\mathbf{f}}$ be a primitive measure generated by a filter $\mathbf{f} = \langle A_1, A_2, ...,$ $A_k \rangle$. Then $\eta_{\mathbf{f}}$ is a lower probability if and only if $\bigcap_{i=1}^{k} A_i \neq \emptyset$.*

4 Algebraic operations on filters

Lemma 4. *Let $\mathbf{f}_1$ and $\mathbf{f}_2$ be filters of algebra $\Im$ then sets $\mathbf{f}_1 \cap \mathbf{f}_2$, $\mathbf{f}_1 \cup \mathbf{f}_2$ are also filters of $\Im$.*

Lemma 5. *Any filter $\mathbf{f} = \langle A_1, A_2, ..., A_k \rangle$ can be represented as a union of principal filters in the following way: $\mathbf{f} = \langle A_1 \rangle \cup \langle A_2 \rangle \cup ... \cup \langle A_k \rangle$.*

Lemma 6. *Let* $\langle A \rangle$, $\langle B \rangle$ *be some principal filters of* $\Im$ *then* $\langle A \rangle \cap \langle B \rangle =$ $\langle A \cup B \rangle$.

One can put a question - how to calculate the intersection of arbitrary filters? The following lemma solves this problem.

Lemma 7. *Let* $\mathbf{f_1} = \langle A_1, A_2, ..., A_k \rangle$, $\mathbf{f_2} = \langle B_1, B_2, ..., B_m \rangle$ *then* $\mathbf{f_1} \cap \mathbf{f_2} =$ $\bigcup_{i,j} \langle A_i \cup B_j \rangle$.

Definition 1. *The filter* $\mathbf{f_1} = \langle A_1, A_2, ..., A_k \rangle$ *is called coherent if* $\bigcap_{i=1}^{k} A_i \neq \emptyset$.

Lemma 8. *Let* $\mathbf{f_1}$ *be an arbitrary coherent filter and* $\mathbf{f_2}$ *be an arbitrary filter, which is non-coherent in general. Then filter* $\mathbf{f_1} \cap \mathbf{f_2}$ *will be coherent.*

5 Ideals

In this section we introduce the operation of fuzzy measure's multiplication. The justification of this operation is given in the following lemma.

Lemma 9. *Let* g_1 *and* g_2 *be fuzzy measures on the algebra* $\Im$, *then the set function constructed by a rule:* $g(A) = g_1(A)g_2(A)$ *for all* $A \in \Im$, *is also fuzzy measure.*

Further we name the rule, described in the lemma, as a rule of fuzzy measure *multiplication*.

Lemma 10. *Let* $\eta_{\mathbf{f_1}}$ *and* $\eta_{\mathbf{f_2}}$ *are primitive fuzzy measures then* $\eta_{\mathbf{f_1}} \eta_{\mathbf{f_2}} = \eta_{\mathbf{f_1} \cap \mathbf{f_2}}$.

Lemma 11. *Let* g_1, g_2, q *be fuzzy measures for arbitrary* $\lambda_1, \lambda_2 \geq 0$, $\lambda_1 + \lambda_2 = 1$. *Then* $(\lambda_1 g_1 + \lambda_2 g_2)q = \lambda_1 g_1 q + \lambda_2 g_2 q$.

Corollary 1. *Let fuzzy measures* g q *have the following representations by convex sums of primitive measures:* $g = \sum_{k=1}^{m} \alpha_k \eta_{\mathbf{f}_{i_k}}$, $q = \sum_{l=1}^{n} \beta_l \eta_{\mathbf{f}_{j_l}}$. *Then*

$$gq = \sum_{k=1}^{m} \sum_{l=1}^{n} \alpha_k \beta_l \eta_{\mathbf{f}_{i_k}} \eta_{\mathbf{f}_{j_l}} = \sum_{k=1}^{m} \sum_{l=1}^{n} \alpha_k \beta_l \eta_{\mathbf{f}_{i_k} \cap \mathbf{f}_{j_l}}.$$

Lemma 12. *Let* g_1 *and* g_2 *are fuzzy measures, in addition,* g_1 *is lower probability. Then* $g_1 g_2$ *is also lower probability.*

Definition 2. *The family* $\mathfrak{g}$ *of fuzzy measures is called an* ideal *if this family is convex and closed relatively multiplication of fuzzy measures, i.e.*
1) if $g_1 \in \mathfrak{g}$, $g_2 \in \mathfrak{g}$ then $\lambda_1 g_1 + \lambda_2 g_2 \in \mathfrak{g}$, $\lambda_1 + \lambda_2 = 1$, $\lambda_1, \lambda_2 \geq 0$;
2) if $g_1 \in g$ $g_2 \in \mathfrak{g}$, then $g_1 g_2 \in \mathfrak{g}$.

Theorem 3. *The following convex families of fuzzy measures are ideals: 1) the set of all fuzzy measures* (M_0); *2) the set of lower probabilities* (M_1); *3) the convex family of fuzzy measures, generated by primitive lower probabilities* (M_3); *4) the set of accurate lower probabilities* (M_4); *5) the set of superadditive measures* (M_5); *6) the set of belief measures* (M_6).

6   Some ways for ideal construction

It is obvious that any ideal $\mathfrak{g}$ is a convex subset of the family of all fuzzy measures. However an arbitrary convex set is not an ideal in general. In particular, (including the most important cases), an ideal has a form of polyhedron in the real vector space $R^{2^{|X|}-2}$. The vertexes of this polyhedron determine a set of fuzzy measures with finite power. The description of such a type of convex families is given in the following definition.

Definition 3. Let M be a convex family of fuzzy measures. Fuzzy measures $\{g_i\} \subseteq M$ are called *generating elements* of M, if any fuzzy measure $g \in M$ can be represented in the form of a convex combination of fuzzy measures from $\{g_i\}$. The system of generating elements is called *minimal* if neither generating element can be represented by a convex sum of other generating elements.

Theorem 4. *Let the convex set M have a finite set of generating elements $\{g_1, g_2, ..., g_n\}$. Then M is an ideal if $g_i g_k \in M$ for any pair of indexes $i, k \in \{1, 2, ..., n\}$.*

Further we shall be interested in closed ideals, which are described with the help of the following definition.

Definition 4. An ideal $\mathfrak{g}$ is called *closed* one, if for any sequence $\{g_n\}_{n=1}^{\infty}$ of fuzzy measures from $\mathfrak{g}$ we have $\lim\limits_{n \to \infty} g_n = g$ and $g \in \mathfrak{g}$.

Remark. One can notice that this definition coincides with the notion of closed set in real vector space $R^{2^{|X|}-2}$ if we consider the values of fuzzy measures as components of such vectors.

Consider an arbitrary fuzzy measure g and construct a closed ideal having minimal properties and containing the fuzzy measure g. It is obvious that this ideal has to contain fuzzy measures g^n, $n = 1, 2, ...$ too.

Theorem 5. *Let g be a fuzzy measure. Then a set M of fuzzy measures of a type*

$$\mu(A) = \lambda_0 g_0(A) + \sum_{i=1}^{\infty} \lambda_i g^i(A),$$

where $\sum\limits_{i=1}^{\infty} \lambda_i = 1$, $\lambda_i \geq 0$, $g_0(A) = \lim\limits_{n \to \infty} g^n(A) = \begin{cases} 1, & g(A) = 1, \\ 0, & g(A) < 1, \end{cases}$ is a closed ideal. Any other closed ideal $\mathfrak{g}$, containing fuzzy measure g, includes also the ideal M, i.e. $M \subseteq \mathfrak{g}$.

Theorem 6. *The minimal ideal M from theorem 5 has a finite system of generating elements if and only if the number of fuzzy measure's g values is not more than 3, i.e. $|\{g(A) \,|\, A \in \Im\}| \leq 3$.*

Show briefly the way of proving this theorem. If $|\{g(A)\,|A \in \mathfrak{S}\}| = 2$ then the fuzzy measure g is primitive. Therefore $g = g_0$ and $g^n = g_0$, $n = 1, 2, ...,$ i.e. theorem 6 is true by theorem 5. If $|\{g(A)\,|A \in \mathfrak{S}\}| = 3$, we can use the following representation for fuzzy measure g: $g = \lambda\eta_{f_1} + (1 - \lambda)\eta_{f_2}$, where $f_1 \supseteq f_2$ and

$$\eta_{f_1}(A) = \begin{cases} 1, & g(A) > 0, \\ 0 & g(A) = 0, \end{cases} \qquad \eta_{f_2} = g_0.$$

Then

$$g^n = [\lambda\eta_{f_1} + (1 - \lambda)\eta_{f_2}]^n = \lambda^n\eta_{f_1} + (1 - \lambda^n)\eta_{f_2} = \lambda^{n-1}g + (1 - \lambda^{n-1})g_0.$$

and we must check only the case, when $|\{g(A)\,|A \in \mathfrak{S}\}| > 3$. To do it, we can use the following lemma.

Lemma 13. *The polynomial $f(x) = x^n - \sum\limits_{i=0}^{n-1} a_i x^i$ has a unique root on the interval $(0, +\infty)$ if $a_i \geq 0$, $i = 1, 2, ..., n - 1$ and $a_0 > 0$.*

Theorem 7. *Let a convex set M of fuzzy measures have a finite set of generating elements and be closed relatively the operation of fuzzy measure's multiplication. Then M is closed ideal.*

Theorem 8. *If a convex set of fuzzy measures is described by a finite system of non-strict linear inequalities then this set has a finite generating system.*

This theorem is the well-known fact from the theory of linear inequalities.

Corollary 2. *It is obvious that lower probabilities, accurate lower probabilities , superadditive measures can be described by a finite system of non-strict linear inequalities. Applying theorems 7 and 8, we get that these convex families are closed ideals.*

7 Algebraic operations on ideals

In this section we consider only closed ideals.

a) Intersection of ideals.

Lemma 14. *Let $\mathfrak{g}_1$ and $\mathfrak{g}_2$ two arbitrary ideals. Then the set $\mathfrak{g} = \mathfrak{g}_1 \cap \mathfrak{g}_2$ is also an ideal.*

Lemma 15. *Let ideals $\mathfrak{g}_1$ and $\mathfrak{g}_2$ have finite generating systems. Then ideal $\mathfrak{g} = \mathfrak{g}_1 \cap \mathfrak{g}_2$ has a finite generating system too.*

Definition 5. Let M be an arbitrary set of fuzzy measures. We denote by $\mathfrak{g}(M)$ a minimal closed ideal, containing the set M. Minimality of $\mathfrak{g}(M)$ means that any other closed ideal $\mathfrak{g}_1$, satisfying $M \subseteq \mathfrak{g}_1$, must contain the ideal $\mathfrak{g}(M)$.

The basic theorem. *For any set M an ideal $\mathfrak{g}(M)$ exists and it is determined uniquely.*

b) Sum of ideals.

Definition 6. The ideal $\mathfrak{g}(\mathfrak{g}_1 \cup \mathfrak{g}_2)$ is called a *sum* of ideals $\mathfrak{g}_1$ and $\mathfrak{g}_2$. This fact we denote $\mathfrak{g} = \mathfrak{g}_1 + \mathfrak{g}_2$.

Theorem 9. *Let ideals $\mathfrak{g}_1$ and $\mathfrak{g}_2$ have finite generating systems $\{g_1, ..., g_n\}$, $\{q_1, ..., q_m\}$ correspondingly. Then the ideal $\mathfrak{g} = \mathfrak{g}_1 + \mathfrak{g}_2$ has the following generating system $\{g_i q_j\}_{\substack{i=1,..,n \\ j=1,..,m}} \cup \{g_i\}_{i=1,..,n} \cup \{q_j\}_{j=1,..,m}$.*

Corollary 3. *Let ideals $\mathfrak{g}_1$ and $\mathfrak{g}_2$ have finite generating systems. Then ideal $\mathfrak{g} = \mathfrak{g}_1 + \mathfrak{g}_2$ has a finite generating system too.*

c) Multiplication of ideals

Definition 7. Let M be a set of all products $\mu_1 \mu_2$ of fuzzy measures μ_1 and μ_2 that belong to ideals $\mathfrak{g}_1$ and $\mathfrak{g}_2$, i.e. $M = \{\mu_1 \mu_2 | \mu_1 \in \mathfrak{g}_1, \mu_2 \in \mathfrak{g}_2\}$. Then the ideal $\mathfrak{g}(M)$ is called a *product* of ideals $\mathfrak{g}_1$ and $\mathfrak{g}_2$. It is used the notation: $\mathfrak{g}(M) = \mathfrak{g}_1 \mathfrak{g}_2$.

Theorem 10. *Let ideals $\mathfrak{g}_1$ and $\mathfrak{g}_2$ have finite generating systems $\{g_1, ..., g_n\}$, $\{q_1, ..., q_m\}$ correspondingly. Then the ideal $\mathfrak{g} = \mathfrak{g}_1 \mathfrak{g}_2$ has the finite generating system $\{g_i q_j\}_{\substack{i=1,..,n \\ j=1,..,m}}$.*

Theorem 11. *Let $\mathfrak{g}$ be an arbitrary ideal and notations of the theorem 3 are used. Then the following relations are true:*
1) $\mathfrak{g}^n \subseteq \mathfrak{g}$, $n = 1, 2, ...$;
2) $M_0 \mathfrak{g} \supseteq \mathfrak{g}$;
3) $M_1 \mathfrak{g} \subseteq M_1$;
4) $M_3 \mathfrak{g} \subseteq M_3$;
5) $M_0 M_1 = M_1$;
6) $M_0 M_3 = M_3$;
7) $M_0 M_6 = M_3$;
8) $M_i M_i = M_i$, $i = 0, 3, 6$.

8 Conclusions

1. The algebraic structure of ideal can be interpreted as an effect of Lukasiewicz logic on the set of fuzzy measures. Actually, consider an ideal $\mathfrak{g} = \{g_i\}$ and the dual convex family of fuzzy measures $\neg \mathfrak{g} = \{\neg g_i\}$. With the help of not complicated computations it is not hard to show that this convex family is closed relatively an operation $f(x, y) = x + y - xy$. This with negation $\neg x = 1 - x$ together gives us justification for such conclusion.

2. Due to introduced ideals we can get classical families of fuzzy measures in a very natural way. In particular, consider the family $\mathcal{P} = \{P_i\}$ of all probability measures, then an ideal $g(\mathcal{P})$, contains any belief measure, i.e. $\mathfrak{g}(\mathcal{P}) = M_6$.

3. We have to imagine what types of relationships among introduced ideals exist. Using results of this paper and [1], one can show that the following inclusions are fulfilled: $M_0 \supset M_1$, $M_3 \supset M_5$, $M_4 \supset M_5$, $M_6 \supset M_5$, however $M_3 \not\supset M_4$, $M_3 \not\subset M_4$. Therefore one can consider an ideal $M_3 \cap M_4$. It is clear that $M_5 \subset M_3 \cap M_4$, but it is easy to show $M_5 \neq M_3 \cap M_4$.

References

1. Bronevich, A.G. and Karkishchenko, A.N. (2002) The structure of fuzzy measure families induced by upper and lower probabilities. *Statistical modeling, analysis and management of fuzzy data.* Heidelberg; New York: Physica-verl., 160-172.
2. Dempster, A.P. (1967) Upper and lower probabilities induced by multivalued mapping. *Ann. Math. Statist.*, **38**, 325–339
3. Shafer, G. (1976) *A mathematical theory of evidence.* Princeton University Press, Princeton, N.J.
4. Walley, P. (1991) *Statistical Reasoning with Imprecise Probabilities*, Chapman and Hall, London.
5. Dubois D., Prade H. (1992) When upper probabilities are possibility measures. *Fuzzy sets and systems*, **24**, 279–300
6. Sugeno, M. (1972) Fuzzy measure and fuzzy integral. *Trans. SICE*, **8**, 95–102
7. Grabisch, M., Nguyen, H.T., and Walker, E.A. (1995). *Fundamentals of Uncertainty Calculi with Applications to Fuzzy Inference.* Kluwer Academic Publishers, New York.
8. Dubois D. and Prade, H. (1988) *Possibility Theory.* Plenum Press, New York.

Operators for Convolution of Fuzzy Measures

Andrew G. Bronevich and Alexander E. Lepskiy

Taganrog State University of Radio-Engineering Nekrasovskij bystreet, 44
Taganrog, RUSSIA

Abstract. In this paper we investigate different mappings of fuzzy measures in the framework of probabilistic approach. Usually these mappings in the decision-making theory are called convolution (aggregation) operators [1,2]. We study what kind of restrictions is required to the convolution operator that one family of fuzzy measures (for example belief measures) is mapped to the same family.

1 Introduction

This paper presents some results devoted to convolution of fuzzy measures. This mathematical tool is applied in decision-making theory for solving the multi-criterion choice problem under fuzzy restrictions [1,2]. This investigation gives us another view to the description of the set of the different convolution operators. In the framework of the probabilistic approach to the fuzzy measure theory we divide the whole set of fuzzy measures into three classes [3–5]:

1) lower probabilities that can be interpreted as lower estimations of probabilities;

2) upper probabilities that can be interpreted as upper estimations of probabilities;

3) contradictory measures.

Taking this into consideration we study the effect of applying convolution operators in a sense how they preserve the properties of fuzzy measures families they belong to, for instance, when lower probabilities are mapped into the similar family of fuzzy measures. This investigation is carried out on the various families of fuzzy measures, considered in [3].

2 Basic fuzzy measures families in the scope of probabilistic approach

The set function g upon the algebra $\mathfrak{A} = 2^X$ of a space $X = \{x_1, x_2, ..., x_N\}$ is called a fuzzy measure [6] if the following conditions holds:

a) $g(\emptyset) = 0$, $g(X) = 1$ (norming);

b) $g(A) \geq 0$ for all $A \in \mathfrak{A}$ (non-negativeness);

c) $g(A) \leq g(B)$ if $A \subseteq B$ (monotonicity).

Let M be the set of all fuzzy measures. Further we shall use a partial order on M by a rule: $g_1 \leq g_2$ if $g_1(A) \leq g_2(A)$ for all $A \in \mathfrak{A}$. We introduce the following families of fuzzy measures [2]:

M_0 is the family of probability measures;

$M_1 = \{g \in M \mid \exists P \in M_0 : g \leq P\}$ is the family of fuzzy measures that are called lower probabilities;

$M_2 = \left\{g \in M \mid \forall B \in A,\ g(B) \neq 0 : \frac{g(A \cap B)}{g(B)} \in M_1\right\}$;

$M_3 = \{g \in M_1 \mid \forall A \in A\ \exists P \in M_0 : g \leq P,\ g(A) = P(A)\}$ is the family of fuzzy measures being called accurate lower probabilities;

M_4 is the family of superadditive fuzzy measures that satisfy the following inequality: $g(A \cup B) \geq g(A) + g(B) - g(A \cap B)$ for all $A, B \in \mathfrak{A}$;

M_5 is the family of belief measures if the following inequality holds

$$\sum_{B \subseteq \{1,\dots,k\}} (-1)^{k-|B|} \mu \left(A \cup \bigcup_{i \in B} A_i\right) \geq 0$$

for an arbitrary system $A, A_1, \dots, A_k$ of mutually exclusive sets [5].

One can prove the following embeddings: $M \supseteq M_1 \supseteq M_2 \supseteq M_3 \supseteq M_4 \supseteq M_5 \supseteq M_0$.

3 Characterizations of monotone functions and fuzzy measures through the use of the difference theory

Here we shall consider different functions $\varphi : R^n \to R$. A first order difference of this function is determined by

$$\Delta\varphi(\mathbf{x}; \Delta\mathbf{x}) = \varphi(\mathbf{x} + \Delta\mathbf{x}) - \varphi(\mathbf{x}), \mathbf{x}, \Delta\mathbf{x} \in R^n.$$

By analogy we can define differences of an arbitrary order:

$$\Delta^2\varphi(\mathbf{x}; \Delta\mathbf{x}_1, \Delta\mathbf{x}_2) = \Delta(\Delta\varphi(\mathbf{x}; \Delta\mathbf{x}_1); \Delta\mathbf{x}_2) =$$

$$\varphi(\mathbf{x} + \Delta\mathbf{x}_1 + \Delta\mathbf{x}_2) - \varphi(\mathbf{x} + \Delta\mathbf{x}_1) - \varphi(\mathbf{x} + \Delta\mathbf{x}_2) + \varphi(\mathbf{x}), \mathbf{x}, \Delta\mathbf{x}_1, \Delta\mathbf{x}_2 \in R^n,$$

$$\Delta^n \varphi(\mathbf{x}; \Delta\mathbf{x}_1, \Delta\mathbf{x}_2, ..., \Delta\mathbf{x}_n) = \Delta(\Delta^{n-1}\varphi(\mathbf{x}; \Delta\mathbf{x}_1, ..., \Delta\mathbf{x}_{n-1}); \Delta\mathbf{x}_n) =$$

$$= \sum_{k=0}^{n} (-1)^{n-k} \sum_{\{i_1, ..., i_k\} \subseteq \{1, ..., n\}} \varphi\left(\mathbf{x} + \sum_{m=1}^{k} \Delta\mathbf{x}_{i_m}\right), \mathbf{x}, \Delta\mathbf{x}_1, ..., \Delta\mathbf{x}_n \in R^n.$$

In the difference theory one of important tools is the Euler's formula [9]. For the sake of brevity introduce the following notations:

$$\Delta^k \varphi(\mathbf{x}; \Delta\mathbf{x}_{i_1}, \Delta\mathbf{x}_{i_2}, ..., \Delta\mathbf{x}_{i_k}) = \Delta^k_{i_1 i_2...i_k}\varphi \Delta^0\varphi = \varphi(\mathbf{x}).$$

Then the Euler's formula is written as follows:

$$\varphi(\mathbf{x} + \Delta\mathbf{x}_1 + \Delta\mathbf{x}_2 + ... + \Delta\mathbf{x}_n) = \sum_{\{i_1 i_2...i_k\} \subseteq \{1, ..., n\}} \Delta^k_{i_1 i_2...i_k}\varphi.$$

In particular, $\varphi(\mathbf{x} + \Delta\mathbf{x}_1) = \Delta^0\varphi + \Delta^1_1\varphi$, $\varphi(\mathbf{x} + \Delta\mathbf{x}_1 + \Delta\mathbf{x}_2) = \Delta^0\varphi + \Delta^1_1\varphi + \Delta^1_2\varphi + \Delta^2_{1,2}\varphi$.

We can consider the Euler's formula as an analog of the inverse Mobius transform [8]. But it has more general sense. This connection can be shown on monotone, superadditive and belief measures. To do this, we must consider each fuzzy measure g on the algebra $\mathfrak{A}$ of a space $X = \{x_1, x_2, ..., x_N\}$ as a function $g(A) = \tilde{g}(\alpha) = \tilde{g}(\alpha_1, ..., \alpha_N)$, where

$$\alpha_i = \begin{cases} 1, & x_i \in A \\ 0, & x_i \notin A \end{cases}, \quad \alpha \in R^n,$$

determined only on the vertices of a binary cube $\{0, 1\}^N$. On the $\{0, 1\}^N$ amy be applied the relation of the partial ordering: $\mathbf{x} \leq \mathbf{y}$ if $x_i \leq y_i$ for all $i = 1, ..., N$. It is clear that the disjunction on the binary cube can be produced as usual sum for disjoint sets. Call the function $\tilde{g}$ to be associated with the fuzzy measure g. Thus we can apply the difference theory to fuzzy measures. In this way one can easily prove the following theorem.

Theorem 1. *Let $\tilde{g}(\mathbf{x})$, $\mathbf{x} \in \{0, 1\}^N$, be a function, determined only on the vertices of the binary cube. Then this function is associated with a fuzzy measure g if and only if*

1) $\tilde{g}(\mathbf{0}) = 0$, $\tilde{g}(\mathbf{1}) = 1$;

2) $\Delta\tilde{g}(\mathbf{x}; \Delta\mathbf{x}) \geq 0$ for all $\mathbf{x}, \Delta\mathbf{x} \geq \mathbf{0}$, $\mathbf{x} + \Delta\mathbf{x} \leq \mathbf{1}$

In addition,

a) the fuzzy measure g is superadditive if and only if $\Delta^2\tilde{g}(\mathbf{x}; \Delta\mathbf{x}_1, \Delta\mathbf{x}_2) \geq 0$, forall $\mathbf{x}, \Delta\mathbf{x}_1, \Delta\mathbf{x}_2 \geq \mathbf{0}$, $\mathbf{x} + \Delta\mathbf{x}_1 + \Delta\mathbf{x}_2 \leq \mathbf{1}$

b) the fuzzy measure is belief measure if and only if $\mathbf{x} + \Delta\mathbf{x}_1 + \Delta\mathbf{x}_2 \leq \mathbf{1}$, for all $\mathbf{x}, \Delta\mathbf{x}_1, ..., \Delta\mathbf{x}_n \geq \mathbf{0}$, $\mathbf{x} + \Delta\mathbf{x}_1 + ... + \Delta\mathbf{x}_n \leq \mathbf{1}$.

Further we shall use the following notations and vector operations. Let $\sigma_1^n = \left\{\alpha = (\alpha_i)_{i=1}^n, \ \alpha_i \geq 0, \ \sum_{i=1}^n \alpha_i = 1\right\}$, $D_n = [0,1]^n$, let $C(D)$ and $C^k(D)$ be sets of continuous and k-differentiable functions correspondingly with the range of definition D. If $\alpha \in \sigma_1^n$, $\mathbf{x}, \mathbf{y} \in D_n$, then we consider the following vector operations $\alpha \cdot \mathbf{x} = \alpha_1 x_1 + \ldots + \alpha_n x_n$, $\mathbf{x}^\alpha = x_1^{\alpha_1} \ldots x_n^{\alpha_n}$, $\mathbf{x} + \mathbf{y} = (x_1 + y_1, \ldots, x_n + y_n)$, $\mathbf{x}/\mathbf{y} = (x_1/y_1, \ldots, x_n/y_n)$, $\mathbf{xy} = (x_1 y_1, \ldots, x_n y_n)$. In addition, we write $\mathbf{x} \leq \mathbf{y}$ if $x_i \leq y_i$ for all $i = 1, \ldots, n$.

Introduce by analogy the following classes of functions that play an important role for describing monotone, superadditive and belief measures. Let $g : D_n \to D$ then $g \in H_m$, $(m = 1, 2, \ldots)$ if $g(\mathbf{0}) = 0$, $g(\mathbf{1}) = 1$, $\Delta^n g(\mathbf{x}; \Delta\mathbf{x}_1, \ldots, \Delta\mathbf{x}_n) \geq 0$, $n = 1, \ldots, m$ for all $\mathbf{x}, \Delta\mathbf{x}_1, \ldots, \Delta\mathbf{x}_n \geq \mathbf{0}$, $\mathbf{x} + \Delta\mathbf{x}_1 + \ldots + \Delta\mathbf{x}_n \leq \mathbf{1}$.

Using the Euler's formula one can prove the following theorem.

Theorem 2. *Let* $\mathbf{g} = (g_1, g_2, \ldots, g_n) \in H_m^n$ *be a Cartesian product of functions* $g_1, g_2, \ldots, g_n \in H_m$ *and* $\varphi \in H_m$, *then* $\varphi(\mathbf{g}) \in H_m$.

Theorem 3. *Let* $g \in C^m(D)$, $g(\mathbf{0}) = 0$, $g(\mathbf{1}) = 1$, *then* $g \in H_m$ *if and only if* $d^n g(\mathbf{x}; d\mathbf{x}) \geq 0$ *($n = 1, \ldots, m$) for all* $\mathbf{x}, d\mathbf{x} \geq \mathbf{0}$.

4 Convolution operators. Their investigation

Let $\mathbf{g} = (g_1, g_2, \ldots, g_n) \in M^n$ be a Cartesian product of fuzzy measures $g_1, g_2, \ldots, g_n \in M$, then the operator of n-convolution is a functional mapping $\varphi : M^n \to M$ defining by a rule: $g = \varphi(\mathbf{g})$ if $g(A) = \varphi(\mathbf{g}(A))$ for all $A \in \mathfrak{A}$. For the sake of simplicity we shall call 2-convolution simply convolution (without 2-) and 1-convolution as a fuzzy measure function. The following theorem is obviously true.

Theorem 4. *The function* $\varphi : [0,1]^n \to [0,1]$ *is an n-convolution if and only if* $\varphi \in H_1$.

Theorem 5. *Let* $\varphi \in C(D_n)$ *then* $\varphi : M_0^n \to M_0$ *if and only if* $\varphi(\mathbf{x}) = \alpha \cdot \mathbf{x}$ *for a certain vector* $\alpha \in \sigma_1^n$.

Proof. Necessity. Let $\varphi : M_0^n \to M_0$ and $A, B \in \mathfrak{A}$ such as $A \cap B = \emptyset$. Then for any probability measure $\mathbf{P} = (P_1, \ldots, P_n) \in M_0^n$ we have $P(A \cup B) = \varphi(\mathbf{P}(A \cup B)) = \varphi(\mathbf{P}(A) + \mathbf{P}(B)) = P(A) + P(B) = \varphi(\mathbf{P}(A)) + \varphi(\mathbf{P}(B))$.

On the other hand, there exists probability measure $\mathbf{P} \in M_0^n$ with the following property: $\mathbf{x} = \mathbf{P}(A)$, $\mathbf{y} = \mathbf{P}(B)$ for any $\mathbf{x}, \mathbf{y} \in D^n$. Thus, $\varphi(\mathbf{x}+\mathbf{y}) = \varphi(\mathbf{x}) + \varphi(\mathbf{y})$ if $\mathbf{x}, \mathbf{y} \in D^n$ and $\mathbf{x}+\mathbf{y} \in D^n$. Last condition and continuity [10] (p.157) of the function together lead to the representation $\varphi(\mathbf{x}) = \alpha \cdot \mathbf{x}$. Because of equality $\varphi(\mathbf{1}) = 1$ we get $\alpha \in \sigma_1^n$.

It is easy to see that sufficiency is also true. $\qquad\square$

Theorem 6. *Let the function φ satisfy all the conditions of the theorem 1 and the condition $\varphi(\mathbf{x}) \leq \alpha \cdot \mathbf{x}$ is fulfilled for a certain vector $\alpha \in \sigma_1^n$ and an arbitrary $\mathbf{x} \in D_n$, then $\varphi : M_1^n \to M_1$.*

Proof. Let $\mathbf{g} \in M_1^n$. Then there exists measure $\mathbf{P} \in M_0^n$ such as $\mathbf{g} \leq \mathbf{P}$. Consider $P = \alpha \cdot \mathbf{P}$. Then $P \in M_0$ and $P(A) = \alpha \cdot \mathbf{P}(A) \geq \alpha \cdot \mathbf{g}(A) \geq \varphi(\mathbf{g}(A)) = g(A)$. Thus, $\varphi(\mathbf{g}) \in M_1$, and the theorem has been proved. $\square$

Example 1. Consider a function $\mathbf{x}^1 = x_1 \cdot \ldots \cdot x_n$. Using the chain of inequalities $\varphi(\mathbf{x}) = \mathbf{x}^1 = x_1 \cdot \ldots \cdot x_n \leq \frac{x_1^n + \ldots + x_n^n}{n} \leq \frac{x_1 + \ldots + x_n}{n}$, $\mathbf{x} \in D^n$, we get $\varphi(\mathbf{x}) \leq \alpha \cdot \mathbf{x}$, where $\alpha = (n^{-1}, \ldots, n^{-1}) \in \sigma_1^n$. Thus, $\varphi : M_1^n \to M_1$.

Notice that in the case $n = 1$ the converse proposition to the theorem 3 is justified, namely

Theorem 7. *Let $\varphi \in C(D_1)$ be a 1-convolution operator then $\varphi : M_1 \to M_1$ if and only if $\varphi(x) \leq x$.*

Proof. Let $\varphi : M_1 \to M_1$. Suppose the contradictory statement that there is a point $x_0 \in (0,1)$, for which $\varphi(x_0) > x_0$. Without putting any restrictions we can assume that $x_0 = m/n$ is a rational quantity. On the space $X = \{x_1, x_2, \ldots, x_n\}$ we can choose the probability measure P such as $P(x_i) = 1/n$ for all $i = 1, \ldots, n$. Consider events $A_{i_1 \ldots i_m} = \{x_{i_1}, \ldots, x_{i_m}\}$. Then $P(A_{i_1 \ldots i_m}) = m/n$. It is easy to see that the number of all such sets with the power m is $\binom{n}{m}$. According to our supposition $\varphi(P) \in M_1$, therefore we can find a probability measure P' such as $P' \geq \varphi(P)$. Thus,

$$P'(A_{i_1 \ldots i_m}) \geq \varphi(P(A_{i_1 \ldots i_m})) = \varphi(m/n) > m/n$$

and

$$\sum_{1 \leq i_1 < \ldots < i_m \leq n} P'(A_{i_1, \ldots, i_m}) > \frac{m}{n} \sum_{1 \leq i_1 < \ldots < i_m \leq n} 1 = \frac{m}{n}\binom{n}{m}. \qquad (1)$$

On the other hand,

$$\sum_{1 \leq i_1 < \ldots < i_m \leq n} P'(A_{i_1, \ldots, i_m}) = \sum_{1 \leq i_1 < \ldots < i_m \leq n} P'(x_{i_1}) + \ldots + P'(x_{i_m}) = \frac{m}{n}\binom{n}{m},$$

but this equality contradicts the estimation (1) and this finishes the proof of our theorem. $\square$

Theorem 8. *Let φ be an n-convolution operator and the following inequality $\varphi(\mathbf{x}) \leq (\alpha \cdot \mathbf{x}/\mathbf{y}) \varphi(\mathbf{y})$ holds for a certain vector $\alpha \in \sigma_1^n$ and any $\mathbf{x}, \mathbf{y} \in D_n$, $\mathbf{x} \leq \mathbf{y}$, then $\varphi : M_2^n \to M_2$.*

Proof. Let $\mathbf{g} \in M_2^n$, i.e. there exist probability measures $P_i \in M_0$ with the following property: $\frac{g_i(A \cap B_i)}{g_i(B_i)} \leq P_i(A)$, $i = 1, ..., n$, for any sets $B_i \in \mathfrak{A}$, $g_i(B_i) \neq 0$. Since $g_i(A \cap B_i) \leq g_i(B_i)$ and $\varphi(\mathbf{x}) \leq (\alpha \cdot \mathbf{x}/\mathbf{y})\varphi(\mathbf{y})$ for any $\mathbf{x}, \mathbf{y} \in D^n$, $\mathbf{x} \leq \mathbf{y}$, then, putting $x_i = g_i(A \cap B_i)$, $y_i = g_i(B_i)$, $i = 1, ..., n$, we get

$$\frac{\varphi(\mathbf{g}(A \cap B))}{\varphi(\mathbf{g}(B))} \leq \alpha_1 \frac{g_1(A \cap B_1)}{g_1(B_1)} + ... + \alpha_n \frac{g_n(A \cap B_n)}{g_n(B_n)} \leq \alpha_1 P_1(A) + ... + \alpha_n P_n(A).$$

Since $\alpha \cdot \mathbf{P} \in M_0$ then $\varphi(\mathbf{g}) \in M_2$, and the required statement is proved. $\square$

Example 2. The function $\varphi(\mathbf{x}) = \mathbf{x}^\alpha = x_1^{\alpha_1} \cdot ... \cdot x_n^{\alpha_n}$, $\alpha \in \sigma_1^n$, satisfies all conditions of the theorem 8.

Theorem 9. *Let φ be an n-convolution operator and $\forall \mathbf{x} \in D_n \ \exists \alpha', \alpha'' \in \sigma_1^n$ $\forall \mathbf{x}', \mathbf{x}'' \in D_n$: $\varphi(\mathbf{x}\mathbf{x}' + (1 - \mathbf{x})\mathbf{x}'') \leq \alpha' \cdot \mathbf{x}'\varphi(\mathbf{x}) + \alpha'' \cdot \mathbf{x}''(1 - \varphi(\mathbf{x}))$. Then $\varphi : M_3^n \to M_3$.*

Proof. It is required to prove that $\varphi(\mathbf{g}) \in M_3$ if $\mathbf{g} \in M_3^n$, i.e. for any $B \in A$ we can find a probability measure P such as $\varphi(\mathbf{g}) \leq P$ and $\varphi(\mathbf{g}(B)) = P(B)$. Show how to choose this measure P. Since $\mathbf{g} \in M_3^n$ then for any $B \in \mathfrak{A}$ there exists $\mathbf{P} \in M_0^n$: $\mathbf{g} \leq \mathbf{P}$ and $\mathbf{g}(B) = \mathbf{P}(B)$. Let $0 < \mathbf{g}(B) < 1$ and denote $\mathbf{x} = \mathbf{P}(B)$, $\mathbf{x}' = \mathbf{P}(A \cap B)/\mathbf{P}(B)$, $\mathbf{x}'' = (\mathbf{P}(A \cup B) - \mathbf{P}(B))/(1 - \mathbf{P}(B))$. Then

$$\varphi(\mathbf{g}(A)) \leq \varphi(\mathbf{P}(A)) = \varphi(\mathbf{P}(A \cap B) + \mathbf{P}(A \cup B) - \mathbf{P}(B)) =$$

$$\varphi(\mathbf{x}\mathbf{x}' + (1 - \mathbf{x})\mathbf{x}'') \leq \alpha' \cdot \mathbf{x}'\varphi(\mathbf{x}) + \alpha'' \cdot \mathbf{x}''(1 - \varphi(\mathbf{x})) \leq$$

$$\alpha' \cdot \mathbf{P}'(A)\varphi(\mathbf{P}(B)) + \alpha'' \cdot \mathbf{P}''(A)(1 - \varphi(\mathbf{P}(B))) = P(A),$$

where $\mathbf{P}'(A) = \mathbf{P}(A \cap B)/\mathbf{P}(B)$, $\mathbf{P}''(A) = (\mathbf{P}(A \cup B) - \mathbf{P}(B))/(1 - \mathbf{P}(B))$ are probability measures and $\alpha'\varphi(\mathbf{P}(B)) + \alpha''(1 - \varphi(\mathbf{P}(B)) \in \sigma_1^n$. Therefore P is also probability measure. $\square$

Theorem 10. *Let φ be an n-convolution operator then $\varphi : M_4^n \to M_4$ if and only if $\varphi \in H_2$.*

Proof. Necessity. Let $\varphi : M_4^n \to M_4$ but the theorem is not fulfilled, i.e. for some $\mathbf{x}, \Delta\mathbf{x}, \Delta\mathbf{y}$ such that $\mathbf{x}, \Delta\mathbf{x}, \Delta\mathbf{y} \geq 0$ and $\mathbf{x} + \Delta\mathbf{x} + \Delta\mathbf{y} \leq 1$ we have

$$\Delta^2(\varphi(\mathbf{x}), \Delta\mathbf{x}, \Delta\mathbf{y}) = \varphi(\mathbf{x} + \Delta\mathbf{x} + \Delta\mathbf{y}) - \varphi(\mathbf{x} + \Delta\mathbf{x}) - \varphi(\mathbf{x} + \Delta\mathbf{y}) + \varphi(\mathbf{x}) < 0. \quad (2)$$

Then we can choose a probability measure $\mathbf{P} \in M_0^n$ as follows: $\mathbf{P}(A \cap B) = \mathbf{x}$, $\mathbf{P}(A) = \mathbf{x} + \Delta\mathbf{x}$, $\mathbf{P}(B) = \mathbf{x} + \Delta\mathbf{y}$, and because of additivity $\mathbf{P}$ we have $\mathbf{P}(A \cup B) = \mathbf{x} + \Delta\mathbf{x} + \Delta\mathbf{y}$. Check the required condition:

$$\varphi(\mathbf{P}(A)) + \varphi(\mathbf{P}(B)) \leq \varphi(\mathbf{P}(A \cup B)) + \varphi(\mathbf{P}(A \cap B)),$$

or

$$\varphi(\mathbf{x} + \Delta \mathbf{x}) + \varphi(\mathbf{x} + \Delta \mathbf{y}) \leq \varphi(\mathbf{x} + \Delta \mathbf{x} + \Delta \mathbf{y}) + \varphi(\mathbf{x}).$$

But last inequality contradicts with our supposition.

Sufficiency. If $\varphi \in H_2$ then $\varphi : M_4^n \to M_4$ (see theorems 1a) and 2). $\quad\square$

Theorem 11. *Let φ be an n-convolution operator, then $\varphi : M_5^n \to M_5$ if and only if $\varphi \in H_\infty$.*

Proof. Necessity. Let $\varphi : M_5^n \to M_5$ but the theorem is not fulfilled, i.e. for some $\mathbf{x}, \Delta \mathbf{x}_1, ..., \Delta \mathbf{x}_n \geq 0$, $\mathbf{x} + \Delta \mathbf{x}_1 + ... + \Delta \mathbf{x}_n \leq 1$ we have $\Delta^n \varphi(\mathbf{x}, \Delta \mathbf{x}_1, ..., \Delta \mathbf{x}_n) < 0$. Then we can choose a probability measure $\mathbf{P} \in M_0^n$ as follows: $\mathbf{P}(A) = \mathbf{x}$, $\mathbf{P}(A_k) = \Delta \mathbf{x}_k$, where $A, A_1, ..., A_n$ is a system of mutually exclusive sets. Check the required condition:

$$\sum_{B \subseteq \{1,...,n\}} (-1)^{k-|B|} \varphi \left(\mathbf{P} \left(A \cup \bigcup_{i \in B} A_i \right) \right) \geq 0,$$

or

$$\sum_{k=0}^{n} (-1)^{n-k} \sum_{\{i_1,...,i_k\} \subseteq \{1,...,n\}} \varphi \left(\mathbf{x} + \sum_{m=1}^{k} \Delta \mathbf{x}_{i_m} \right)$$
$$= \Delta^n \varphi(\mathbf{x}, \Delta \mathbf{x}_1, ..., \Delta \mathbf{x}_n) \geq 0.$$

But last inequality contradicts with our supposition.

Sufficiency. If $\varphi \in H_\infty$ then $\varphi : M_5^n \to M_5$ (see theorems 1b) and 2). $\quad\square$

5   Conclusions

In a view of convolution operators and presented results we can get another understanding of fuzzy measure theory. Actually, if we shall consider introduced convolution operators $\varphi(\mathbf{x})$, $\mathbf{x} \in R^n$ on the vertices of a binary cube (i.e. $\mathbf{x} = (x_1, x_2, ..., x_n)$, $x_i \in \{0, 1\}$), then this operator obviously is fuzzy measure, in addition, if this operator maps lower probabilities to the similar class of fuzzy measures then the pointed measure φ will be a lower probability. The same property refers to all considered families of fuzzy measures. From this fact we can put a question. How to construct convolution operators from fuzzy measures, keeping desirable properties? In this case we obviously know the values of the function only in the vertices of a binary cube and it is necessary to extend the range of definition to the whole cube. To do this, we can use different integrals [6,11,12] upon fuzzy measure; perhaps, it is required to apply another technique.

There are many problems how to interpret convolution operators in the framework of the probabilistic approach. For instance, how to combine independent or correlated imprecise random variables [13,14] keeping required accuracy? Here we have to extend the notion of convolution operator to the following context. For example we must also consider 2-convolution $\varphi(x,y)$ as a function, which transfers two fuzzy spaces $(X_1, \mathfrak{A}_1, g_1)$, $(X_2, \mathfrak{A}_2, g_2)$ to the fuzzy space $(X_1 \times X_2, \mathfrak{A}_1 \times \mathfrak{A}_2, g)$ according to a rule: $g(A) = \varphi(g_1(A_1), g_1(A_1))$, $A = A_1 \times A_2$, $A_1 \in \mathfrak{A}_1$, $A_2 \in \mathfrak{A}_2$.

References

1. Dubois D. and Prade, H. (1988) *Possibility Theory*. Plenum Press, New York.
2. Grabisch, M., Nguyen, H.T., and Walker, E.A. (1995). *Fundamentals of Uncertainty Calculi with Applications to Fuzzy Inference*. Kluwer Academic Publishers, New York.
3. Bronevich, A.G. and Karkishchenko, A.N. (2002) The structure of fuzzy measure families induced by upper and lower probabilities. *Statistical modeling, analysis and management of fuzzy data*. Heidelberg; New York: Physica-verl., 160-172.
4. Dempster, A.P. (1967) Upper and lower probabilities induced by multivalued mapping. *Ann. Math. Statist.*, **38**, 325–339
5. Walley, P. (1991) *Statistical Reasoning with Imprecise Probabilities*, Chapman and Hall, London.
6. Sugeno, M. (1972) Fuzzy measure and fuzzy integral. *Trans. SICE*, **8**, 95–102
7. Shafer, G. (1976) *A mathematical theory of evidence*. Princeton University Press, Princeton, N.J.
8. Chateauneuf, A. and Jaffray, J.Y. (1989) Some characterizations of lower probabilities and other monotone capacities through the use of Mobius inversion, *Mathematical Social Sciences*, **17**, 263-283.
9. Gelfond, A.O. (1967) *Calculi of finite differences*. "Nauka" Press, Moscow. (In Russian)
10. Fihtengoltz, G. (1969) *Course of the differential and integral calculi, V.1*, "Nauka" Press, Moscow. (In Russian)
11. Murofushi, T., Sugeno, M. (1989) An interpretation of fuzzy measure and Choquet integral as an integral with respect to fuzzy measure. *Fuzzy sets and Systems*, **29**, 229-235.
12. Mesiar R. (1995) Choquet-like integrals, *J. Math. Anal. Appl.*, **194**, 477-488.
13. Dubois, D., Farinas del Cerro, Herzig, A., Prade, H. (1994) An original view of independence with application to plausible reasoning. *Proceedings of 10-th conference "Uncertainty in Artificial reasoning"*, 195-203.
14. De Cooman, G. Possibility Theory I-III. *International journal of General systems*, **25**, 291-371.

Probabilities of Fuzzy Events Based on Scalar Cardinalities*

Jaume Casasnovas and Francesc Rosselló

Departament de Matemàtiques i Informàtica, Universitat de les Illes Balears,
E-07071 Palma de Mallorca, Spain.
E-mail: {dmijcc0,dmifrl0}@clust.uib.es

Abstract. The standard set of Bell-type inequalities is satisfied by the extension of probabilities to fuzzy events based on the axiomatic definition of scalar cardinality of a fuzzy set, even though the lattice defined by the intersection, union and negation of fuzzy sets in the sense of Zadeh is not a boolean algebra.

1 Introduction

Fuzzy probability theory allows to consider non-sharp, fuzzy random events like "The number will be much more than 26" or "The winner will be young". This kind of events can be described by the theory of fuzzy sets [9], where a fuzzy subset A of a universe X can be identified with a membership function $\mu_A : X \longrightarrow [0,1]$, and in this paper we shall understand a *fuzzy probability* as an extension of a measure of probability defined on subsets of X to these fuzzy events.

If a classical Kolmogorov probability space $(X, \mathcal{B}(X), \mathcal{P})$ can be defined, where $\mathcal{B}(X)$ is a σ-algebra of subsets of X, $\mathcal{P}$ is a traditional probability measure over $\mathcal{B}(X)$ and the elements of $\mathcal{B}(X)$ are the $\mathcal{P}$-measurable sets, then we can consider as fuzzy random events those fuzzy subsets of X whose membership functions are $\mathcal{P}$-measurable, i.e., those fuzzy subsets A of X such that $\mu_A^{-1}(\beta)$ is a $\mathcal{P}$-measurable set for every Borel set $\beta \subseteq [0,1]$. The probability measure of such a (measurable) fuzzy random event A is defined in [10] as

$$m(A) = \int_X \mu_A(x)\, dP(x),$$

where $\int_X$ stands for the Lebesgue-Stieljes integral.

If A is finite crisp subset of X and $\mathcal{B}(X)$ is the whole σ-algebra of parts of X, then $m(A) = \sum_{x \in A} \mathcal{P}(x)$. Therefore, if $\mathcal{P}(x) = p$ for every $x \in A$, then $m(A) = p|A|$. If A is a finite fuzzy subset of X, i.e., if $\mathrm{Supp}(A) = \{x \in X \mid \mu_A(x) > 0\}$ is a finite crisp subset of X, then $m(A) = \sum_{x \in \mathrm{Supp}(A)} \mu_A(x)\mathcal{P}(x)$, and if $\mathcal{P}(x) = p$ for every $x \in \mathrm{Supp}(A)$, then $m(A) = p \sum_{x \in \mathrm{Supp}(A)} \mu_A(x)$. The value $\sum_{x \in \mathrm{Supp}(A)} \mu_A(x)$ is defined by Zadeh and other authors [2,10,7]

---

* This work has been partially supported by the Spanish DGES, grants BFM2000-1113-C02-01 and BFM2000-1114.

as the *cardinality* of the fuzzy subset A, but we can find several definitions of cardinality of a finite fuzzy subset in the literature [2,7,8].

Cardinalities for fuzzy sets are a generalization of classical cardinality theory for crisp sets. They are defined with the aim of being used in areas such as modelling the meaning of imprecise quantifiers, probabilities of fuzzy events, analysis of grey images, or in algebraic and topological structures on fuzzy sets. Among the different ways of defining the concept of cardinality of a finite fuzzy set that can be found in the literature on this subject, the *scalar cardinality*, which associates a natural or real quantity to each finite fuzzy set, is a relevant one. Recently, an axiomatic approach to scalar cardinalities of finite fuzzy sets has been presented by M. Wygralak [6]. This approach provides an infinite family of possible scalar cardinalities including all standard ones like the sigma count of a fuzzy set, the cardinality of its core or its support and also the cardinality of its α-level set; see Section 2 below. Namely, these are cardinalities of the form $sc_f(A) = \sum_{x \in \text{Supp}(A)} f(\mu_A(x))$, where $f : [0,1] \to [0,1]$ is a nondecreasing function with $f(0) = 0$ and $f(1) = 1$

Therefore, if we define a generalization of a probability measure

$$m_f(A) = \int_X f(\mu_A(x)) \, dP(x)$$

for those fuzzy events A such that the function $f \circ \mu_A$ is $\mathcal{P}$-measurable, we have, for finite fuzzy sets and if $\mathcal{B}(X)$ is the whole σ-algebra of the parts of X, that

$$m_f(A) = \sum_{x \in \text{Supp}(A)} f(\mu_A(x))\mathcal{P}(x).$$

Moreover, if $\mathcal{P}(x) = p$ for every $x \in A$, then

$$m_f(A) = p \sum_{x \in \text{Supp}(A)} f(\mu_A(x)) = p \cdot sc_f(A),$$

which is a result that also generalizes the crisp case.

Bell's inequalities have been used in mathematical physics as a way to check whether a situation in quantum probability calculus, i.e., a measure of probability defined over an orthomodular orthocomplemented lattice or poset, allows a description in terms of the classical probability calculus of Kolmogorov or not [1,6,4,5]. Thus, according to the original idea of Bell, the first approach to his inequalities was a physical one, but there exists also a mathematical way of looking at these inequalities according to which their violation means first of all that the considered situation does not allow a description in terms of Kolmogorov probability calculus. This mathematical way of looking at Bell-type inequalities allows to apply them in any situation in which numerical data may be interpreted as probabilities, even if this situation does not concern quantum physics or even physics at all. In particular, J. Pykacz and B. d'Hooghe studied in [5] the satisfaction of these inequalities in several models of probability of fuzzy events, namely for Zadeh's definition of probability measure in a theory of fuzzy sets using Frank's t-norms and t-conorms, and they showed that the most popular model of fuzzy probability calculus based on the original Zadeh operations [9] cannot be distinguished from Kolmogorov's model by any one of Bell's inequalities.

In this paper we investigate these Bell-type inequalities for the definition of generalized "scalar" probability measure m_f given above.

2 Scalar cardinalities of finite fuzzy sets

Recall that, with the following operations, the class of fuzzy subsets of the universe X is a distributive complemented lattice:

$$\mu_{A \cup B}(x) = \max\{\mu_A(x), \mu_B(x)\} = \mu_A(x) \vee \mu_B(x),$$
$$\mu_{A \cap B}(x) = \min\{\mu_A(x), \mu_B(x)\} = \mu_A(x) \wedge \mu_B(x),$$
$$\mu_{\bar{A}}(x) = 1 - \mu_A(x)$$

Definition 1. ([7]) Let $\mathrm{FFS}(X)$ be the class of finite fuzzy subsets of a universe X. A function sc : $\mathrm{FFS}(X) \to [0, +\infty]$ is called a *scalar cardinality* if the following postulates are satisfied:

1. For every $a, b \in [0, 1]$ and $x \in X$, if $a \leq b$, then $\mathrm{sc}(a/x) \leq \mathrm{sc}(b/x)$, where $a/x \in \mathrm{FFS}(X)$ is defined by $a/x(x) = a$ and $a/x(z) = 0$ for every $z \neq x$.
2. For every finite family $\{A_i\}_{i \in J}$ of elements of $\mathrm{FFS}(X)$ such that $A_i \cap A_{i'} = T$ whenever $i \neq i'$ (where T is the constant 0 fuzzy set), it happens that $\mathrm{sc}(\bigcup_{i \in J} A_i) = \sum_{i \in J} \mathrm{sc}(A_i)$.
3. If A is a finite crisp set, then $\mathrm{sc}(A) = |\mathrm{Supp}(A)|$.

Proposition 1. ([7]) *A function* sc : $\mathrm{FFS}(X) \to [0, +\infty]$ *is a scalar cardinality if and only if there exists a non-decreasing function* $f : [0, 1] \to [0, 1]$ *with* $f(0) = 0$ *and* $f(1) = 1$, *such that*

$$\mathrm{sc}(A) = \sum_{x \in \mathrm{Supp}(A)} f(\mu_A(x)).$$

We have, among others, the following examples of scalar cardinalities:

- If $f(t) = t$ for every $t \in [0, 1]$, then $\mathrm{sc}_f(A) = \sum_{x \in \mathrm{Supp}(A)} \mu_A(x)$.
- If $f(t) = 1$ for every $t > 0$, then $\mathrm{sc}_f(A) = |\mathrm{Supp}(A)|$.
- If $f(t) = 1$ for every $t \geq \alpha$ and $f(t) = 0$ for every $t < \alpha$, then $\mathrm{sc}_f(A) = |A_\alpha|$, where $A_\alpha = \{x \in X | \mu_A(x) \geq \alpha\}$.
- If $f(t) = t^p, p \in \mathbb{Q}$, for every $t \in [0, 1]$, then $\mathrm{sc}_f(A) = \sum_{x \in \mathrm{Supp}(A)} (\mu_A(x)^p)$.

3 The Bell-Type inequalities

Let m be a measure defined over a lattice (or simply a poset) $(L, \leq, \wedge, \vee, \tilde{}, \mathbf{0}, \mathbf{1})$ with $m(\mathbf{1}) = 1$, and let $\{p_l, p_{ls}\}_{1 \leq l, s \leq n} \subseteq [0, 1]$ be a collection of numbers such that there exists a collection $\{e_l\}_{1 \leq l \leq n}$ of elements of L with $m(e_l) = p_l$ and $m(e_l \wedge e_s) = p_{ls}$ for every l, s. If m is a probability measure in the sense of Kolmogorov, as well as in some other cases, the following inequalities must be satisfied (see [1,5,6]): for every i, j, k, t

$$0 \leq p_i + p_j - p_{ij} \leq 1 \tag{1}$$
$$0 \leq p_i - p_{ij} - p_{ik} + p_{jk} \tag{2}$$
$$0 \leq p_j - p_{jk} - p_{ij} + p_{ik} \tag{3}$$
$$0 \leq p_k - p_{ik} - p_{jk} + p_{ik} \tag{4}$$
$$p_i + p_j + p_k - p_{ij} - p_{ik} + p_{jk} \leq 1 \tag{5}$$
$$-1 \leq p_{ik} + p_{ir} + p_{jr} - p_{jk} - p_i - p_r \leq 0 \tag{6}$$
$$-1 \leq p_{jk} + p_{jr} + p_{ir} - p_{ik} - p_j - p_r \leq 0 \tag{7}$$
$$-1 \leq p_{ir} + p_{ik} + p_{jk} - p_{jr} - p_i - p_k \leq 0 \tag{8}$$
$$-1 \leq p_{jr} + p_{jk} + p_{ik} - p_{ir} - p_j - p_k \leq 0. \tag{9}$$

4 Main results

Let $\mathrm{FFS}(X)$ be the lattice of fuzzy subsets of X. If a classical Kolmogorov probability space $(X, \mathcal{B}(X), \mathcal{P})$ is defined, where $\mathcal{B}(X)$ is the σ-algebra of $\mathcal{P}$-measurable subsets of X and $\mathcal{P}$ is a traditional probability measure over $\mathcal{B}(X)$, i.e. $\mathcal{P}(X) = 1$ and if $B_i \cap B_{i'} = T$ whenever $i \neq i'$, then $\mathcal{P}(\bigcup_{i \in J} A_i) = \sum_{i \in J} \mathcal{P}(A_i)$, then we can define a generalization of a probability measure

$$m_f(A) = \int_X f(\mu_A(x)) \, dP(x)$$

for those fuzzy events A such that the function $f \circ \mu_A$ is $\mathcal{P}$-measurable, where $\int_X$ denotes, here and in the sequel, the Lebesgue-Stieljes integral [3,5].

Proposition 2. *The inequalities of Bell are satisfied for every set of values $\{p_l, p_{ls}\}_{1 \leq l, s \leq n} \subseteq [0, 1]$ for which there exists a collection $\{A_l\}_{1 \leq l \leq n}$ of fuzzy subsets of X such that $m_f(A_l) = p_l$ and $m_f(A_l \cap A_s) = p_{ls}$ for every l, s, if and only if for every $u, v, w, t \in f([0, 1])$ we have that:*

$$0 \leq u + v - u \wedge v \leq 1 \tag{10}$$
$$0 \leq u - u \wedge v - u \wedge w + v \wedge w \tag{11}$$
$$0 \leq v - v \wedge w - u \wedge v + u \wedge w \tag{12}$$
$$0 \leq w - u \wedge w - v \wedge w + u \wedge v \tag{13}$$
$$u + v + w - u \wedge v - u \wedge w + v \wedge w \leq 1 \tag{14}$$

$$-1 \le u \wedge w + u \wedge t + v \wedge t - v \wedge w - u - t \le 0 \tag{15}$$
$$-1 \le v \wedge w + v \wedge t + u \wedge t - u \wedge w - v - t \le 0 \tag{16}$$
$$-1 \le u \wedge t + u \wedge w + v \wedge w - v \wedge t - u - w \le 0 \tag{17}$$
$$-1 \le v \wedge t + v \wedge w + u \wedge w - u \wedge t - v - w \le 0 \tag{18}$$

(Notice that inequalities (10) to (18) correspond to Bell's inequalities for u, v, w, t and their binary minimums.)

Proof. Assume that inequalities (10) ... (18) are satisfied for any set of values $\{u, v, w, t\} \subseteq f([0,1])$, and let $\{C_s\}_{s \in J}$ be a finite family of fuzzy sets whose membership mappings, which we will represent by C_s instead μ_{C_s} for the sake of simplicity, satisfy that $f \circ C_s$ is $\mathcal{P}$-measurable for every $s \in J$. Then we have that the values $f(C_i(x)), f(C_j(x)), f(C_k(x)), f(C_t(x))$ satisfy inequalities (10)—(18) for every $x \in X$ and for every i, j, k, t. And since f is nondecreasing, for any pair of values $p, q \in [0, 1]$ we have that $f(p \wedge q) = f(p) \wedge f(q)$. Therefore, if we take the values $f(C_i(x)), f(C_j(x)), f(C_k(x)), f(C_t(x))$ and, instead of their binary minimums, the values $f(C_i(x) \wedge C_j(x)), f(C_i(x) \wedge C_k(x)), \ldots$, then the corresponding inequalities are obtained. Let us denote these inequalities by

$$c_h \le D_h(f(C_i(x)), \ldots, f(C_t(x)), f(C_i(x) \wedge C_j(x)), \ldots) \le c_h' \quad h = 1, \ldots, 9$$

where every D_h is a linear function with coefficients in $\{-1, 0, 1\}$ and all c_h, c_h' can be understood as constant functions taking values in $\{-1, 0, 1\}$.

The functions c_h, c_h', D_h, for $l = 1, \ldots, 9$ are $\mathcal{P}$-measurable and Lebesgue-integrable, because they are either constant functions or additions and subtractions of $\mathcal{P}$-measurable and Lebesgue-integrable functions. Moreover, for every pair G, G' of them, if $G(x) \le G'(x)$ for every $x \in X$, then $\int_X G \, dP(x) \le \int_X G' \, dP(x)$.

Therefore, we have that

$$\int_X c_h \, dP(x) \le \int_X D_h(f(C_i(x)), \ldots, f(C_t(x)), f(C_i(x) \wedge C_j(x)), \ldots) \, dP(x)$$
$$\le \int_X c_h' \, dP(x),$$

from where we deduce that

$$c_h \int_X dP(x) \le D_h(\int_X f(C_i(x)) \, dP(x), \ldots, \int_X f(C_i(x) \wedge C_j(x)) \, dP(x), \ldots)$$
$$\le c_h' \int_X dP(x),$$

and, finally,

$$c_h \le D_h(m_f(C_i), \ldots, m_f(C_t), m_f(C_i \cap C_j), \ldots) \le c_h'$$

for every $h = 1, \ldots, 9$, which are Bell's inequalities (1)—(9).

Conversely, if there exists a collection $\{u_1, u_2, u_3, u_4\} \subseteq f([0,1])$ that violates inequality (h_0), for some $10 \le h_0 \le 18$, we can take a collection of fuzzy sets C_1, C_2, C_3, C_4 such that $f(C_i(x)) = u_i$ for every $x \in X$ and for every $i = 1, \ldots, 4$, and then

$$m_f(C_i) = \int_X f(C_i(x)) \, dP(x) = \int_X u_i \, dP(x) = u_i \int_X dP(x) = u_i.$$

Therefore, the values $m_f(C_i)$, $i = 1, \ldots, 4$, violate the inequality (h_0).

Corollary 1. *Let m_f be the measure over the fuzzy events*

$$m_f(A) = \int_X f(\mu_A(x)) \, dP(x),$$

where f is a nondecreasing function $f : [0,1] \to [0,1]$ with $f(0) = 0$ and $f(1) = 1$. Then, Bell's inequalities are satisfied for every collection of numbers $\{p_l, p_{ls}\}_{1 \leq l, s \leq n} \subseteq [0,1]$ such that there exists a collection $\{A_l\}_{1 \leq l \leq n}$ of fuzzy subsets of X with $m_f(A_l) = p_l$ and $m_f(A_l \cap A_s) = p_{ls}$ for every l, s.

Proof: By Proposition 2, we only need to prove that inequalities (10) to (18) are satisfied by any set of values $\{u, v, w, t\} \subseteq f([0,1])$. But it is an straightforward consequence of the fact, proved in [4], that Bell's inequalities hold for any $\{u, v, w, t\} \subseteq [0,1]$.

References

1. E.G. Beltrametti and M.J. Maczinski. On a characterization of classical and nonclassical probabilities. *Journal of Mathematical Physics* 32 (1991), 1280–1286
2. D. Dubois. A new definition of the fuzzy cardinality of finite sets preserving the classical additivity property. *Bull. Stud. Ecxch. Fuzziness Appl. (BUSEFAL)* 5 (1981), 11–12.
3. R. Mesiar and M. Navara. T_s-tribes and T_s-measures. *J. Math. Anal. Appl.* 201 (1996), 91–102.
4. J. Pycacz and E. Santos. Hidden variables in quantum logic approach reexamined. *Journal of Mathematical Physics* 32 (1991), 1287–1292.
5. J. Pycacz and B. d'Hooghe. Bell-Type inequalities in fuzzy probability calculus. *International Journal of Uncertainty, Fuzziness and Knowledge–Based Systems* 9 (2000), 263–275.
6. S. Pulmanova and V. Majernik. Bell inequalities on quantim logics. *Journal of Mathematical Physics* 33 (1992), 2173–2178.
7. M. Wygralak. Questions of cardinality of finite fuzzy sets. *Fuzzy Sets and Systems* 102 (1999), 185–210.
8. M. Wygralak. An axiomatic approach to scalar cardinalities of fuzzy sets. *Fuzzy Sets and Systems* 110 (2000), 175–179.
9. L.A. Zadeh. Fuzzy sets. *Information and Control* 8 (1965), 338–353.
10. L.A. Zadeh. Probability measures of fuzzy events. *J. Math. Anal. Appl.* 23 (1968), 421–427.

On the Variation of the f-Inequality of a Random Variable*

Cascos-Fernández, I., López-Díaz, M., and Gil-Álvarez, M.A.

Dpto. de Estadística e I.O. y D.M., Universidad de Oviedo, 33071 Oviedo, Spain

Abstract. In this paper we recall the definition of a generalized inequality index of a real-valued random variable, and present some new useful properties in which we analyze how this index varies in terms of the variable variation. This will serve us in a future to state set-valued inequality indices for certain random sets.

1 The f-inequality index

The inequality index associated with a positive random variable is commonly intended to be a measure of its relative dispersion. The measurement of inequality is a topic that has received attention from diverse fields, such as Economics, Industry and Social Sciences. Our study concentrates on the family of f-inequality indices (see for instance [1,3]) (based on Csiszar's generalized directed divergence between two probability distributions [5]), and we are primarily concerned on finding some properties of these indices that will help us to approximate their value when we lack of an accurate knowledge of the random variable whose inequality we are studying.

Definition 1. Let $(\Omega, \mathcal{A}, P)$ be a probability space, let $\xi : \Omega \longrightarrow \mathbb{R}^+$ be a random variable and let $f : (0, +\infty) \longrightarrow \mathbb{R}$ be a (strictly) convex function such that $f(1) = 0$, its f-**inequality index** is defined by

$$I_f(\xi) = E\left(f\left(\frac{\xi}{E(\xi)} \right) \right),$$

whenever this expectation exists.

Besides generalizing the additively decomposable inequality indices (defined for $\alpha \in \mathbb{R}$ as $I^\alpha(\xi) = I_{f_\alpha}(\xi)$ with $f_\alpha(x) = \frac{x^\alpha}{\alpha(\alpha-1)}$ if $\alpha \notin \{0,1\}$, $f_0(x) = -\log(x)$ and $f_1(x) = x\log(x)$, see for example [2]), we can mention that this family fulfills the standard desirable properties for inequality indices and some additional ones (see for instance [4,8]). Thus

- $I_f(\xi) \geq 0$, and for f strictly convex $I_f(\xi) = 0$ iff ξ is degenerated.
- $I_f(k\xi) = I_f(\xi)$ for all $k > 0$.

* The research in this paper has been partially supported by MCYT Grant DGE-99-PB98-1534. Its financial support is gratefully acknowledged

- Schur-convexity.
- Meets Pigou-Dalton Principle of Transfers.
- Meets Lorenz Criterium.
- Grouping effects.

In addition to these properties, some other ones are now presented. These new properties concern mainly to the bounding of the Euclidean distance between the f-inequality indices of two positive random variables in terms of the appropriate distance between them.

2   More properties on the f-inequality index

First, one can prove that whenever we add a positive constant to a random variable, its inequality decreases (i.e., positive constant translations applied to variable values entail a decreasing in inequality).

Lemma 1. *Given ξ a positive random variable, $k > 0$ and $f : (0, +\infty) \longrightarrow \mathbb{R}$ satisfying the conditions stated on Definition 1, monotonic and such that $I_f(\xi)$ exists, then*

$$I_f(\xi) \geq I_f(\xi + k).$$

Proof: Let us suppose that f is monotone increasing. As the domain of f, $\mathrm{dom} f = (0, +\infty)$ is non-empty and f takes on finite values over all its domain, then f has left- and right-hand derivatives (see for instance [10]), let us take the left-hand derivative at $x = 1$, denoted by $f'_-(1)$. We define $g(x) = f'_-(1)(x-1)$, whose graph is a straight line intersecting the graph of f at the point of abscissa 1 and such that $g(x) \leq f(x)$ in the whole of the domain of f. We know

$$I_f(\xi) - I_f(\xi + k) = \int \left(f\left(\frac{x}{E(\xi)}\right) - f\left(\frac{x+k}{E(\xi)+k}\right) \right) dF_\xi(x).$$

We can compute the last integral in two regions: first on $(0, E(\xi)]$ and then on $[E(\xi), +\infty)$.

As f is increasing and $x/E(\xi) < (x+k)/(E(\xi)+k)$ at the left of $E(\xi)$, then

$$\int_{\{z: z < E(\xi)\}} \left(f\left(\frac{x}{E(\xi)}\right) - f\left(\frac{x+k}{E(\xi)+k}\right) \right) dF_\xi(x) \leq 0,$$

we also get a non-positive result if we substitute f for the new function g, also increasing. For values lower than $x = 1$ the slope of g is greater or equal than the one of f, because this last one is convex, then in absolute value this number is greater or equal for g, that is,

$$\int_{\{z:z<E(\xi)\}} (g(\frac{x}{E(\xi)}) - g(\frac{x+k}{E(\xi)+k}))dF_\xi(x)$$

$$\leq \int_{\{z:z<E(\xi)\}} (f(\frac{x}{E(\xi)}) - f(\frac{x+k}{E(\xi)+k}))dF_\xi(x).$$

With the other member of the sum it happens the other way round; it is positive, and in absolute value greater than when we substitute f and write g, because for values greater than $x = 1$ the slope of g is lower than the one of f. Therefore both are greater than for g, and we have

$$I_f(\xi) - I_f(\xi + k) \geq \int \left(g\left(\frac{x}{E(\xi)}\right) - g\left(\frac{x+k}{E(\xi)+k}\right) \right) dF_\xi(x)$$

$$= f'_-(1) \int \frac{k(x - E(\xi))}{E(\xi)(E(\xi)+k)} dF_\xi(x) = 0.$$

It is clear that if f were a decreasing function the proof would follow the same structure. $\qquad\square$

Proposition 1. *For any f-inequality index I_f, any positive random variable ξ such that $I_f(\xi)$ exists and any $k > 0$, we have $I_f(\xi) \geq I_f(\xi + k)$.*

Proof: If f were monotonic we would have proved this in the previous Lemma; otherwise based on the function f that determines the f-inequality index, we define a new function g sharing its fundamental properties, but monotonic. If f is convex, but not monotonic, it would reach a minimum for some $c \in \mathbb{R}^+$. If $c \leq 1$, we define

$$g(x) = \begin{cases} f(c) & \text{if } x \leq c \\ f(x) & \text{if } x > c. \end{cases}$$

If $c > 1$, then we define g equal to f on the left-hand of point c and equal to $f(c)$ on its right-hand.

The new function g is clearly monotonic and convex. And also, if $c < 1$ (resp. $c > 1$) $(f - g)(x) = 0$, for $x > c$ (resp. $x < c$) and $f - g$ decreasing if $x < c$ (resp. increasing if $x > c$) and convex, and then $f - g$ is also monotonic and convex, so we can apply the previous result $I_{f-g}(\xi) \geq I_{f-g}(\xi + k)$. Due to the linearity of the expectation functional and to the possibility of defining both I_f and I_g, we have $I_{f-g}(\xi) = E\left(\frac{(f-g)(\xi)}{E(\xi)}\right) = I_f(\xi) - I_g(\xi)$. Joining the last two results we get that $I_f(\xi) - I_g(\xi) \geq I_f(\xi + k) - I_g(\xi + k)$; we can rewrite this and apply the previous proposition on the inequality defined by g to get

$$I_f(\xi) - I_f(\xi + k) \geq I_g(\xi) - I_g(\xi + k) \geq 0. \qquad\square$$

There is a well-known result on the classical concept of the Lorenz order (see for instance [9]) that states that two positive random variables ξ, η are ordered according to the Lorenz order $\xi \leq_{Lorenz} \eta$ if and only if $Eh\left(\frac{\xi}{E(\xi)}\right) \leq Eh\left(\frac{\eta}{E(\eta)}\right)$ for every convex function h. Now $\xi + k \leq_{Lorenz} \xi$ for every $k > 0$ can be obtained as a corollary from Proposition 1.

3 Setting bounds to the f-inequality

By finding a bound to the distance between two realizations of a general convex function, we can also set bounds to the distance between the inequality indices of two different random variables. Note that every convex function is continuous and has both, left and right-hand derivatives on all the points of its domain.

Lemma 2. *Given f convex, and given $x, h \in \mathbb{R}$, $f(x) - f(x+h) \leq f'_+(x)(-h)$.*

Proof:
If $h = 0$ it clearly holds.
If $h > 0$, let's see $f'_+(x) \leq \frac{f(x+h)-f(x)}{h}$,

$$f'_+(x) = \lim_{t\downarrow 0} \frac{f(x+t) - f(x)}{t} = \lim_{t\downarrow 0} \frac{f((1-\frac{t}{h})x + \frac{t}{h}(x+h)) - f(x)}{t}$$

for $t < h$, by convexity, we have:

$$f'_+(x) \leq \lim_{t\downarrow 0} \frac{(1-\frac{t}{h})f(x) + \frac{t}{h}f(x+h) - f(x)}{t} = \frac{f(x+h) - f(x)}{h}$$

and from there $f(x) - f(x + h) \leq f'_+(x)(-h)$.

If $h < 0$ we can show $f'_-(x) \geq \frac{f(x+h)-f(x)}{h}$, from where $f(x) - f(x+h) \leq f'_-(x)(-h) \leq f'_+(x)(-h)$. This last step is true due to the convexity of f, thanks to which the right-hand derivative of f at some given point is always bigger or equal than the left-hand derivative. $\square$

Using the latter result, it is easy to get the next two theorems.

Theorem 1. *Given a positive random variable ξ and $k > 0$, if $I_f(\xi)$ exists, we have that*

$$I_f(\xi) - I_f(\xi + k) \leq \frac{k\sqrt{\operatorname{var}(\xi)}}{E(\xi)(E(\xi) + k)} \sqrt{E\left(f'_+\left(\frac{\xi}{E(\xi)}\right)^2\right)}.$$

Proof: Just by making a small computation with the normalized random variables we get:

$$\frac{\xi}{E(\xi)} - \frac{\xi + k}{E(\xi) + k} = k\frac{\xi - E(\xi)}{E(\xi)(E(\xi) + k)}$$

and as f is convex we can apply the previous Lemma and the Cauchy-Schwartz's inequality.

$$I_f(\xi) - I_f(\xi + k) \le E\left(f'_+\left(\frac{\xi}{E(\xi)}\right) k \frac{\xi - E(\xi)}{E(\xi)(E(\xi) + k)}\right)$$

$$\le \frac{k\sqrt{\text{var}(\xi)}}{E(\xi)(E(\xi) + k)} \sqrt{E\left(f'_+\left(\frac{\xi}{E(\xi)}\right)^2\right)}. \qquad \square$$

Theorem 2. *Let $\xi, \eta \in L^p$ for some $p \ge 1$. If $\max\left\{ \left| f'_+\left(\frac{\xi}{E(\xi)}\right)\right|, \left| f'_+\left(\frac{\eta}{E(\eta)}\right)\right|\right\}$ $\in L^q$ for q such that $\frac{1}{p} + \frac{1}{q} = 1$, we have that*

$$|I_f(\xi) - I_f(\eta)|$$

$$\le \frac{\left\| \max\left\{ \left| f'_+\left(\frac{\xi}{E(\xi)}\right)\right|, \left| f'_+\left(\frac{\eta}{E(\eta)}\right)\right|\right\}\right\|_q}{\|\xi\|_1 \|\eta\|_1} (\|\xi - \eta\|_1 \|\eta\|_p + \|\xi - \eta\|_p \|\eta\|_1).$$

Proof: Let us define $\zeta = \xi - \eta$. As $(L^p, +, \cdot)$ is a vector space on the body of the real numbers, $\zeta \in L^p$.

Now, we have that

$$\frac{\xi}{E(\xi)} - \frac{\eta}{E(\eta)} = \frac{\xi E(\eta) - \eta E(\xi)}{E(\xi)E(\eta)} = \frac{\xi E(\eta) - \eta E(\zeta + \eta)}{E(\xi)E(\eta)}$$

$$= \frac{(\xi - \eta)E(\eta) - \eta E(\zeta)}{E(\xi)E(\eta)} = \frac{\zeta E(\eta) - \eta E(\zeta)}{E(\xi)E(\eta)}.$$

In accordance with Lemma 2 for every $\omega \in \Omega$

$$f\left(\frac{\xi(\omega)}{E(\xi)}\right) - f\left(\frac{\eta(\omega)}{E(\eta)}\right) \le \left(\frac{\zeta(\omega)E(\eta) - \eta(\omega)E(\zeta)}{E(\xi)E(\eta)}\right) f'_+\left(\frac{\xi(\omega)}{E(\xi)}\right).$$

By exchanging the roles played by ξ and η, we get

$$\frac{\eta}{E(\eta)} - \frac{\xi}{E(\xi)} = \frac{\eta E(\zeta) - \zeta E(\eta)}{E(\xi)E(\eta)}$$

whence for every $\omega \in \Omega$

$$f\left(\frac{\eta(\omega)}{E(\eta)}\right) - f\left(\frac{\xi(\omega)}{E(\xi)}\right) \le \left(\frac{\eta(\omega)E(\zeta) - \zeta(\omega)E(\eta)}{E(\xi)E(\eta)}\right) f'_+\left(\frac{\eta(\omega)}{E(\eta)}\right),$$

and hence

$$|I_f(\xi) - I_f(\eta)| \leq \frac{E\big(|\eta E(\zeta) - \zeta E(\eta)| \max\{|f'_+(\frac{\xi}{E(\xi)})|, |f'_+(\frac{\eta}{E(\eta)})|\}\big)}{E(\xi)E(\eta)}$$

$$\leq \frac{E|\zeta|E\big(\eta \max\{|f'_+(\frac{\xi}{E(\xi)})|, |f'_+(\frac{\eta}{E(\eta)})|\}\big)}{E(\xi)E(\eta)}$$

$$+ \frac{E(\eta)E\big(|\zeta| \max\{|f'_+(\frac{\xi}{E(\xi)})|, |f'_+(\frac{\eta}{E(\eta)})|\}\big)}{E(\xi)E(\eta)}.$$

Now, by Hölder's inequality

$$|I_f(\xi) - I_f(\eta)| \leq \frac{E|\zeta| \|\eta\|_p \| \max\{|f'_+(\frac{\xi}{E(\xi)})|, |f'_+(\frac{\eta}{E(\eta)})|\}\|_q}{E(\xi)E(\eta)}$$

$$+ \frac{E(\eta)\|\zeta\|_p \| \max\{|f'_+(\frac{\xi}{E(\xi)})|, |f'_+(\frac{\eta}{E(\eta)})|\}\|_q}{E(\xi)E(\eta)}$$

with $\frac{1}{p} + \frac{1}{q} = 1 \leq p, q \leq \infty$. Consequently,

$$|I_f(\xi) - I_f(\eta)|$$

$$\leq \frac{\| \max\{|f'_+(\frac{\xi}{E(\xi)})|, |f'_+(\frac{\eta}{E(\eta)})|\}\|_q}{\|\xi\|_1 \|\eta\|_1} \big(\|\xi - \eta\|_1 \|\eta\|_p\| + \|\xi - \eta\|_p \|\eta\|_1 \big). \quad \square$$

Finally, we are going to formalize the fact that when the available information on a random variable corresponds to data grouped in intervals, the maximal inequality is achieved when its realizations are concentrated on the boundary of the intervals, and the minimal inequality is achieved for a random variable concentrating all of its mass on one point for each interval.

Lemma 3. *Given h a convex function and being ξ a random variable with bounded support, $\text{supp}(\xi) \subseteq [a, b]$, then*

$$E(h(\xi)) \leq h(a)\frac{b - E(\xi)}{b - a} + h(b)\frac{E(\xi) - a}{b - a}.$$

We will give a proof of this Lemma simpler than the one by [6].

Proof: We define $g(x) = h(a)\frac{b-x}{b-a} + h(b)\frac{x-a}{b-a}$, due to the convexity of h it is clear that for $x \in [a, b], h(x) \leq g(x)$, and then $E(h(\xi)) \leq E(g(\xi)) = h(a)\frac{b-E(\xi)}{b-a} + h(b)\frac{E(\xi)-a}{b-a}$. $\quad \square$

Proposition 2. *Given a random variable ξ with finite support and assuming that the probabilities of ξ lying on each one of the sets of a partition of its support $(p_i = P(\xi \in [a_{i-1}, a_i]), \quad i = 1, \ldots, n)$ are know to us, then*

$$\inf_{x_i \in [a_{i-1}, a_i]} \sum_{i=1}^{n} p_i f\left(\frac{x_i}{\sum p_j x_j}\right) \le I_f(\xi)$$

$$\le \sup_{x_i \in [a_{i-1}, a_i]} \sum_{i=1}^{n} p_i \left[\left(\frac{a_i - x_i}{a_i - a_{i-1}}\right) f\left(\frac{a_{i-1}}{\sum p_j x_j}\right) + \left(\frac{x_i - a_{i-1}}{a_i - a_{i-1}}\right) f\left(\frac{a_i}{\sum p_j x_j}\right)\right]$$

The previous extreme values are reachable, because f is convex, therefore continuous and then the functions involving linear combinations of f are also continuous and are maximized on a compact set.

4 Conclusions

The properties on the last section bounding the distance between inequality indices will be used in the future to define inequality indices of set-valued random events.

References

1. Alonso, M.C., Brezmes, T., Lubiano, M.A. and Bertoluzza C. (2001) A generalized real-valued measure of the inequality associated with a fuzzy random variable. Int. J. Approx. Reason. **26**, 47–66.
2. Bourguignon, F. (1979) Decomposable income inequality measures. Econometrica **47**, 901–920.
3. Cascos, I. (2001) Asymptotic Results on Inequality Indices for Radom Sets. Master Thesis. Universidad de Oviedo.
4. Colubi, A. (1997) The fuzzy f-inequality indices associated to a fuzzy random variable *(in Spanish)*. Master Thesis. Universidad de Oviedo.
5. Csiszár, I. (1967) Information-type mesaures of difference of probability distributions and indirect observations. Studia Scient. Math. Hung. **2**, 299–318.
6. Gastwirth, J.L. (1975) The Estimation of a Family of Measures of Economic Inequality. Journal of Econometrics **3**, 61–70.
7. Gastwirth, J.L., Nayak, T.K. and Krieger, A.M. (1986) Large Sample Theory for the Bounds on the Gini and Related Indices of Inequality Estimated From Grouped Data. Journal of Business & Economic Statistics **4**, 269–273.
8. Lubiano Gómez, M.A. (1998) Variation measures for imprecise random elements. PhD Thesis. Universidad de Oviedo.
9. Shaked, M. and Shanthikumar, J.G. (1994) Stochastic Orders and Their Applications. Academic Press. Boston.
10. Rockafellar, R.T. (1970) Convex Analysis. Princeton University Press.

Probability of Intuitionistic Fuzzy Events

Przemysław Grzegorzewski and Edyta Mrówka

Systems Research Institute, Polish Academy of Sciences,
Newelska 6, 01-447 Warsaw, Poland,
E-mail: e-mail: pgrzeg@ibspan.waw.pl, mrowka@ibspan.waw.pl

Abstract. The notion of probability measure of intuitionistic fuzzy events is investigated and basic properties of that concept are examined.

1 Introduction

Probability theory provides very powerful tools for dealing with uncertainty. However, in classical theory all random events should be precisely defined. Unfortunately this assumption appears too rigid in many real life problems. Very often people deal with imprecisely defined notions, like: "high income", "cloudy sky", "low temperature", etc. For the traditional probability theory such expressions are ill defined and they are beyond the scope of that theory.

To handle situations like described above Zadeh [9] introduced the concept of a fuzzy set. In a conventional fuzzy set a membership function assigns to each element of the universe of discourse a number from the unit interval to indicate the degree of belongingness to the set under consideration. The degree of nonbelongingness is just automatically the complement to 1 of the membership degree. However, in real life the linguistic negation not always identifies with logical negation. This situation is very common in natural language processing, computing with words, etc. Therefore Atanassov [1] suggested a generalization of the classical fuzzy set, called an intuitionistic fuzzy set, which is characterized by two functions expressing the degree of belongingness and the degree of nonbelongingness, respectively. This idea, which is a natural generalization of the classical fuzzy set, seems to be useful in modeling many real life situations, like negotiation processes, etc. (see [3], [5], [6]).

Zadeh [10] was the first who defined a fuzzy event and suggested how to compute probabilities of such events. Gerstenkorn and Mańko [4] and Szmidt and Kacprzyk ([7], [8]) discussed how to define the probability of an intuitionistic fuzzy event in a finite universe of discourse. In this paper we give a more general definition and we examine basic properties of the probability of intuitionistic fuzzy events. Our approach is consistent with Zadeh's approach when considered intuitionistic fuzzy events become fuzzy events.

2 Intuitionistic fuzzy sets

Let X denote a universe of discourse. Then a fuzzy set A in X is defined as a set of ordered pairs, i.e.

$$A = \{\langle x, \mu_A(x) \rangle : x \in X\}, \tag{1}$$

where $\mu_A : X \to [0,1]$ is the membership function of A and $\mu_A(x)$ is the grade of belongingness of x to A. Thus automatically the grade of nonbelongingness of x to A is equal to $1 - \mu_A(x)$. An intuitionistic fuzzy set A in X is given by an ordered triple

$$A = \{\langle x, \mu_A(x), \nu_A(x) \rangle : x \in X\}, \tag{2}$$

where $\mu_A, \nu_A : X \to [0,1]$ such that

$$0 \le \mu_A(x) + \nu_A(x) \le 1 \qquad \forall x \in X. \tag{3}$$

For each x the numbers $\mu_A(x)$ and $\nu_A(x)$ represent the degree of membership and degree of nonmembership of the element $x \in X$ to $A \subset X$, respectively. It is easily seen that a $\{\langle x, \mu_A(x), 1 - \mu_A(x)\rangle : x \in X\}$ is equivalent to (1), i.e. each fuzzy set is a particular case of the intuitionistic fuzzy set. For each element $x \in X$ we can compute, so called, the intuitionistic fuzzy index (hesitation margin) of x in A defined as follows

$$\pi_A(x) = 1 - \mu_A(x) - \nu_A(x), \tag{4}$$

which measures the degree of hesitation of whether x belongs to A.

In his papers [1], [2] Atanassov defined basic operations on intuitionistic fuzzy sets corresponding to classical and algebraic t−norms and s−norms, respectively. For intuitionistic fuzzy sets A, B we have

$$A \cup B = \{\langle x, \mu_A(x) \vee \mu_B(x), \nu_A(x) \wedge \nu_B(x) \rangle : x \in X\}, \tag{5}$$

$$A \cap B = \{\langle x, \mu_A(x) \wedge \mu_B(x), \nu_A(x) \vee \nu_B(x) \rangle : x \in X\}, \tag{6}$$

$$A \oplus B = \{\langle x, \mu_A(x) + \mu_B(x) - \mu_A(x)\mu_B(x), \nu_A(x)\nu_B(x) \rangle : x \in X\}, \tag{7}$$

$$A \cdot B = \{\langle x, \mu_A(x)\mu_B(x), \nu_A(x) + \nu_B(x) - \nu_A(x)\nu_B(x) \rangle : x \in X\}, \tag{8}$$

The complement of an intuitionistic fuzzy set A is an intuitionistic fuzzy set A^c such that

$$A^c = \{\langle x, \nu_A(x), \mu_A(x) \rangle : x \in X\}. \tag{9}$$

Two intuitionistic fuzzy sets A and B are equal if and only if their membership and nonmembership function are equal, i.e.

$$A = B \Leftrightarrow (\mu_A(x) = \mu_B(x) \quad \text{and} \quad v_A(x) = v_B(x) \quad \forall x \in X), \tag{10}$$

while

$$A \subset B \iff (\mu_A(x) \le \mu_B(x) \quad \text{and} \quad v_A(x) \ge v_B(x) \quad \forall x \in X). \tag{11}$$

We say that intuitionistic fuzzy sets A and B are disjoint if their intersection is empty. Therefore, we get different definitions according to different t-norms, e.g.

$$A \text{ and } B \text{ are disjoint} \iff A \cap B = \emptyset \tag{12}$$

$$A \text{ and } B \text{ are disjoint} \iff A \cdot B = \emptyset. \tag{13}$$

It is easily seen that definitions (12) and (13) are equivalent.

3   Intuitionistic fuzzy events

The basic concept of probability theory is a probability space $(X, \mathcal{A}, P)$, where X is a sample space, $\mathcal{A}$ is a $\sigma-$ field of subsets of X and P is a real-valued function which assigns to every event A in $\mathcal{A}$ its probability $P(A)$. Zadeh [10] extended the notions of an event and its probability to fuzzy context. Further on, we shall assume for simplicity, that X is a set in $\mathcal{R}^n$, $\mathcal{A}$ is the smallest Borel $\sigma-$field on X and P is a probability distribution on $(X, \mathcal{A})$. Then, according to Zadeh's definition, a fuzzy event in X is a fuzzy set A in X whose membership function μ_A is Borel measurable and the probability of such fuzzy event is given by

$$P(A) = \int_X \mu_A(x) dP. \tag{14}$$

The existence of that Lebesgue-Stielties integral is assured by the assumption that μ_A is Borel measurable. In his paper Zadeh also showed some properties of the probabilities defined on fuzzy events.

With reference to Zadeh's paper Szmidt and Kacprzyk ([7], [8]) proposed the definition of intuitionistic fuzzy event and its probability. By an intuitionistic fuzzy event A they mean an intuitionistic fuzzy subset of the universe of discourse whose membership function μ_A and hesitation margin π_A are Borel measurable. Then they defined the notion of the probability of an intuitionistic fuzzy set. However, they considered finite universe of discourse only. Below we present a more general definition formulated in the spirit of the definition given by Szmidt and Kacprzyk.

Definition *Let A denote an intuitionistic fuzzy event in X whose membership function μ_A and nonmembership function v_A are Borel measurable, where X is a set in $\mathcal{R}^n$, and let P denote a probability measure over X. Then the probability of the intuitionistic fuzzy event A is a number $p(A)$ from the interval*

$$\mathcal{P}(A) = [p_{\min}(A), p_{\max}(A)], \tag{15}$$

where

$$p_{\min}(A) = \int_X \mu_A(x)dP, \tag{16}$$

$$p_{\max}(A) = \int_X (\mu_A(x) + \pi_A(x))dP = 1 - \int_X \nu_A(x)dP. \tag{17}$$

$p_{\min}(A)$ gives "sure" probability that the event A will occur while $p_{\max}(A)$ gives the highest possible probability that the event A will occur. It is achieved only if the hesitation margin function support occurrence of the event A. The difference between the maximal and minimal probabilities, i.e. $p_{\max}(A) - p_{\min}(A)$, reflects the unsureness of occurrence of the intuitionistic fuzzy event A. Therefore, if A is a classical fuzzy set, then (15) obviously reduces to (14), i.e. to the probability of a fuzzy set in the sense of Zadeh's definition. We will denote the family of intuitionistic fuzzy events in the universe of discourse X by $IFE(X)$.

Since probabilities of intuitionistic fuzzy events are described by intervals let us recall well known formulae for arithmetic addition, subtraction, multiplication and division of closed intervals. In particular, assuming that $I_1 = [a, b]$ and $I_2 = [c, d]$, we get

$$[a, b] + [c, d] = [a + c, b + d], \tag{18}$$
$$[a, b] - [c, d] = [a - d, b - c], \tag{19}$$
$$[a, b] \circ [c, d] = [\min\{ac, ad, bc, bd\}, \max\{ac, ad, bc, bd\}], \tag{20}$$
$$[a, b]/[c, d] = [\min\{a/c, a/d, b/c, b/d\}, \max\{a/c, a/d, b/c, b/d\}], \tag{21}$$

provided that $c \neq 0$ and $d \neq 0$ in (21). This way we may also consider operations between interval and a real number $\lambda \in \mathcal{R}$ treating λ as a degenerated interval, i.e. $[\lambda, \lambda]$.

We say that two intervals are equal if

$$[a, b] = [c, d] \Leftrightarrow (a = c \quad \text{and} \quad b = d) \tag{22}$$

while

$$[a, b] \geq [c, d] \Leftrightarrow (a \geq c \quad \text{and} \quad b \geq d). \tag{23}$$

A sequence of intervals $I_1, I_2, \ldots$ such that $I_n = [a_n, b_n], n = 1, 2, \ldots$ tends to interval $I = [a, b]$ if the interval borders tend to a and b, respectively, i.e.

$$\lim_{n \to \infty} I_n = I \Leftrightarrow (\lim_{n \to \infty} a_n = a \quad \text{and} \quad \lim_{n \to \infty} b_n = b). \tag{24}$$

In the sequel we also use a following notation: $\mathbb{O} = [0, 0]$ and $\mathbf{I} = [1, 1]$.

Last of all we have to mention that Gerstenkorn and Mańko [4] were the first who considered the concept of a probability of intuitionistic fuzzy events.

One can easily see that their definition of the probability of intuitionistic fuzzy subset A of a finite universe of discourse, is a real number that coincide with the average $\frac{1}{2}(p_{\min}(A) + p_{\max}(A))$, where $p_{\min}(A)$ and $p_{\max}(A)$ are given by (16) and (17), respectively, and P is a counting measure.

4 Properties

In this section we briefly examine all basic properties of the probability of intuitionistic fuzzy events. We are interested especially in their comparison with the properties of the classical crisp events and fuzzy events. Let X denote a set in $\mathcal{R}^n$ and let P be a probability measure over X. Then we have

Proposition 1. *The following relations are valid for each $A \in IFE(X)$:*

$$\mathcal{P}(A) \geq \mathbb{0} \tag{25}$$

$$\mathcal{P}(X) = \mathbb{I} \tag{26}$$

$$\mathcal{P}(\emptyset) = \mathbb{0} \tag{27}$$

$$if \quad A \subset B \quad then \quad \mathcal{P}(A) \leq \mathcal{P}(B) \tag{28}$$

$$\mathcal{P}(A) \leq \mathbb{I} \tag{29}$$

$$p_{\min}(A) + p_{\max}(A^c) = 1 \tag{30}$$

$$p_{\max}(A) + p_{\min}(A^c) = 1 \tag{31}$$

$$1 \in \mathcal{P}(A^c) + \mathcal{P}(A). \tag{32}$$

Proof:
We see at once that $p_{\min}(A) = \int_X \mu_A(x)dP \geq 0$ hence (25) holds. Similarly, $p_{\min}(X) = \int_X \mu_X(x)dP = \int_X 1dP = 1$ and (26) holds. Since $p_{\max}(\emptyset) = 1 - \int_X \nu_\emptyset(x)dP = 1 - \int_X 1dP = 0$ which gives (27).

By (11) $\mu_A(x) \leq \mu_B(x)$ and $\nu_A(x) \geq \nu_B(x)$ for all $x \in X$. Therefore $p_{\min}(A) = \int_X \mu_A(x)dP \leq \int_X \mu_B(x)dP = p_{\min}(B)$ and $p_{\max}(A) = 1 - \int_X \nu_A(x)dP \leq 1 - \int_X \nu_B(x)dP = p_{\max}(B)$, hence $\mathcal{P}(A) \leq \mathcal{P}(B)$ and (28) is proved. Since $A \subseteq X$, (28) and (26) shows that $\mathcal{P}(A) \leq \mathcal{P}(X) = \mathbb{I}$ and (29) holds.

By definition (9) of the compliment A^c of A we have

$$p_{\min}(A) + p_{\max}(A^c) = \int_X \mu_A(x)dP + \int_X (1 - \mu_A(x))dP = 1$$

$$p_{\max}(A) + p_{\min}(A^c) = \int_X (1 - \nu_A(x))dP + \int_X \nu_A(x)dP = 1$$

which proves (30) and (31). Moreover, by (3) and (4)

$$\mathcal{P}(A^c) + \mathcal{P}(A) = [p_{\min}(A), p_{\max}(A)] + [p_{\min}(A^c), p_{\max}(A^c)]$$
$$= [p_{\min}(A) + p_{\min}(A^c),\ p_{\max}(A) + p_{\max}(A^c)]$$
$$= \left[1 - \int_X \pi_A(x)dP,\ 1 + \int_X \pi_A(x)dP\right] \ni 1 \qquad (33)$$

and (32) holds. □

Remark:
Properties (25)–(29) correspond to those known from the classical probability theory, namely: nonnegativity, norming, zero probability of the impossible event, monotonicity. However, (33) shows us that if $\int_X \pi_A(x)dP > 0$ then

$$\mathcal{P}(A^c) + \mathcal{P}(A) \neq \mathbf{I} \qquad (34)$$

while both for fuzzy events and for crisp events the complementation property $P(A) + P(A^c) = 1$ holds.

Proposition 2. *The following equalities hold for all* $A, B \in IFE(X)$:

$$\mathcal{P}(A \cup B) + \mathcal{P}(A \cap B) = \mathcal{P}(A) + \mathcal{P}(B) \qquad (35)$$
$$\mathcal{P}(A \oplus B) + \mathcal{P}(A \cdot B) = \mathcal{P}(A) + \mathcal{P}(B). \qquad (36)$$

The proof follows immediately from (5)–(8). From (12) and (13) we obtain

Corollary 1. *If* $A, B \in IFE(X)$ *are disjoint then*

$$\mathcal{P}(A \cup B) = \mathcal{P}(A) + \mathcal{P}(B) \qquad (37)$$
$$\mathcal{P}(A \oplus B) = \mathcal{P}(A) + \mathcal{P}(B). \qquad (38)$$

In this paper we consider two definitions of the union of intuitionistic fuzzy sets and two definitions of the intersection. Below we show the relation between corresponding probabilities.

Proposition 3. *The following inequalities hold for all* $A, B \in IFE(X)$:

$$\mathcal{P}(A \cup B) \leq \mathcal{P}(A \oplus B) \qquad (39)$$
$$\mathcal{P}(A \cap B) \geq \mathcal{P}(A \cdot B). \qquad (40)$$

Proof:
We first recall that $a \vee b \leq a + b - ab$ and $a \wedge b \geq ab$ for any $a, b \in [0, 1]$. Hence we get $\mu_A(x) \vee \mu_B(x) \leq \mu_A(x) + \mu_B(x) - \mu_A(x)\mu_B(x)$ and $\nu_A(x) \wedge \nu_B(x) \geq \nu_A(x)\nu_B(x) \ \forall x \in X$ which means by (5) and (7) that $\mu_{A \cup B}(x) \leq \mu_{A \oplus B}(x) \ \forall x \in X$ and therefore by (11) $A \cup B \subseteq A \oplus B$. Now according to (28) we have $\mathcal{P}(A \cup B) \leq \mathcal{P}(A \oplus B)$.

Similarly, $\mu_A(x) \wedge \mu_B(x) \geq \mu_A(x)\mu_B(x)$ and $\nu_A(x) \vee \nu_B(x) \leq \nu_A(x) + \nu_B(x) - \nu_A(x)\nu_B(x) \ \forall x \in X$ which means by (6) and (8) that $\mu_{A \cap B}(x) \geq \mu_{A \cdot B}(x) \ \forall x \in X$ and therefore by (11) $A \cap B \subseteq A \cdot B$. Hence according to (28) we get $\mathcal{P}(A \cap B) \geq \mathcal{P}(A \cdot B)$ which completes the proof. □

Proposition 4. *Let $A_1, \ldots, A_n \in IFE(X)$. Then following equalities hold*

$$P(\bigcup_{k=1}^{n} A_k) = \sum_{k=1}^{n} (-1)^{k-1} \sum_{1 \leq i_1 < i_2 < \ldots < i_k \leq n} P(A_{i_1} \cap A_{i_2} \cap \ldots \cap A_{i_k}) \quad (41)$$

$$P(\bigoplus_{k=1}^{n} A_k) = \sum_{k=1}^{n} (-1)^{k-1} \sum_{1 \leq i_1 < i_2 < \ldots < i_k \leq n} P(A_{i_1} \cdot A_{i_2} \cdot \ldots \cdot A_{i_k}). \quad (42)$$

Proof:
Let us first prove (41). It is easy to check that

$$P(\bigcup_{k=1}^{n} A_k) = \left[\int_X (\mu_{A_1} \vee \ldots \vee \mu_{A_n}) dP, 1 - \int_X (\nu_{A_1} \wedge \ldots \wedge \nu_{A_n}) dP \right].$$

From (35) it follows that (41) holds for $n = 2$. Assuming (41) holds for n, we will prove it for $n + 1$

$$P(\bigcup_{k=1}^{n+1} A_k) = P(\bigcup_{k=1}^{n} A_k \cup A_{n+1}).$$

By (35) we have

$$p_{\min}(\bigcup_{k=1}^{n+1} A_k) = p_{\min}(\bigcup_{k=1}^{n} A_k) + p_{\min}(A_{n+1}) - p_{\min}(\bigcup_{k=1}^{n} A_k \cap A_{n+1})$$

$$= \int_X (\mu_{A_1} \vee \ldots \vee \mu_{A_n}) dP + \int_X \mu_{A_{n+1}} dP - \int_X ((\mu_{A_1} \vee \ldots \vee \mu_{A_n}) \wedge \mu_{A_{n+1}}) dP$$

$$= \int_X (\mu_{A_1} \vee \ldots \vee \mu_{A_n} \vee \mu_{A_{n+1}}) dP$$

and in a similar way we get

$$p_{\max}(\bigcup_{k=1}^{n+1} A_k) = 1 - \int_X (\nu_{A_1} \wedge \ldots \wedge \nu_{A_n} \wedge \nu_{A_{n+1}}) dP.$$

Hence

$$\left[\int_X (\mu_{A_1} \vee \ldots \vee \mu_{A_{n+1}}) dP, 1 - \int_X (\nu_{A_1} \wedge \ldots \wedge \nu_{A_{n+1}}) dP \right] = P(\bigcup_{k=1}^{n+1} A_k)$$

which is the desired conclusion.

Now let us turn to (42). It follows easily that

$$\mathcal{P}(\bigoplus_{k=1}^{n} A_k) = \left[\int_X \left(\sum_{i=1}^{n} \mu_{A_i} - \sum_{1 \le i < j \le n} \mu_{A_i}\mu_{A_j} + \sum_{1 \le i < j < k \le n} \mu_{A_i}\mu_{A_j}\mu_{A_k}\right.\right.$$

$$\left.\left.- \ldots + (-1)^{n-1}\mu_{A_1} \cdot \ldots \cdot \mu_{A_n}\right) dP, 1 - \int_X \nu_{A_1} \cdot \ldots \cdot \nu_{A_n} dP\right].$$

From (36) it follows that (42) holds for $n = 2$. Assuming (42) holds for n, we will prove it for $n + 1$

$$\mathcal{P}(\bigoplus_{k=1}^{n+1} A_k) = \mathcal{P}(\bigoplus_{k=1}^{n} A_k \oplus A_{n+1}).$$

By (36) we have

$$p_{\min}(\bigoplus_{k=1}^{n+1} A_k) = p_{\min}(\bigoplus_{k=1}^{n} A_k) + p_{\min}(A_{n+1}) - p_{\min}(\bigoplus_{k=1}^{n} A_k \cdot A_{n+1})$$

$$= \int_X \left(\sum_{i=1}^{n} \mu_{A_i} - \sum_{1 \le i < j \le n} \mu_{A_i}\mu_{A_j} + \sum_{1 \le i < j < k \le n} \mu_{A_i}\mu_{A_j}\mu_{A_k}\right.$$

$$\left.- \ldots + (-1)^{n-1}\mu_{A_1} \cdot \ldots \cdot \mu_{A_n}\right) dP + \int_X \mu_{A_{n+1}} dP$$

$$- \int_X \left(\sum_{i=1}^{n} \mu_{A_i} - \sum_{1 \le i < j \le n} \mu_{A_i}\mu_{A_j} + \sum_{1 \le i < j < k \le n} \mu_{A_i}\mu_{A_j}\mu_{A_k}\right.$$

$$\left.- \ldots + (-1)^{n-1}\mu_{A_1} \cdot \ldots \cdot \mu_{A_n}\right) \cdot \mu_{A_{n+1}} dP$$

$$= \int_X \left(\sum_{i=1}^{n+1} \mu_{A_i} - \sum_{1 \le i < j \le n+1} \mu_{A_i}\mu_{A_j} + \sum_{1 \le i < j < k \le n+1} \mu_{A_i}\mu_{A_j}\mu_{A_k}\right.$$

$$\left.- \ldots + (-1)^{n}\mu_{A_1} \cdot \ldots \cdot \mu_{A_n} \cdot \mu_{A_{n+1}}\right) dP$$

and similarly we get

$$p_{\max}(\bigoplus_{k=1}^{n+1} A_k) = 1 - \int_X \nu_{A_1} \cdot \ldots \cdot \nu_{A_n} \cdot \nu_{A_{n+1}} dP.$$

Hence

$$\left[\int_X \left(\sum_{i=1}^{n+1} \mu_{A_i} - \sum_{1 \le i < j \le n+1} \mu_{A_i}\mu_{A_j} + \sum_{1 \le i < j < k \le n+1} \mu_{A_i}\mu_{A_j}\mu_{A_k}\right.\right.$$

$$-\ldots+(-1)^n \mu_{A_1} \cdot \ldots \cdot \mu_{A_n} \cdot \mu_{A_{n+1}}\Big)\, dP, 1-\int_X \nu_{A_1} \cdot \ldots \cdot \nu_{A_n} \cdot \nu_{A_{n+1}}\, dP\Bigg]$$

$$= \mathcal{P}\Big(\bigoplus_{k=1}^{n+1} A_k\Big)$$

which completes the proof. □

According to the above proposition we get immediately finite additivity and subadditivity.

Corollary 2. *If* $A_1, \ldots, A_n \in IFE(X)$ *are pairwise disjoint intuitionistic fuzzy events then:*

$$\mathcal{P}(A_1 \cup \ldots \cup A_n) = \mathcal{P}(A_1) + \mathcal{P}(A_2) + \ldots + \mathcal{P}(A_n) \qquad (43)$$
$$\mathcal{P}(A_1 \oplus \ldots \oplus A_n) = \mathcal{P}(A_1) + \mathcal{P}(A_2) + \ldots + \mathcal{P}(A_n). \qquad (44)$$

Corollary 3. *If* $A_1, \ldots, A_n, \ldots \in IFE(X)$ *then*

$$\mathcal{P}(A_1 \cup \ldots \cup A_n) \leq \mathcal{P}(A_1) + \mathcal{P}(A_2) + \ldots + \mathcal{P}(A_n) \qquad (45)$$
$$\mathcal{P}(A_1 \oplus \ldots \oplus A_n) \leq \mathcal{P}(A_1) + \mathcal{P}(A_2) + \ldots + \mathcal{P}(A_n). \qquad (46)$$

Next proposition shows that the probability of intuitionistic fuzzy events is continuous from above and from below.

Proposition 5. *Let* $A_1, \ldots, A_n, \ldots \in IFE(X)$ *such that* $A_n \subseteq A_{n+1}$ *and* $A = \bigcup_{n=1}^{\infty} A_n \in IFE(X)$. *Then*

$$\lim_{n \to \infty} \mathcal{P}(A_n) = \mathcal{P}(A), \qquad (47)$$

while for $A_1, \ldots, A_n, \ldots \in IFE(X)$ *such that* $A_n \supseteq A_{n+1}$ *and* $A = \bigcap_{n=1}^{\infty} A_n \in IFE(X)$ *we have*

$$\lim_{n \to \infty} \mathcal{P}(A_n) = \mathcal{P}(A). \qquad (48)$$

Proof:
Let $A_n = \{\langle x, \mu_{A_n}(x), \nu_{A_n}(x)\rangle : x \in X\}$, $A = \{\langle x, \mu_A(x), \nu_A(x)\rangle : x \in X\}$ and $A = \bigcup_{n=1}^{\infty} A_n$ where $\mu_A(x) = \sup_n \mu_{A_n}(x)$ and $\nu_A(x) = \inf_n \nu_{A_n}(x)$ $\forall x \in X$. Since $|\mu_{A_n}(x)| \leq 1$ and $|\nu_{A_n}(x)| \leq 1$ $\forall x \in X$ then by the Lebesgue theorem on the dominated convergence we get

$$p_{\min}(A_n) = \int_X \mu_{A_n}(x) dP \to \int_X \mu_A(x) dP = p_{\min}(A)$$

$$p_{\max}(A_n) = 1 - \int_X \nu_{A_n}(x) dP \to 1 - \int_X \nu_A(x) dP = p_{\max}(A)$$

and thus $P(A_n) \longrightarrow P(A)$ which proves (47). The proof for (48) is similar. $\square$

Now we show the countable additivity of the probability measure for intuitionistic fuzzy events.

Proposition 6. *If* $A_1, \ldots, A_n, \ldots \in IFE(X)$ *are pairwise disjoint and* $\bigcup\limits_{i=1}^{\infty} A_i$, $\bigoplus\limits_{i=1}^{\infty} A_i \in IFE(X)$ *then*

$$P(\bigcup_{i=1}^{\infty} A_i) = \sum_{i=1}^{\infty} P(A_i), \tag{49}$$

$$P(\bigoplus_{i=1}^{\infty} A_i) = \sum_{i=1}^{\infty} P(A_i). \tag{50}$$

Proof:
Since our intuitionistic fuzzy events are disjoint, (37) shows that

$$P(\bigcup_{i=1}^{\infty} A_i) = P(\bigcup_{i=1}^{n} A_i) + P(\bigcup_{i=n+1}^{\infty} A_i).$$

From (43) we conclude that

$$\sum_{i=1}^{\infty} P(A_i) = \lim_{n\to\infty} \sum_{i=1}^{n} P(A_i) = \lim_{n\to\infty} P(\bigcup_{i=1}^{n} A_i)$$

$$= \lim_{n\to\infty} \left[P(\bigcup_{i=1}^{\infty} A_i) + (-1)P(\bigcup_{i=n+1}^{\infty} A_i) \right]$$

$$= P(\bigcup_{i=1}^{\infty} A_i) + \lim_{n\to\infty} (-1)P(\bigcup_{i=n+1}^{\infty} A_i) = P(\bigcup_{i=1}^{\infty} A_i) + (-1)\lim_{n\to\infty} P(\bigcup_{i=n+1}^{\infty} A_i)$$

Since $\bigcup\limits_{i=n+1}^{\infty} A_i \to \emptyset$ as $n \to \infty$, then by (48)) $\lim_{n\to\infty} P(\bigcup\limits_{i=n+1}^{\infty} A_i) = 0$. Hence $\sum\limits_{i=1}^{\infty} P(A_i) = P(\bigcup\limits_{i=1}^{\infty} A_i)$. The similar proof works for (50). $\square$

Therefore, we get at once subadditivity:

Corollary 4. *If* $A_1, \ldots, A_n, \ldots \in IFE(X)$ *then*

$$P(\bigcup_{i=1}^{\infty} A_i) \le \sum_{i=1}^{\infty} P(A_i) \tag{51}$$

$$P(\bigoplus_{i=1}^{\infty} A_i) \le \sum_{i=1}^{\infty} P(A_i). \tag{52}$$

5   Conclusions

In the paper we try to handle two types of uncertainty: randomness - described by the probability theory, and imprecision - expressed here by intuitionistic fuzzy set theory. Both sources of uncertainty play a central role in decision making. We have investigated the fundamental notion of that approach – the probability of intuitionistic fuzzy event. We have examined basic properties of the probability of an intuitionistic fuzzy events. As it was shown, many properties are equivalent to those known from the classical probability theory of crisp events or fuzzy events. However, there are also some dissimilarities. This is due to the fact that the probability of intuitionistic fuzzy event is described by an interval, contrary to the classical probability theory and to fuzzy set theory where it is given by a single real number.

References

1. Atanassov K. (1986), *Intuitionistic Fuzzy Sets*, Fuzzy Sets and Systems **20**, 87–96.
2. Atanassov K. (1989), *More on Intuitionistic Fuzzy Sets*, Fuzzy Sets and Systems **33**, 37–46.
3. Atanassov K. (1999), *Intuitionistic Fuzzy Sets: Theory and Applications*, Physica-Verlag.
4. Gerstenkorn T., Mańko J., (1991), Probability of Fuzzy Intuitionistic Sets, BUSEFAL 45, 128–136.
5. Szmidt E., Kacprzyk J. (1998a), *Group Decision Making under Intuitionistic Fuzzy Preference Relations*, Proceedings of the 7th Int. Conference IMPU'98, Paris, pp. 172–178.
6. Szmidt E., Kacprzyk J. (1998b), *Applications of Intuitionistic Fuzzy Sets in Decision Making*, Proceedings of the 8th Congreso EUSFLAT'98, Pampelona, pp. 150–158.
7. Szmidt E., Kacprzyk J. (1999a), *A Concept of a Probability of an Intuitionistic Fuzzy Event*, Proceedings of the 1999 IEEE International Fuzzy Systems Conference, Seoul, pp. 1346–1349.
8. Szmidt E., Kacprzyk J. (1999c), *Probability of Intuitionistic Fuzzy Events and Their Applications in Decision Making*, Proceedings of the 9th Congreso EUSFLAT'99, pp. 457–460.
9. Zadeh L.A. (1965), *Fuzzy Sets*, Inf. and Control **8**, 338–353.
10. Zadeh L.A. (1968), *Probability Measures of Fuzzy Events*, J. Math. Anal. Appl. **23**, 421–427.

A New Calculus for Linguistic Prototypes in Data Analysis

Jonathan Lawry

Department of Engineering Mathematics, University of Bristol, Bristol, UK,
E-mail: e-mail: j.lawry@bris.ac.uk

Abstract. A random set semantics for imprecise concepts is introduced. It is then demonstrated how label descriptions of data sets can be learnt in this framework. These descriptions take the form of linguistic prototypes representing amalgams of elements. The potential of this approach for classification and query evaluation is then investigated.

1 Introduction

The area of automated learning from data is becoming increasingly important in an age of almost continuous data collection. From data we must be able to learn models which are flexible enough to facilitate a wide range of queries and which allow for insight into the underlying nature of the system under consideration. A principal requirement of such models is that they should be transparent and in order to achieve such clarity, ideally the representation framework should be high-level and capture certain aspects of natural language. In this paper we shall focus on modelling the imprecision associated with adjective labels that describe a quantity or the value of a measurement. Typical examples of this type of label occur in expressions of the form 'the diastolic blood pressure is *high*' or 'The sodium concentration is *quite low*'. Specifically, we aim to provide label descriptions of attributes values for sets of similar elements contained in a database. In a sense such descriptions can be viewed as imprecise prototype definitions. These prototypes can then be used for clustering as well as classification and prediction tasks. In the sequel we introduce a random set based calculus [4] for attribute labels with a clear underlying semantics.

2 Label Semantics

For an attribute (or variable) x into a domain of discourse Ω we identify a finite set of words LA with which to label the values of x. Then for a specific value $a \in \Omega$ an individual I identifies a subset of LA, denoted $\mathcal{D}_a^I$ to stand for the description of a given by I, as the set of words with which it is appropriate to label a. Within this framework then, an expression such as 'the diastolic blood pressure is *high*', as asserted by I, is interpreted to mean *high* $\in \mathcal{D}_{bp}^I$ where bp denotes the value of the variable blood pressure. If we allow I to vary across a population of individuals V then we naturally obtain a random set $\mathcal{D}_x$ from V into the power set of LA where $\mathcal{D}_x(I) = \mathcal{D}_x^I$. A probability distribution (or mass assignment) associated with this random set can be defined and is dependent on the prior distribution over the population V. We can view the random set $\mathcal{D}_x$ as a description of the variable x in terms of the labels in LA.

Definition 1. (Value Description) For $x \in \Omega$ the label description of x is a random set from V into the power set of LA, denoted $\mathcal{D}_x$, with associated distribution $m_{\mathcal{D}_x}$, given by

$$\forall S \subseteq LA \ m_{\mathcal{D}_x}(S) = Pr(\{I \in V : \mathcal{D}_x^I = S\})$$

Another high level measure associated with $m_{\mathcal{D}_x}$ is the following quantification of the degree of appropriateness of a particular word $L \in LA$ as a label of x.

Definition 2. (Appropriateness Degrees)

$$\forall x \in \Omega, \ \forall L \in LA \ \mu_L(x) = \sum_{S \subseteq LA : L \in S} m_{\mathcal{D}_x}(S)$$

Now clearly μ_L is a function from Ω into $[0,1]$ and therefore can technically be viewed as a fuzzy set. However, we shall use the term 'appropriateness degree' partly because it more accurately reflects the underlying semantics and partly to highlight the quite distinct calculus for these functions that will be introduced in the sequel. We now make the additional assumption that value descriptions are consonant random sets (see [2]). In the current context consonance simply requires the restriction that individuals in V differ regarding what labels are appropriate for a value only in terms of generality or specificity. Certainly, given that the meaning of the labels in LA must be sufficiently invariant across V to allow for effective communication then some strong restriction on $\mathcal{D}_x$ should be expected. The consonance restriction could be justified by the idea that all individuals share a common ordering on the appropriateness of labels for a value and that the composition of $\mathcal{D}_x^I$ is consistent with this ordering for each I. The consonance assumption means that $m_{\mathcal{D}_x}$ can be completely determined from the values of $\mu_L(x)$ for $L \in LA$ as follows [2]: If $\{\mu_L(x) : L \in LA\} = \{y_1, \ldots, y_n\}$ ordered such that $y_i > y_{i+1}$ for $i = 1, \ldots, n-1$ then for $S_i = \{L \in LA : \mu_L(x) \geq y_i\}$,

$$m_{\mathcal{D}_x}(S_i) = y_i - y_{i+1} \text{ for } i = 1, \ldots, n-1$$
$$m_{\mathcal{D}_x}(S_n) = y_n \text{ and } m_{\mathcal{D}_x}(\emptyset) = 1 - y_1$$

This has considerable practical advantages since we no longer need to have any knowledge of the underlying population of individuals V in order to determine $m_{\mathcal{D}_x}$. Rather, for reasoning with label semantics in practice we need only define appropriateness degrees μ_L for $L \in LA$ corresponding to the imprecise definition of each label.

For more general linguistic reasoning a mechanism is required for evaluating compound label expressions. For example, we may wish to know whether or not expressions such as $medium \wedge low$, $medium \vee low$ and $\neg high$ can be applied to a value $x \in \Omega$. In the context of this assertion-based framework we interpret the main logical connectives in the following manner: $L_1 \wedge L_2$ means that both L_1 and L_2 are appropriate labels, $L_1 \vee L_2$ means that either L_1 or L_2 are appropriate labels and $\neg L$ means that L is not an appropriate label. More generally, if we consider label expressions formed from LA by recursive application of the connectives then an expression θ identifies a set of possible label sets $\lambda(\theta)$ as follows:

Definition 3. Possible Label Sets

- For $L \in LA$ $\lambda(L) = \{S \subseteq LA : L \in S\}$

- For label expressions θ and φ $\lambda(\theta \wedge \varphi) = \lambda(\theta) \cap \lambda(\varphi)$

- For label expressions θ and φ $\lambda(\theta \vee \varphi) = \lambda(\theta) \cup \lambda(\varphi)$

- For label expression θ $\lambda(\neg\theta) = \overline{\lambda(\theta)}$

The notion of appropriateness measure given above can now be extended so that it applies to compound label expressions. The intuitive idea here is that $\mu_\theta(x)$ quantifies the degree to which expression θ is appropriate as a description of x.

Definition 4. (Compound Appropriateness Degrees) For θ a label expression and $x \in \Omega$ the appropriateness of θ to x is given by:

$$\mu_\theta(x) = \sum_{S \in \lambda(\theta)} m_{\mathcal{D}_x}(S)$$

3 Label Descriptions of Data Sets

Suppose we have a database DB of N elements, associated with each of which are n measurements $x_1, \ldots, x_n$ so that $DB = \{\langle x_1(i), \ldots, x_n(i)\rangle : i = 1, \ldots, n\}$ where $x_j(i)$ denotes the value of x_j for object i. Further, suppose that we select a set of labels LA_j for each attribute x_j for $j = 1, \ldots, n$ where each label is defined by an appropriateness measure. The label description of DB is now defined to be a vector of mass assignments as follows:

Definition 5. (Label Description of DB) The label description of DB is a vector $\mathcal{L}(DB) = \langle m_1, \ldots, m_n \rangle$ where

$$\forall S \subseteq LA \ m_j(S) = \frac{1}{N} \sum_{i=1}^{N} m_{\mathcal{D}_{x_j(i)}}(S)$$

Given a label description of DB we now evaluate a joint mass assignment on $2^{LA_1} \times \ldots \times 2^{LA_n}$ so that:

$$\forall S_j \in 2^{LA_j}, j = 1, \ldots, n \ m_{DB}(S_1, \ldots, S_n) = \prod_{j=1}^{n} m_j(S_j)$$

Clearly, we are making an independence assumption here and in some cases this may not be appropriate. In order to overcome this problem one approach is to partition DB into a number of disjoint sets $P_1, \ldots, P_c$, perhaps according to some standard clustering algorithm, where the elements contained in each partition set are assumed to be sufficiently similar to allow an independence assumption. We can then learn label descriptions $\mathcal{L}(P_k)$ for $k = 1, \ldots, c$ and combine them to form an overall mass assignment for DB as follows: Let $\mathcal{L}(P_k) = \langle m_{1,k}, \ldots, m_{n,k} \rangle$ then

$$\forall S_j \in 2^{LA_j}, j = 1, \ldots, n \ m_{DB}(S_1, \ldots, S_n) = \sum_{k=1}^{c} \frac{|P_k|}{N} \prod_{j=1}^{n} m_{j,k}(S_j)$$

Given a joint mass assignment on DB and tuple of label expressions $\boldsymbol{\theta} = \langle \theta_1, \ldots, \theta_n \rangle$ where θ_j is an expressions based on labels LA_j, we can now use labels semantic to evaluate the appropriateness of $\boldsymbol{\theta}$ for describing DB in the following way.

$$\mu_{\boldsymbol{\theta}}(DB) = \sum_{S_1 \in \lambda(\theta_1)} \cdots \sum_{S_n \in \lambda(\theta_n)} m_{DB}(S_1, \ldots, S_n)$$

For many types of data analysis it is useful to be able to estimate the distribution on underlying variables given the information contained in DB. In the current context our knowledge of DB is represented by the mass assignment m_{DB} and hence we need to be able to evaluate a distribution on the base variables $x_1, \ldots, x_n$ conditional on m_{DB}. For simplicity, we now assume that all variables are continuous with domains of discourse comprising of bounded closed intervals of the real line. Furthermore, we assume a prior joint distribution $p(x_1, \ldots, x_n)$ for the base variables. In the case that p is unknown we will assume it to be the uniform distribution. Furthermore, to simplify the following definition we will at least assume that $x_1, \ldots, x_n$ are *a priori* independent so that $p(x_1, \ldots, x_n) = \prod_{j=1}^{n} p_j(x_j)$, p_j being the marginal prior on x_j. The following definition is based on a Bayesian argument the details of which are given in [3].

Definition 6. (Conditional Density given a Mass Assignment) Let x be a variable into Ω with prior distribution $p(x)$, LA be a set of labels for x and m be a posterior mass assignment for the set of appropriate labels of x (i.e. $\mathcal{D}_x$) inferred from some database DB. Then the posterior distribution of x conditional on m is given by:

$$\forall x \in \Omega \ p(x|m) = p(x) \sum_{S \subseteq LA} \frac{m(S)}{pm(S)} m_{\mathcal{D}_x}(S)$$

where pm is the prior mass assignment generated by the prior distribution p according to

$$pm(S) = \int_{\Omega} m_{\mathcal{D}_x}(S) p(x)$$

This definition is motivated by the following argument based on the theorem of total probability:

$$p(x|m) = \sum_{S \subseteq LA} p(x|\mathcal{D}_x = S) Pr(\mathcal{D}_x = S) = \sum_{S \subseteq LA} p(x|\mathcal{D}_x = S) m(S)$$

Also

$$p(x|\mathcal{D}_x = S) = \frac{Pr(\mathcal{D}_x = S|x) p(x)}{Pr(\mathcal{D}_x = S)} = \frac{m_{\mathcal{D}_x}(S) p(x)}{pm(S)}$$

Making the relevant substitutions and then simplifying gives the expression from definition 6. This definition is then extended to the case where the posterior knowledge consists of a set of prototype descriptions of DB.

Definition 7. (Conditional Densities from Prototype Descriptions) For m_{DB} generated from label descriptions $\mathcal{L}(P_k), k = 1, \ldots, c$

$$\forall \boldsymbol{x} \in \Omega_1 \times \ldots \times \Omega_n \ p(\boldsymbol{x}|m_{DB}) = \sum_{k=1}^{c} \frac{|P_k|}{N} \prod_{j=1}^{n} p(x_j|m_{j,k})$$

4 Label Models and Linguistic Queries

To illustrate the potential of this framework we shall briefly describe how it can be applied to classification problems. In principle, however, the approach can also be applied to prediction and cluster analysis. Suppose then that the objects of DB can be categorised as belonging to one of the classes $C_1, \ldots, C_t$ and let DB_j denote the subset of DB containing only the elements with class C_j. We can now determine m_{DB_j} on the basis of some partition and evaluate $p(\boldsymbol{x}|m_{DB_j})$ as described above. If we now take $p(\boldsymbol{x}|m_{DB_j})$ as an approximation for $p(\boldsymbol{x}|C_j)$ then from Bayes theorem we have $Pr(C_j|\boldsymbol{x}) \propto p(\boldsymbol{x}|m_{DB_j})|DB_j|$. Given this estimate for each class probability, classification can be then carried out in the normal way. In the limit case when the partition of DB_j has only one set (i.e. DB_j) then this method corresponds to a version of the well known Naive Bayes algorithm [5].

Also, in the context of classification problems we extend the vector notation for linguistic queries as follows:

$$\langle \theta_1, \ldots, \theta_n \rangle : C_j$$

This represents the question: Do elements of class C_j satisfy θ (i.e. x_1 is θ_1 and x_2 is θ_2 and ... and x_n is θ_n)? The support for this query is given by $Pr(\theta|C_j) = \mu_\theta(DB_j)$.

$$\langle \theta_1, \ldots, \theta_n \rangle$$

This represents the question: Do elements of DB satisfy θ? The support for this query is given by $Pr(\theta) = \mu_\theta(DB) = \sum_{k=1}^{t} Pr(DB_k)\mu_\theta(DB_k)$.

$$C_j : \langle \theta_1, \ldots, \theta_n \rangle$$

This represents the question: Do elements satisfying θ belong to class C_j? The support for this query is given by

$$Pr(C_j|\theta) = \frac{\mu_\theta(DB_j)Pr(DB_j)}{\mu_\theta(DB)}$$

Example 1. The Naive Bayes version of the above algorithm was applied to the UCI repository problem on glass categorisation. The problem has 6 classes and 9 continuous attributes. 5 labels were define for attributes 1-2 and 4-7. Attribute 3 was allocated 4 labels and attributes 8 and 9 were not used as their variance across DB was too low for effective labelling. For all attributes the labels were defined by trapezoidal appropriateness degrees positioned according to a simple percentile method (see figure 1). The database of 214 elements was randomly split into a test and training set of 107 elements each and a classification accuracy of 78.5% on the training set and 72.9% on the test set was obtained. This is comparable with other approaches; for example a feedforward Neural Network with architecture 9-6-6 gives 72% on a smaller test set where the network was trained on 50% of the data, validated on 25% and tested on 25%. The density function for attribute 1 generated from the label description of class 1 according to definition 6 is shown in figure 2.

Now suppose for the attributes with five labels that these correspond to domain specific versions of *very low(vl)*, *low(l)*, *medium(m)*, *high(h)* and *very high (vh)* (see figure 1). Consider the following queries:

Query 1

What is the probability that float processed building window glass (class 1) has a medium to low or high refractive index (att. 1) and a very low or low sodium concentration (att. 2)?

To answer this query we note that $\mathcal{L}(DB_1) = \langle m_{1,1}, \ldots, m_{n,1} \rangle$ where $m_{1,1} = \{vl\} : 0.01373, \{l, vl\} : 0.04342, \{l\} : 0.02804, \{l, m\} : 0.37391,$

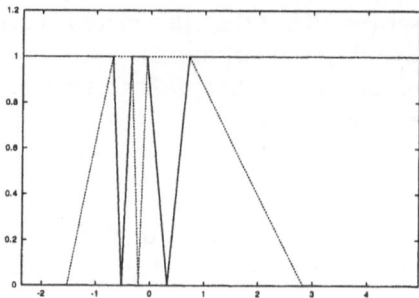

Fig. 1. Non uniform appropriateness degrees for, from left to right, *verysmall*, *small*, *medium*, *large* and *verylarge* for attribute 1 generated using a percentile algorithm.

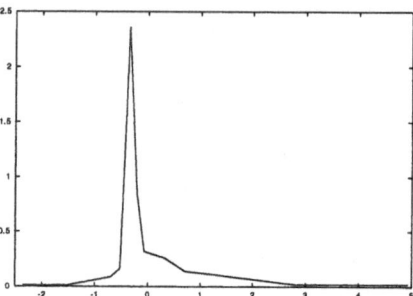

Fig. 2. Density function for attribute 1 conditional on class 1 generated from the mass assignment for attribute 1 in the label description of class 1.

$\{m\} : 0.12208, \{m, h\} : 0.08424, \{h\} : 0.10059, \{h, vh\} : 0.17233, \{vh\} : 0.06167.$

In this case the vector representation of the query is given by

$$\boldsymbol{\theta} : C_1 = \langle \theta_1, \theta_2, \top, \ldots, \top \rangle : C_1$$

where $\top$ denotes a tautology, $\theta_1 \equiv (medium \wedge low) \vee high$ and $\theta_2 \equiv very\ low \vee low$.

For this query we have that

$$\mu_{\boldsymbol{\theta}}(DB_1) = \sum_{S_1 \in \lambda(\theta_1)} \sum_{\lambda \in (\theta_2)} \sum_{S_3 \subseteq LA_3} \cdots \sum_{S_n \subseteq LA_n} \prod_{i=1}^{n} m_{i,1}(S_i)$$

$$= \left(\sum_{S_1 \in \lambda(\theta_1)} m_{1,1}(S_1) \right) \times \left(\sum_{S_1 \in \lambda(\theta_1)} m_{2,1}(S_2) \right)$$

Now from the training database we have

$$\sum_{S \in \lambda(\theta_1)} m_{1,1}(S_1)$$

$$= m_{1,1}(\{l, m\}) + m_{1,1}(\{m, h\}) + m_{1,1}(\{h\}) + m_{1,1}(\{h, vh\}) = 0.73107$$

Similarly, $\sum_{S \in \lambda(\theta_2)} m_2(S) = 0.92923$ so that $\mu_\theta(DB_1) = 0.73107 \times 0.92923 = 0.67933$ this being the required probability.

Query 2

What is the probability that a glass fragment has a medium to low or high refractive index (att. 1) and a very low or low sodium concentration (att. 2)?

This query has vector representation

$$\langle \theta_1, \theta_2, \top, \ldots, \top \rangle$$

and the required probability is given by $\mu_\theta(DB)$. To evaluate this we note that for classes $C_1, \ldots, C_6$ the probabilities of each class satisfying θ are given by

$$\mu_\theta(DB_1) = 0.67933, \quad \mu_\theta(DB_2) = 0.18429, \quad \mu_\theta(DB_3) = 0.09036,$$
$$\mu_\theta(DB_4) = 0.42108, \quad \mu_\theta(DB_5) = 0, \quad \mu_\theta(DB_6) = 0.02173.$$

Also the number of data elements in $DB_1, \ldots, DB_6$ are respectively 35, 38, 8, 7, 4 and 15. From this we can evaluate:

$$\mu_\theta(DB) = \frac{1}{107}(0.0.67933(35) + 0.18429(38) + 0.09036(8) + 0.42108(7)$$

$$+ 0.02173(15)) = \frac{34.7760}{107} = 0.32501.$$

Query 3

What is the probability that a glass fragment with a medium to low or high refractive index (att. 1) and a very low or low sodium concentration (att. 2) is a fragment of float processed building window glass (class 1)?

For this query the vector representation is

$$C_1 : \langle \theta_1, \theta_2, \top, \ldots, \top \rangle$$

and the required probability $Pr(C_1|\theta)$ is given by:

$$Pr(C_1|\theta) = \frac{\mu_\theta(DB_1)Pr(DB_1)}{\mu_\theta(DB)} = \frac{0.67933\frac{35}{107}}{0.32501} = 0.6837$$

Example 2. In this problem a figure eight shape (see figure 4) was generated according to the parametric equation $x = 2^{-0.5}(\sin 2t - \sin t)$, $y = 2^{-0.5}(\sin 2t + \sin t)$ where $t \in [0, 2\pi]$. Points in $[-1.6, 1.6]^2$ are classified as legal if they lie within the figure and illegal if they lie outside. The training database consisted of a regular grid of points on $[-1.6, 1.6]^2$. Clearly, a Naive Bayes independence assumption is inappropriate for this problem and will lead to significant decomposition error. Instead, c-means [1], was used to partition both the legal and illegal sub-databases into four. A joint mass assignment and joint distribution (figure 3) were then generated for each class, as described in section 3 and based on six labels for each variable. A classification accuracy of 95.4% on the training set and 95.8% on a denser test set of 2116 elements was obtained (see figure 4). This compares with an accuracy for the Naive Bayes model of 85% on the training set and 85.1% on the test set.

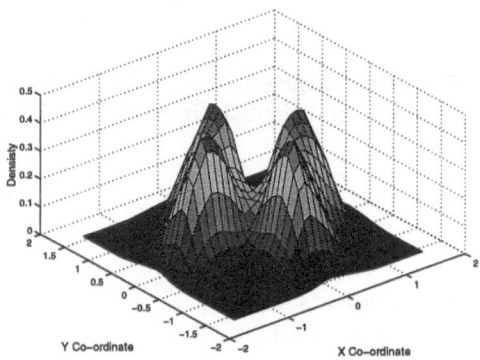

Fig. 3. Density function for legal class generated from the label model consisting of four legal prototypes

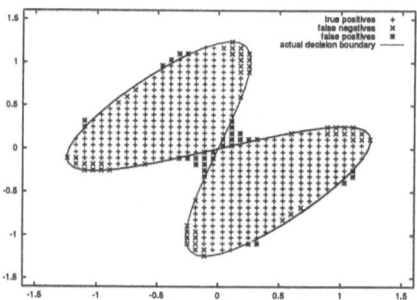

Fig. 4. Scatter plot showing true positives, false negative and false positive points for the figure eight test set

5 Conclusions

A framework for evaluating label descriptions of a database has been introduced. This involves a new random set based semantics for imprecise concepts. The potential of this approach has been demonstrated by its application to classification problems and to the evaluation of linguistic queries.

References

1. Duda R., Hart P.: Pattern Classification and Scene Analysis, Wiley, New York (1973).
2. Goodman I.R., Nguyen H.T.: Uncertainty Models for Knowledge Based Systems, North Holland (1985).
3. Lawry, J.: Label Prototypes for Modelling with Words, Proceedings of The North American Fuzzy Information Processing Society 2001 Conference, (2001) .
4. Lawry, J.: Label Semantics: A Formal Framework for Modelling with Words. Lecture Notes in Artificial Intelligence, **2143** (ed. S. Benferhat, P. Besnard), (2001) 374–384.
5. Lewis, D,D.: Naive Bayes at Forty: The independence Assumption in Information Retrieval, Lecture Notes in Artificial Intelligence, **1398** (1998) 4–15.

Upper Probabilities and Selectors of Random Sets

Enrique Miranda, Inés Couso and Pedro Gil

University of Oviedo, Department of Statistics and Operations Research
C-Calvo Sotelo s/n, 33007 Oviedo, Spain
e-mail: (alu426,couso,pedro)@pinon.ccu.uniovi.es

Abstract. We investigate the probabilistic information given by a random set when it represents the imprecise observation of a random variable. We compare the information given by the distributions of the selectors with that provided by the upper and lower probabilities induced by the random set. In particular, we model the knowledge on both the probability of an event and the probability distribution of the original random variable. Some characterizations and examples are given for the case of a finite final space, and the main difficulties for the infinite case are commented.

The theory of random sets has been successfully applied in many different contexts. It has been used in connection with fuzzy set theory ([4]), as a tool for dealing with imprecise information ([3,9]), but also in the context of stochastic geometry ([8,10]). Formally, a random set is a multi-valued mapping satisfying a certain measurability condition. Different conditions have been defined ([7]). Most of them depend on the notion of upper and lower inverse.

Definition 1. Consider a probabilistic space $(\Omega, \mathcal{A}, P)$, a measurable space $(X, \mathcal{A}')$ and a multi-valued mapping $\Gamma : \Omega \to \mathcal{P}(X)$. Given $A \in \mathcal{A}'$, the *upper inverse* of A by Γ is $\Gamma^*(A) \equiv A^* = \{\omega \in \Omega : \Gamma(\omega) \cap A \neq \emptyset\}$, and the *lower inverse* is given by $\Gamma_*(A) \equiv A_* = \{\omega \in \Omega : \Gamma(\omega) \neq \emptyset, \Gamma(\omega) \subseteq A\}$.

Definition 2. A multi-valued mapping $\Gamma : \Omega \to \mathcal{P}(X)$ is said to be *strongly measurable* when $A^* \in \mathcal{A} \ \forall A \in \mathcal{A}'$.

Other conditions, such as the measurability, the weak-measurability or the $\mathcal{C}$-measurability, only require the upper inverse of closed (resp. open, compact) sets to belong to $\mathcal{A}$. Our choice of the strong measurability is due to our interpretation of a random set: we will regard ([9]) Γ as an imprecise observation of a random variable U_0, in the sense that $U_0(\omega) \in \Gamma(\omega) \ \forall \omega$. Then, we have $A_* \subset U_0^{-1}(A) \subset A^* \ \forall A \in \mathcal{A}'$. That is, the upper and lower inverse of a set A are respectively a superset and a subset of the anti-image of A by the random variable U_0.

Dempster defined the upper and lower probabilities induced by Γ in the following form:

Definition 3. [3] Let $\Gamma : \Omega \to \mathcal{P}(X)$ be a strongly measurable multi-valued mapping, and let $A \in \mathcal{A}'$. Then, the *upper probability* of A is given by $P^*(A) = \frac{P(A^*)}{P(X^*)}$, and the *lower probability* is given by $P_*(A) = \frac{P(A_*)}{P(X^*)}$.

If Γ is the imprecise observation of U_0, it must be $\Gamma(\omega) \neq \emptyset \; \forall \omega$ (for $U_0(\omega) \in \Gamma(\omega)$), and hence $P^*(A) = P(A^*)$, $P_*(A) = P(A_*) \; \forall A \in \mathcal{A}'$. We deduce that $P^*(A)$ and $P_*(A)$ are an upper and a lower bound of $P_{U_0}(A)$. The strong measurability of Γ is necessary in order for P^*, P_* to be defined. Note that $A_* = [(A^c)^*]^c$, whence if Γ is strongly measurable, $A_* \in \mathcal{A} \; \forall A \in \mathcal{A}'$. Moreover, $P_*(A) = 1 - P^*(A^c) \; \forall A \in \mathcal{A}'$. This shows that the studies concerning the upper and the lower probability will be dual of each other.

As we have said, we will regard a random set Γ as the result of the imprecise observation of a random variable U_0, in the sense that, given an element ω in the initial space, we only know that $U_0(\omega)$ belongs to the set $\Gamma(\omega)$. Therefore, all we know about U_0 is that it belongs to the class

$$S(\Gamma) := \{U : \Omega \to X \;\; \text{r.v. such that } U(\omega) \in \Gamma(\omega) \; \forall \omega\},$$

and the probability measure induced by U_o will belong to the class

$$P(\Gamma) := \{P_U : U \in S(\Gamma)\}.$$

The elements of $S(\Gamma)$ are called *selectors* of the random set Γ. Note that because of the relationship $A_* \subseteq U^{-1}(A) \subseteq A^*$, valid for any $U \in S(\Gamma)$, it is $P_*(A) \leq P_U(A) \leq P^*(A) \; \forall U \in S(\Gamma), \forall A \in \mathcal{A}'$. That is, if we define the family

$$M(P^*) := \{Q \text{ probability} : Q(A) \leq P^*(A) \; \forall A\},$$

it is $P(\Gamma) \subseteq M(P^*)$. Moreover, because of the duality between the upper and lower probability, it will be $Q(A) \leq P^*(A) \forall A \Leftrightarrow Q(A) \geq P_*(A) \forall A$. There are several studies in the literature concerning convex sets of probabilities dominated by a Choquet capacity ([1,13,15]). However, studies for non-convex sets of probabilities are much more rare. Because of this fact, many authors model the information given by Γ about P_{U_0} with the set $M(P^*)$, instead of working with $P(\Gamma)$, which is the most precise set of probabilities we can consider. As, moreover, convex sets provide some advantages respect to non-convex sets from an operational point of view, it will be preferable to work with $M(P^*)$ when no significant information is lost. Our aim in this paper is to make a thorough study of the relationships between the classes $M(P^*)$ and $P(\Gamma)$, determining which of them is more appropriate to work with in different situations.

We are first going to consider the case of a finite referential X. In that case, the measurability conditions introduced above are equivalent. We will show through different examples that $P(\Gamma)$ is not convex in general. Nevertheless, the replacement of $P(\Gamma)$ by $M(P^*)$ will not cause any problem in most cases: if our goal is to obtain upper and lower bounds for some parameter of the distribution P_{U_0}, $f(P_{U_0})$, the substitution of $P(\Gamma)$ will not alter our information when f satisfies the condition

$$\{f(Q) \mid Q \in \mathrm{Conv}\,(\mathcal{C})\} \subseteq \mathrm{Conv}(\{f(Q) \mid Q \in \mathcal{C}\}). \qquad (1)$$

for any class $\mathcal{C}$ of probability distributions, where $\mathrm{Conv}(A)$ denotes the convex hull of A. This is for instance the case for any linear operator, such as the mean, but also for operators like the median or the variance. However, condition (1) does not always hold, and this means an important loss of information in the case of parameters like the entropy or the modal value, as we show next:

Example 1. Consider the set $\Omega = \{1, 2, 3\}$ with the uniform distribution, and the multi-valued mapping $\Gamma : \Omega \to \mathcal{P}(\{1, 2\})$ given by $\Gamma(\omega) = \{1, 2\}\ \forall \omega$. Γ is trivially strongly measurable, because for any non-empty $A \in \mathcal{P}(\{1, 2\})$, it is $A^* = \Omega$.

Then, $P(\Gamma) = \{(0, 1), (1/3, 2/3), (2/3, 1/3), (1, 0)\}$, whereas it is $P^*(A) = 1\ \forall A \in \mathcal{P}(\{1, 2\})$, and hence $M(P^*)$ is given by $\{(\alpha, 1 - \alpha) : \alpha \in [0, 1]\}$.

Now, consider f the Shannon entropy on the probability distributions on $\{1, 2\}$: it is given by $f((\alpha, 1 - \alpha)) = -[\alpha \log_2 \alpha + (1 - \alpha) \log_2(1 - \alpha)]$. Then, $f(P_{U_0})$ will belong to the set $f(P(\Gamma)) = \{0, f(1/3, 2/3)\} = \{0, 0.92\}$. On the other hand, $f(M(P^*)) = [0, 1]$. Therefore, our estimation is more imprecise when we work with the class $M(P^*)$.♦

Example 2. Take $\Omega = \{1, 2\}$, and a probability P on $\mathcal{P}(\Omega)$ given by $P(\{1\}) = 0.4, P(\{2\}) = 0.6$. Consider $X = \mathbb{R}$, and the multi-valued mapping $\Gamma(1) = \mathbb{R} \setminus \{1, 3\}, \Gamma(2) = \{1, 3\}$. It is clearly strongly measurable.

Now, $P(\Gamma) := \{0.4\delta_x + 0.6\delta_y : x \in \mathbb{R} \setminus \{1, 3\}, y \in \{1, 3\}\}$, where δ_x is the probability distribution given by $\delta_x(A) = 1$ if and only if $x \in A$; on the other hand, $M(P^*) := \{Q \text{ probability} : Q(\{1, 3\}) = 0.6\}$. Then, $P(\Gamma) \subsetneq M(P^*)$: a probability Q satisfying $Q(\{1\}) = Q(\{3\}) = 0.3$ belongs to $M(P^*) \setminus P(\Gamma)$.

We can see that the modal point of the distribution P_{U_0} can only be 1 or 3; however, if we consider the set of distributions given by $M(P^*)$, any other point x of $\mathbb{R}$ can be the modal point: take for instance the distribution Q given by $Q(\{x\}) = 0.4, Q(\{1\}) = Q(\{3\}) = 0.3$. Thus, we are losing some essential information with this bigger set of probabilities.♦

The reader can find other relevant examples in [2]. This serves as a motivation for a study of the relationship between both sets: it is clear that working with the upper probability is much easier in practice than with the set of distributions of the selectors; one advantage is for instance that the convex set $M(P^*)$ is characterised by its finite number of extreme points, which are in correspondence with the permutations on $S_{|X|}$, as will later show. Nevertheless, the use of the results on $M(P^*)$ is only justified when we have the equality between both sets or when no relevant information is lost, as we argued for equation 1.

Let us define the class

$$P(\Gamma)(A) := \{Q(A) : Q \in P(\Gamma)\}.$$

It represents the information provided by Γ on the probability of a set A. The convexity of this class and its relationship with the upper probability is characterized through the following theorem:

Theorem 1. *Let $(\Omega, \mathcal{A}, P)$ be a probabilistic space, $(X, \mathcal{P}(X))$ be a finite space, with $|X| = n$, and $\Gamma : \Omega \to \mathcal{P}(X)$ a random set.*

1. *$P^*(A) = \max P(\Gamma)(A), P_*(A) = \min P(\Gamma)(A)$ for any $A \subseteq X$.*
2. *$P(\Gamma)(A) = [P_*(A), P^*(A)] \Leftrightarrow A^* \setminus A_*$ is not an atom.*
3. *$Conv(P(\Gamma)) = M(P^*)$, and $P(\Gamma) = M(P^*)$ iff $P(\Gamma)$ is convex.*

Proof:

1. Take $A = \{x_1, \ldots, x_m\} \subset X$; then, we can denote without loss of generality $X = \{x_1, \ldots, x_m, x_{m+1}, \ldots, x_n\}$. Let us prove for instance that $\max P(\Gamma)(A) = P^*(A)$. Define $U : \Omega \to X$ by

$$U := x_1 I_{\{x_1\}^*} + \sum_{i=2}^{n} x_i I_{\{x_i\}^* \setminus \{x_1, \ldots, x_{i-1}\}^*}$$

(i.e., $U(\omega) = x_i$ if and only if ω belongs to $\{x_i\}^* \setminus \{x_1, \ldots, x_{i-1}\}^*$). We deduce from the strong measurability of Γ that U is measurable, and from its definition we have $U(\omega) \in \Gamma(\omega)$ $\forall\omega$. Moreover, it is $U^{-1}(A) = A^*$, whence $P_U(A) = P^*(A)$.
We can show similarly that $P_*(A) = \min P(\Gamma)(A)$ $\forall A$.

2. From the previous point, we can consider $U_1, U_2 \in S(\Gamma)$ s.t. $U_1^{-1}(A) = A^*, U_2^{-1}(A) = A_*$. We deduce that $P(\Gamma)(A) = [P_*(A), P^*(A)]$ if and only if $P(\Gamma)(A)$ is convex.
($\Rightarrow$) Consider $\alpha \in (0, 1)$, and let us find some $B \in \mathcal{A}, B \subset A^* \setminus A_*$ with $P(B) = \alpha P(A^* \setminus A_*)$; take $x = P_*(A) + \alpha P(A^* \setminus A_*) \in [P_*(A), P^*(A)] = P(\Gamma)(A)$. Then, $\exists U \in S(\Gamma)$ s.t. $x = P_U(A) = P(U^{-1}(A))$. Now, take $B = U^{-1}(A) \setminus A_*$. It is $B \subset A^* \setminus A_*$, and $P(B) = P_U(A) - P_*(A) = \alpha P(A^* \setminus A_*)$.
($\Leftarrow$) Conversely, let $x \in [P_*(A), P^*(A)]$; then, $\exists \alpha \in [0, 1]$ s.t. $x = P_*(A) + \alpha P(A^* \setminus A_*)$. As $A^* \setminus A_*$ is not an atom, there exists $B \subset A^* \setminus A_*, B \in \mathcal{A}$ s.t. $P(B) = \alpha P(A^* \setminus A_*)$. Define $U = U_1 I_B + U_2 I_{B^c}$. Then, U is a selector of Γ, because $U_1, U_2 \in S(\Gamma)$, and $P_U(A) = P(U^{-1}(A)) = P(U_1^{-1}(A) \cap B) + P(U_2^{-1}(A) \cap B^c) = P(B) + P_{U_2}(A) = x$.

3. As $M(P^*), P(\Gamma)$ belong to a finite-dimensional space, $M(P^*)$ is the convex hull of its class of extreme points. These extreme points ([1,3,12]) are related to the permutations of $|X|$ elements, in the following way:
Take $\pi = (i_1, \ldots, i_n) \in S^n$, and consider the probability distribution P_π satisfying $P_\pi(\cup_{j=1}^{k}\{x_{i_j}\}) = P^*(\cup_{j=1}^{k}\{x_{i_j}\})$ $\forall k = 1, \ldots, n$. Then, P_π is an

extreme point of $M(P^*)$, and all the extreme points of $M(P^*)$ are of this type.

The result follows immediately if we show that these extreme points belong to $P(\Gamma)$. Fix $\pi = (i_1, \ldots, i_n) \in S^n$, and define

$$U_\pi := x_{i_1} I_{\{x_{i_1}\}^*} + \sum_{j=2}^{n} x_{i_j} I_{\{x_{i_j}\}^* \setminus \{x_{i_1}, \ldots, x_{i_{j-1}}\}^*}.$$

Then, it is $U_\pi(\omega) \in \Gamma(\omega) \; \forall \omega$, and we deduce from the strong measurability of Γ that U_π is a random variable. Moreover, it is $U_\pi^{-1}(\cup_{j=1}^{k}\{x_{i_j}\}) = \cup_{j=1}^{k}\{x_{i_j}\}^* \; \forall k$, whence $P_{U_\pi} = P_\pi$.∎

Consider now the class of probability distributions

$$\Delta(\Gamma) := \{Q \text{ probability} : Q(A) \in P(\Gamma)(A) \; \forall A \in \mathcal{A}'\}.$$

It is the class of distributions determined by the sets $P(\Gamma)(A)$. When $P(\Gamma) = \Delta(\Gamma)$, the probabilistic information of the random set will be determined by the information on the sets of $\mathcal{A}'$. Hence, it is interesting to see whether this equality holds or not, from an operational point of view. It is also of interest to see when $\Delta(\Gamma)$ coincides with $M(P^*)$. In this sense, we have proven the following:

Theorem 2. *Let $(\Omega, \mathcal{A}, P)$ be a probabilistic space, $(X, \mathcal{P}(X))$ be a finite space, and $\Gamma : \Omega \to \mathcal{P}(X)$ a random set. The following statements hold:*

1. $P(\Gamma) \subset \Delta(\Gamma) \subset M(P^*)$.
2. *The equality $P(\Gamma) = \Delta(\Gamma)$ is implied by the equality $P(\Pi \circ \Gamma) = \Delta(\Pi \circ \Gamma)$ for every projection $\Pi : X \to \{1, 2, 3\}$.*
3. *If $(\Omega, \mathcal{A}, P)$ is non-atomic, then $P(\Gamma) = M(P^*)$ holds.*

Proof:

1. It is clear that any $Q \in P(\Gamma)$ satisfies $Q(A) \in P(\Gamma)(A) \; \forall A$, whence $P(\Gamma) \subset \Delta(\Gamma)$. On the other hand, from theorem 1 it is $P(\Gamma)(A) \subseteq [P_*(A), P^*(A)] \; \forall A$ and consequently $\Delta(\Gamma) \subset M(P^*)$.
2. (sketch of proof:) The result follows applying induction on $|X|$. If $|X| = 3$, it is trivial. For $|X| = n$, we consider $(p_1, \ldots, p_n) \in \Delta(\Gamma)$, and we decompose it through certain projections into $(p_1 + p_2, \ldots, p_n), (p_1, \ldots, p_{n-1} + p_n) \in \Delta(\Pi_1 \circ \Gamma), \Delta(\Pi_2 \circ \Gamma)$, respectively. Applying the hypothesis of induction, there are $U_1 \in S(\Pi_1 \circ \Gamma), U_2 \in S(\Pi_2 \circ \Gamma)$ inducing these probabilities. Combining these selections in a proper way, we get a selector $U \in S(\Gamma)$ inducing the distribution $(p_1, \ldots, p_n)$.
3. From theorem 1, if $(\Omega, \mathcal{A}, P)$ is non-atomic, then $A^* \setminus A_*$ is not an atom for every A, whence $\Delta(\Gamma) = M(P^*)$. It can also be checked that under the non-atomicity of $(\Omega, \mathcal{A}, P)$, the hypothesis of the previous point is satisfied. Hence, $P(\Gamma) = M(P^*)$.∎

Remark 1. Let us give a few examples concerning this theorem:

- The converse of (3) is not true: Consider for instance the probabilistic space $(\mathbb{N}, \mathcal{P}(\mathbb{N}), P)$, with $P(\{n\}) = \frac{1}{2^n}$, the finite space $X = \{1, 2\}$, and take $\Gamma : \mathbb{N} \to \mathcal{P}(X)$ given by $\Gamma(n) = X \ \forall n$. Then, it is $P(\Gamma) = M(P^*) = \{(\alpha, 1 - \alpha) : \alpha \in [0, 1]\}$. However, the initial probabilistic space is atomic.
- Let us show that the equality $P(\Gamma) = M(P^*)$ is not implied by any of the equalities $P(\Gamma) = \Delta(\Gamma)$ and $\Delta(\Gamma) = M(P^*)$ separately: Consider $(\mathbb{N}, \mathcal{P}(\mathbb{N}), P)$ as in the previous example, and take $X = \{1, 2, 3\}$. Define $\Gamma : \Omega \to \mathcal{P}(X)$ by $\Gamma(n) = X$ for all n. Then, $P(\Gamma)(A) = [0, 1]$ for all A, whence $\Delta(\Gamma) = M(P^*) = \{(x_1, x_2, x_3) : x_1, x_2, x_3 \geq 0, x_1 + x_2 + x_3 = 1\}$. However, $P(\Gamma)$ is given by figure 1 (we represent the $P(\Gamma)$ on black over the simplex $x + y + z = 1$ of all probabilities on X).

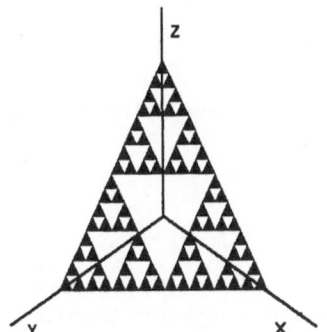

Fig. 1. The class $P(\Gamma)$

Conversely, consider the random set from example 1. We can easily see that it satisfies $P(\Gamma) = \Delta(\Gamma)$, and we saw there that it is $P(\Gamma) \subsetneq M(P^*)$.◆

Theorem 2 characterizes the behaviour of a random set in the case where X is finite. As we have showed, the non-atomicity of $(\Omega, \mathcal{A}, P)$ is sufficient for the equality $P(\Gamma) = M(P^*)$, but it is not necessary. It holds for instance when there exists a random variable $U : (\Omega, \mathcal{A}, P) \to ([0, 1], \beta)$ with absolutely continuous distribution.

In the case of an infinite referential, a number of technical complications arise. Concerning the extreme points of $M(P^*)$, we cannot extend trivially the result for the finite case. Nevertheless, in [11] some studies are carried out for the case where X is a separable metric space, and for the more general situation where P^* only needs to be 2-alternating. We must remark nonetheless that $M(P^*)$ is not in general the convex hull of its extreme points, and these do not belong necessarily to $P(\Gamma)$.

Although the result on the convexity of $P(\Gamma)(A)$ can be extended for the infinite case, the bounds $P^*(A), P_*(A)$ are not necessarily attained. This

132

problem is related to the existence of selectors of a random set. Although in the finite case this existence is trivial, it is not the same in general (see a review on the subject on [14]). Also, the result on the equality between $P(\Gamma)$ and $\Delta(\Gamma)$ cannot be immediately extended to the infinite case.

On the other hand, when we provide the final set with a certain topological structure, we get some additional results in the infinite case. For instance, we have proven that, when a compact-valued random set on a Polish space is considered, and the initial probability space is non-atomic, the equality $M(P^*) = P(\Gamma)$ holds if and only if the set $P(\Gamma)$ is closed with respect to the weak topology. This is related to other studies on the subject, such as those on [5,6]. In the near future, we intend to complete the work we have outlined here with results for the general case where the final space is not necessarily finite.

Acknowledgements

The research in this paper has been partially supported by FEDER-MCYT, grant number BFM2001-3515.

References

1. Chateauneuf, A. and Jaffray, J.-Y. (1989) Some characterizations of lower probabilities and other monotone capacities through the use of Möbius inversion. *Math. Soc. Sci.*, **17**, 263–283.
2. Couso, I. (1999) *Teora de la probabilidad para datos imprecisos. Algunos aspectos.* PhD Thesis, Universidad de Oviedo.
3. Dempster, A. P. (1967) Upper and lower probabilities induced by a multivalued mapping. *Ann. of Math. Stat.*, **38**, 325–339.
4. Dubois, D. and Prade, H. (1987) The mean value of a fuzzy number. *Fuz. Sets and Syst.*, **24**, 279–300.
5. Hart, S. and Kohlberg, E. (1974) Equally distributed correspondences. *J. of Math. Econ.*, 1, 167–174.
6. Hess, C. (1999) The distribution of unbounded random sets and the multivalued strong law of large numbers in nonreflexive banach spaces. *J. of Conv. Anal.*, **6**, 163–182.
7. Himmelberg, C. J. (1975) Measurable relations. *Fund. Math.*, **87**, 53–72.
8. Kendall, D. G. (1974) Foundations of a theory of random sets. In E. F. Harding and D. G. Kendall, editors, *Stochastic Geometry*, pages 322–376. Wiley, New York.
9. Kruse, R. and Meyer, K. D. (1987) *Statistics with vague data.* D. Reidel Publishing Company, Dordretch.
10. Matheron, G. (1975) *Random sets and integral geometry.* Wiley, New York.
11. Miranda, E., Couso, I., and Gil, P. (2001) On the probabilities dominated by a 2-alternating capacity on a separable metric space. In *Proc. of AGOP'01*, Oviedo (Spain).
12. Shapley, L. S. (1971) Cores of convex games. *Int. J. of Game Theory*, 1, 11–26.

13. Verdegay, J. L. and Moral, S. (2000) Network of probabilities associated with a capacity of order-2. *Inf. Sci.*, **125**, 187–206.
14. Wagner, D. H. (1977) Survey of measurable selection theorems. *SIAM J. Cont. and Opt.*, **15**, 859–903.
15. Walley, P. (1981) Coherent lower (and upper) probabilities. *Statistics Research Report, University of Warwick (Coventry)*, **22**.

On Multivalued Logic and Probability Theory

Beloslav Riečan

Department of Mathematics M. Bel University
Tajovského 40,
SK-97401 Banská Bystrica, Slovakia
riecan@fpv.umb.sk
and
Mathematical Institute,
Slovak Academy of Sciences,
Štefánikova 49,
SK-81473 Bratislava, Slovakia
riecan@mat.savba.sk

Abstract. The idea to built probability theory on the families of fuzzy sets belongs to the first ideas of the fuzzy sets theory (see [18]). In the paper we consider the Lukasiewicz connectives ([2,7,15,16]) in the corresponding family of fuzzy sets as a base of the probability. First we present some typical methods of the theory on multivalued logics. Then we mention recent development of the theory.

1 Probability on fuzzy sets

For simplicity we shall consider the family $\mathcal{F}$ of all $\mathcal{S}$-measurable functions $f : \Omega \to [0,1]$, where $\mathcal{S}$ is a σ-algebra of subsets of a non-empty set Ω. Of course,the set $\mathcal{F}$ would be considered together with two binary operations $\oplus, \odot$, where

$$f \oplus g = min(f + g, 1),$$
$$f \odot g = max(f + g - 1, 0),$$

one unary operation $\neg$, where

$$\neg f = 1 - f,$$

and two fixed elements

$$1_\Omega, 0_\Omega.$$

The natural algebraic generalization of the notion is an MV-algebra (see [2,15,16]), what is an algebraic system $(M, \oplus, \odot, \neg, 1, 0)$ satisfying some axioms. MV-algebras plays a similar role in multivalued logics as Boolean algebras in two valued logics.

Kolmogorov's probability model is based on Boolean algebras. It works with two basic notions: probability $P : \mathcal{S} \to [0,1]$ and random variable ($\mathcal{S}$-measurable map) $\xi : \Omega \to R$. Inspiring by quantum logics ([3]) we consider states instead of probabilities and observables instead of random variables.

State is a function $m : \mathcal{F} \to [0,1]$ satisfying the following conditions:

(i) $m(1_\Omega) = 1$;

(ii) if $f \odot g = 0_\Omega$, (i.e. $f + g \leq 1$), then $m(f) + m(g) = m(f \oplus g)(= m(f + g))$;

(iii) if $f_n \nearrow f$, then $m(f_n) \nearrow m(f)$.

Denote by $\mathcal{B}(R)$ the family of all Borel subsets of R. An observable is a mapping $x : \mathcal{B}(R) \to \mathcal{F}$ such that

(i) $x(R) = 1_\Omega$;

(ii) if $A, B \in \mathcal{B}(R), A \cap B = \emptyset$, then $x(A) \odot x(B) = 0_\Omega$ and $x(A \cup B) = x(A) \oplus x(B)(= x(A) + x(B))$;

(iii) if $A_n \in \mathcal{B}(R), A_n \nearrow A$, then $x(A_n) \nearrow x(A)$.

If $m : \mathcal{F} \to [0,1]$ is a state, and $x : \mathcal{B}(R) \to \mathcal{F}$ is an observable, then the composite map $m_x = m \circ x : \mathcal{B}(R) \to \mathcal{F}$ is a probability measure. It plays the same role as the probability distribution P_ξ of a random variable ξ in the Kolmogorovian theory. Therefore we define the mean value $E(x)$ of an observable x by the formula

$$\int_R t \, dm_x(t)$$

and we say that x is integrable, if the integral exists.

2 Joint observable

If $T = (\xi, \eta) : \Omega \to R^2$ is a random vector then the mapping $A \mapsto T^{-1}(A)$ is a morphism from $\mathcal{B}(R^2)$ to S such that

$$T^{-1}(C \times D) = \xi^{-1}(C) \cap \eta^{-1}(D).$$

Motivated by this example we define the notion of the joint observable.

Definition 1. If $x, y : \mathcal{B}(R) \to \mathcal{F}$ are two observables, then its joint observable is a mapping $h : \mathcal{B}(R^2) \to \mathcal{F}$ satisfying the following conditions:

(i) $h(R^2) = 1_\Omega$;

(ii) if $A \cap B = \emptyset$, then $h(A \cup B) = h(A) + h(B)$;

(iii) if $A_n \nearrow A$, then $h(A_n) \nearrow h(A)$;

(iv) $h(C \times D) = x(C) \cdot y(D)$ for any $C, D \in \mathcal{B}(R)$.

If a map $h : \mathcal{B}(R^2) \to \mathcal{F}$ satisfies only (i), (ii) and (iii), then it is called the 2-dimensional observable.

The following theorem has been proved in [9] (see also [15,16]).

Theorem 1. *The joint observable exists for any observables $x, y : \mathcal{B}(R) \to \mathcal{F}$.*

Proof:

Fix $\omega \in \Omega$ and for any $A \in \mathcal{B}(R)$ put $\mu_\omega(A) = x(A)(\omega), \nu_\omega(A) = y(A)(\omega)$. Then μ_ω, ν_ω are probability measures on $\mathcal{B}(R)$, hence their product $\lambda_\omega = \mu_\omega \times \nu_\omega$ is a probability measure on $\mathcal{B}(R^2)$. Define $h(A)(\omega) = \lambda_\omega(A)$. It is not difficult to see that h is actually a mapping from $\mathcal{B}(R^2)$ to $\mathcal{F}$ and that h has all properties of the joint observable. $\square$

By using the joint observable, some arithmetical operations with observables can be introduced. Let us motivate again by a random vector $T = (\xi, \eta)$. If we define $g : R^2 \to R$ by $g(u, v) = u + v$, then

$$(\xi + \eta)^{-1}(A) = (g \circ T)^{-1}(A) = T^{-1}(g^{-1}(A)).$$

Therefore the sum $x + y$ of observables x, y can be defined by the formula

$$(x + y)(A) = h(g^{-1}(A)),$$

where $g(u, v) = u + v$, and h is the joint observable of x, y. Generally, if h is the joint observable of $x_1, x_2, ..., x_n$, and $g : R^n \to R$ is a Borel measurable function, then we define the observable

$$g(x_1, , x_2, ..., x_n)$$

by the formula

$$g(x_1, x_2, ..., x_n)(A) = h(g^{-1}(A)).$$

E.g.

$$\frac{1}{n} \sum_{i=1}^{n} x_i(A) = h(g^{-1}(A)),$$

where

$$g(u_1, ..., u_n) = \frac{1}{n} \sum_{i=1}^{n} u_i.$$

Everybody knows the importance of the arithmetic means for probability theory. The question arises what to do in a general MV-algebra M. Namely, how to substitute the product sign in the formula $h(C \times D) = x(C) \cdot y(D)$. We see at least two ways.

The first one was completely realised in [16]. We say that two observables x, y are *independent*, if there exists such 2-dimensional observable $h :$ $\mathcal{B}(R^2) \to M$ such that

$$m(h(C \times D)) = m(x(C)).m(y(D))$$

for any $C, D \in \mathcal{B}(R)$. This is the same as the characterization of independency in the boolean case: two random variables ξ, η are independent if and only if

$$P_T = P_\xi \times P_\eta.$$

The second possibility is to define the product axiomatically. It was suggestes first in [12] (see [15] for a complete explanation). Recall only that the notion of an MV-algebra with product has been introduced independently in [11] with respect to some questions of multivalued logics.

3   Local representation

There exists nice probability theory on Boolean algebras due to Kolmogorov. Evidently, there is no possibility to represent MV-algebras by Boolean algebras. Of course, to a given sequence of observables can be constructed a Boolean probability space corresponding to this sequence.

Theorem 2. *Let (x_n) be a sequence of observables from $\mathcal{B}(R)$ to $\mathcal{F}$. Then there exists a probability space (X, σ, P) and a sequence (ξ_n) of random variables with the following property. If g is a Borel measurable function $g : R^n \to R$, $y = g(x_1, ..., x_n)$, and $\eta = g(\xi_1, ..., \xi_n)$, then y and η have the same probability distribution, i.e. $P_\eta = m_y$.*

Proof:
Put $X = R^N$ and consider the σ-algebra σ generated by the family $\mathcal{C}$ of all cylinders. Put $P_n = m \circ h : \mathcal{B}(R^n) \to \mathcal{F}$, where h is the joint observable of $x_1, ..., x_n$. By the Kolmogorov consistency theorem there exists exactly one measure $P : \sigma \to [0, 1]$ such that $P(\pi_n^{-1}(D)) = P_n(D)$ whenever $D \in \mathcal{B}(R^n)$. We have obtained a probability space (X, σ, P).

Let ξ_n be the n-th coordinate of R^N, i.e. $\xi_n((u_i)_{i=1}^\infty) = u_i$. Then $\eta = g \circ \pi_n$, where π_n is the projection from R^N to the first n coordinates. We obtain

$$P_\eta(A) = P(\eta^{-1}(A)) = P(\pi_n^{-1}(g^{-1}(A))) = P_n(g^{-1}(A)) =$$
$$= m(h(g^{-1}(A))) = m(y(A)) = m_y(A).$$

This finishes the proof. □

The second crucial problem for probability theory is the problem of convergences. Analogously with the boolean case we say that a sequence (y_n) of observables converges in distribution to a function $F : R \to [0, 1]$, if for each $t \in R$

$$lim_{n \to \infty} m(y_n((-\infty, t))) = F(t).$$

It converges in measure m to 0, if for each $\varepsilon > 0, \varepsilon \in R$

$$lim_{n \to \infty} m(y_n((-\varepsilon, \varepsilon))) = 1.$$

Theorem 3. *Let $g_n : R^n \to R (n = 1, 2, ...)$ be Borel measurable functions, (x_n) and (ξ_n) given as in Theorem 2, $y_n = g_n(x_1, ..., x_n), \eta_n = g_n(\xi_1, ..., \xi_n)$. Then*

(i) *(y_n) converges to F if and only if so does (η_n);*

(ii) (y_n) *converges in measure* m *to* 0 *if and only if* (η_n) *converges in measure* P *to* 0.

A little more complicated is the situation with respect to almost everywhere convergence.For MV-algebras of fuzzy sets (tribes) it was described in [16], for general MV-algebras in [13]. Recall only that in the case only one implication holds: from the a.e. convergence of (η_n) follows the a.e. convergence of (y_n). Of course, it suffices for our purposes, because from the known results for boolean case (applied to the sequence (η_n)) follow the corresponding results for the MV-algebra case (applied to the sequence (y_n)).

Here we list some achieved results ([16,15,5,6,10,17]): weak and strong laws of large numbers, central limit theorem, martingale convergence theorem, individual ergodic theorem, Kolmogorov - Sinai entropy convergence theorem.

References

1. Ban. A.:Ergodic transformations, Soft Computing **5** (2001), 327 - 222.
2. Cignoli, R., D'Ottaviano, I.M.L., Mundici, D.: Algebraic Foundations of Many-valued reasoning, Kluwer, Dordrecht 2000.
3. Dvurečenskij, A., Pulmannová, S.: New Trenda in Quantum Structures. Kluwer, Dordrecht 2000.
4. Jakubík, J.: On the product MV-algebras. Czech. Math. J. (to appear).
5. Jurečková, M.: On the conditional expectation on probability MV-algebras with product. Soft Computing **5** (2001), 381 - 385.
6. Jurečková, M.: A note on the individual ergodic theorem on product MV-algebras. Internat. J. Theor. Physics **39** (2000), 753 - 760.
7. Klement, E.P., Mesiar, R., Pap. E.: Triangular Norms, Kluwer, Dordrecht 2000.
8. Maličký, P., Riečan, B.: On the entropy of dynamical systems. In: Proc. Conf. Ergodic theory and Related Topics II (H. Michel ed.), Teubner, Leipzig 1986, 135 - 138.
9. Mesiar, R., Riečan, B.: On the joint observable in some quantum structures. Tatra Mt. Math. Publ. **9** (1993), 183 - 190.
10. Petrovičová, J.: On the entropy of dynamical systems. Fuzzy Sets and systems **121** (2001), 347 - 351.
11. Montagna, F.: An algebraic approach to propositional fuzzy logic. J. of Logic, Language and Information **9** (2000), 91 - 124.
12. Riečan, B.: On the product MV-algebras. Tatra Mt. Math. Publ. **16** (1999), 143 - 149.
13. Riečan, B.: Almost everywhere convergence in probability MV-algebras with product. Soft Computing **5** (2001), 396 - 399.
14. Riečan, B.: Free products of probability MV-algebras. Atti Sem. Mat. Fis. Univ. Modena **50** (2002), 173 - 186.
15. Riečan, B., Mundici, D.: Probability on MV-algebras. In: Handbook on Measure theory (E. Pap ed.), North Holland, Amsterdam 2002.
16. Riečan, B., Neubrunn, T.: Integral, Measure, and Ordering, Kluwer, Dordrecht 1997.

17. Vrábelová, M.: On the conditional probability in product MV-algebras. Soft Computing **4** (2000), 58 - 61.
18. Zadeh, L.: Probability measures for fuzzy events. J.Math. Anal. App. **23** (1968), 421 - 427.

Reversing the Order of Integration in Iterated Expectations of Compact Convex Random Sets

Luis J. Rodríguez-Muñiz, Miguel López-Díaz, and María Ángeles Gil

Universidad de Oviedo, Departamento de Estadística e I.O. y D.M., Facultad de Ciencias, c/ Calvo Sotelo s/n, 33007 Oviedo (Asturias), Spain

Abstract. In this paper we state some results about the differentiability under the integral sign of compact convex random sets in the special case of working with probability distributions depending on a family of parameters. As a consequence of this fact we obtain some results regarding the exchange of iterated expectacions of compact convex random sets.

1 Introduction

The purpose of this paper is to obtain a result which allows us to reverse two iterated integrals of the random set in those frameworks we are dealing with random sets depending on a parameter, and the involved probability distributions are also depending on a familiy of parameteres (sometimes called conditional probabilities). The originality of the present work arises in the fact that we will not use a Fubini's theorem to get the reversing result. We will obtain a result that will be useful in the case of mappings that are not product measurable.

Thus, our problem is the following: suppose a compact convex random set depending on a real valued parameter. So, we can think of it as a mapping defined on a product space. We want to obtain sufficient conditions to reverse the order of two iterated integrals: one is the expected value of the random set, for a fixed value of the parameter, and the other one is the integral over the values of the parameter itself, for a fixed value of the sample. Since we are working with compact convex subsets of $\mathbb{R}$, this expected values will be calculate by the Kudō-Aumann integral (see Kudō, 1954 and Aumann, 1965).

In order to get our goal, we will have to state and prove some intermediate and supporting results, regarding the measurability of a certain class of integrals, and the Hukuhara derivability under the integral sign of a compact convex random set when working with conditional probabilities. We will also use some supporting results about the similar problem without conditional probabilities. Some of this results and their proofs can be found in Rodríguez-Muñiz and López-Díaz (2002).

2 Preliminaries

We will denote by $\mathcal{K}_c$ the class of non-empty compact convex subsets of $\mathbb{R}$. On this class we can define the well-known Hausdorff distance, d_H (see, for instance, Hiai and Umegaki, 1977). $\mathcal{K}_c$ can be endowed by a semilinear structure by means of Minkowski's addition and the product by a scalar.

Given a probability space $(\Omega, \mathcal{A}, P)$, a (compact convex) random set is a mapping $X : \Omega \to \mathcal{K}_c$ such that it is $\mathcal{A}|\mathcal{B}(\mathcal{K}_c)$-measurable, where $\mathcal{B}(\mathcal{K}_c)$ is the Borel σ-field generated by Hausdorff distance on $\mathcal{K}_c$. A random set is said to be integrably bounded if there exists $h \in L^1(P)$ such that $d_H(X(\omega), \{0\}) \leq h(\omega)$ a.s. $[P]$.

For an integrably bounded random set X we can define its expected value by means of Kudō–Aumann integral, that is, $EX = \{Ef : f \in L^1(P), f(\omega) \in X(\omega) \, a.s. \, [P]\}$, where Ef is the Lebesgue integral of the random variable f. We will also denote EX by $\displaystyle\int_\Omega X(\omega) \, dP(\omega)$, and if $\Omega = [a, b] \subset \mathbb{R}$ then

$$EX = \int_a^b X(\omega) \, dP(\omega).$$

Previous definitions hold in the case of using measures instead of probabilites (see Hiai and Umegaki, 1977).

3 Supporting results

In this section we will state some preliminary results. The first one regards the Radon-Nikodym derivative and the expected value of the random set.

Proposition 1. *Let $([a, b], \mathcal{B}_{[a,b]}, m_1)$ be a σ-finite measure space, being $\mathcal{B}_{[a,b]}$ the Borel σ-field on $[a, b]$ and let $F : [a, b] \to \mathcal{K}_c$ be an integrably bounded random set with respect to m_1. Let m_2 be another σ-finite measure on $([a, b], \mathcal{B}_{[a,b]})$ such that m_1 is absolutely continuous with respect to m_2. Let $\dfrac{dm_1}{dm_2}$ denote a Radon-Nikodym derivative of m_1 with respect to m_2. Then:*

$$\int_a^b F(s) \, dm_1(s) = \int_a^b F(s) \cdot \frac{dm_1}{dm_2}(s) \, dm_2(s).$$

The following result extends First Fundamental Calculus Theorem in the case of random sets with respect to conditioned probabilites.

Proposition 2. *Let $F : [a, b] \to \mathcal{K}_c$ be a continuous mapping at $[a, b]$ and let $([a, b], \mathcal{B}_{[a,b]}, P)$ be a probability space such that P is absolutely continuous with respect to Lebesgue measure on $[a, b]$ –denoted by m– and there exists a continuous Radon-Nikodym derivative, $\dfrac{dP}{dm}$. If we define*

$$G(t) = \int_a^t F(s) \, dP(s),$$

142

then it holds:

i) $G(t)$ is Hukuhara derivable for all $t \in (a, b)$,

ii) $G'(t) = F(t) \cdot \dfrac{dP}{dm}(t)$ for all $t \in (a, b)$.

Now, we state a result regarding differentiability under the integral sign, which will be used in the proof of the main result:

Proposition 3. *Let $(\Omega, \mathcal{A}, P)$ be a probability space with $\Omega \subseteq \mathbf{R}^k$ and $[a, b] \subset \mathbf{R}$. For a given $\omega \in \Omega$, let P_ω be a probability on $([a, b], \mathcal{B}_{[a,b]})$ such that $P_\omega \ll m$ and there exists a continuous Radon-Nikodym derivative, $\dfrac{dP_\omega}{dm}$. If $X : \Omega \times [a, b] \to \mathcal{K}_c$ satisfies that:*

i) for every $t \in [a, b]$, $X_t : \Omega \to \mathcal{K}_c$, given by $X_t(\omega) = X(\omega, t)$, is a random set,

ii) for every $\omega \in \Omega$, $X_\omega : [a, b] \to \mathcal{K}_c$, given by $X_\omega(t) = X(\omega, t)$, is an integrably bounded random set with respect to P_ω and, moreover, for almost every $\omega \in \Omega$ $[P]$ it is continuous at $[a, b]$,

iii) there exists $h \in L^1(P)$ such that

$$\left\| X(\omega, t) \cdot \frac{dP_\omega}{dm}(t) \right\| \leq h(\omega) \quad a.s. \, [P]$$

for every $t \in [a, b]$ and the mapping

$$\omega \mapsto X(\omega, t) \cdot \frac{dP_\omega}{dm}(t)$$

is continuous at ω, a.e. $[m]$,

then

$$t \in [a, b] \mapsto \int_\Omega \left(\int_a^t X(\omega, s) \, dP_\omega(s) \right) dP(\omega)$$

is Hukuhara derivable at (a, b) and:

$$\frac{\partial}{\partial t} \int_\Omega \left(\int_a^t X(\omega, s) \, dP_\omega(s) \right) dP(\omega) = \int_\Omega X(\omega, t) \cdot \frac{dP_\omega}{dm}(t) \, dP(\omega)$$

for every $t \in (a, b)$.

The next result regards continuity of the expected values.

Proposition 4. *Let $(\Omega, \mathcal{B}_\Omega, P)$ be a probability space with $\Omega \subseteq \mathbf{R}^k$. Let $T = [a, b] \subset \mathbf{R}$ and $t_0 \in T$. For every $t \in T$, let us consider P_t a probability on $(\Omega, \mathcal{B}_\Omega)$ such that $P_t \ll P$ and there exists a Radon-Nikodym derivative, $\dfrac{dP_t}{dP}$. If $Y : \Omega \times T \to \mathcal{K}_c$ satisfies that:*

i) *for every* $t \in T$, $Y_t : \Omega \rightarrow \mathcal{K}_c$ *is an integrably bounded random set with respect to* P_t,

ii) *for almost every* $\omega \in \Omega$ $[P]$, $t \mapsto Y(\omega, t) \cdot \dfrac{dP_t}{dP}(\omega)$ *is continuous*,

iii) *there exists* $h \in L^1(P)$ *such that*

$$\|Y(\omega, t) \cdot \frac{dP_t}{dP}(\omega)\| \leq h(\omega) \ a.s. \ [P]$$

for every $t \in N \cap T$, *being* N *a neighbourhood of* t_0,

then

$$t \mapsto \int_\Omega Y(\omega, t) \, dP_t(\omega)$$

is continuous at t_0.

4 Main results

Once we have explained all the basics for the paper, now we state the main result regarding the reverse in the order of integration of two iterated integrals of random sets. Since our main interest is to apply this result to Statistical Decision Theory we have checked that the imposed conditions are not as constraining as they could appear at the first look, moreover, they usually hold in the majority of the decision problems.

Theorem 1. *Let* $(\Omega, \mathcal{B}_\Omega, P)$ *be a probability space with* $\Omega \subseteq \mathbb{R}^k$. *Let us consider* $T = [a, b] \subset \mathbb{R}$ *and the inheritated measurable space* $(T, \mathcal{B}_T, m)$. *For every* $t \in T$, *let* P_t *be a probability on* $(\Omega, \mathcal{B}_\Omega)$ *such that* $P_t \ll P$ *and there exists a continuous Radon-Nikodym derivative,* $\dfrac{dP_t}{dP}$.

For every $\omega \in \Omega$, *let* P_ω *be probability on* $(T, \mathcal{B}_T)$ *such that* $P_\omega \ll m$ *and there exists a continuous Radon-Nikodym derivative,* $\dfrac{dP_\omega}{dm}$,

Let us consider the mapping $X : \Omega \times T \rightarrow \mathcal{K}_c$ *such that*

i) *for every* $t \in T$, $X_t : \Omega \rightarrow \mathcal{K}_c$ *is an integrably bounded random set with respect to* P_t,

ii) *for every* $\omega \in \Omega$, $X_\omega : T \rightarrow \mathcal{K}_c$ *is an integrably bounded random set with respect to* P_ω *and, for almost every* $\omega \in \Omega$ $[P]$ *it is continuous at* T,

iii) *there exists* $h \in L^1(P)$ *such that*

$$\|X(\omega, t) \cdot \frac{dP_\omega}{dm}(t)\| \leq h(\omega) \ a.s. \ [P]$$

for every $t \in T$ *and*

$$\omega \mapsto X(\omega, t) \cdot \frac{dP_\omega}{dm}(t)$$

is continuous at ω, *a.e.* $[m]$,

144

iv) $t \mapsto X(\omega, t) \cdot \dfrac{dP_t}{dP}$ *is continuous at* T, *a.s.* $[P]$,

v) there exists $h_1 \in L^1(P)$ *such that:*

$$\|X(\omega, t) \cdot \frac{dP_t}{dP}(\omega)\| \leq h_1(\omega) \ c.s. \ [P],$$

for every $t \in T$.

Let m' *be a probability on* $(T, \mathcal{B}_T)$ *such that* $m' \ll m$ *and there exists a continuous Radon-Nikodym derivative* $\dfrac{dm'}{dm}$. *If the following equation holds for every* $t \in T$:

$$\frac{dP_\omega}{dm}(t) = \frac{dP_t}{dP}(\omega) \cdot \frac{dm'}{dm}(t) \quad a.s. \ [P],$$

then

$$\int_\Omega \left(\int_a^t X(\omega, s) \, dP_\omega(s) \right) dP(\omega) = \int_a^t \left(\int_\Omega X(\omega, s) \, dP_s(\omega) \right) dm'(s)$$

for every $t \in T$.

Theorem 1 can be easily generalized in the following sense

Proposition 5. *Under conditions in* Theorem 1, *if* T *is a nonbounded interval and there exists* $g \in L^1(P)$ *such that*

$$\int_T \|X(\omega, s)\| dP_\omega(s) \leq g(\omega) \ a.s. \ [P],$$

then it holds

$$\int_\Omega \left(\int_T X(\omega, s) \, dP_\omega(s) \right) dP(\omega) = \int_T \left(\int_\Omega X(\omega, s) \, dP_s(\omega) \right) dm'(s).$$

5 Concluding remarks

The results in this paper guarantee the exchange in the order of integration under quite general conditions and without using Fubini's Theorem, moreover, we do not use product measure spaces.

Our main purpose is to apply these results to the analysis of the Statistical Decision Theory with imprecise utilities, which can be set-valued as well as fuzzy set-valued.

Acknowledgements

Authors want to sincerely acknowledge the financial support from the Spanish Ministry of Education, Culture and Sports (Grant **DGES PB98-1534**).

References and related literature

Aumann, R.J. (1965). Integrals of set-valued functions. J Math Anal Appl, 12:1–12.

Banks, H.T. and Jacobs, M.Q. (1970). A differential calculus for multifunctions. J Math Anal Appl, 29:246–272.

Breiman, L. (1968). Probability. Addison-Wesley Publishing Company, Reading London.

Debreu, G. (1967). Integration of correspondences. In: Proc. Fifth Berkeley Symp. Math. Stat. Prob., 351–372.

Hiai, F. and Umegaki, H. (1977). Integrals, conditional expectations and martingales of multivalued functions. Trans Amer Math Soc, 291:613–627.

Hukuhara, M. (1967). Intégration des applications mesurables dont la valeur est un compact convexe. Funkcial Ekvac, 10:205–223.

Kudō, H. (1954). Dependent experiments and sufficient statistics. Nat Sci Rep Ochanomizu Univ, 4:151–163.

Puri, M.L. and Ralescu, D.A. (1985). The concept of normality for fuzzy random variables. Ann Prob, 13:1373–1379.

Puri, M.L. and Ralescu, D.A. (1986). Fuzzy random variables. J Math Anal Appl, 114:409–422.

Rodríguez-Muñiz, L.J. and López-Díaz, M. (2002). Hukuhara derivative of the fuzzy expected value. Submitted.

Lower Previsions for Unbounded Random Variables

Matthias C. M. Troffaes and Gert de Cooman

Ghent University, SYSTeMS Research Group
Technologiepark 9, 9052 Zwijnaarde, Belgium

Abstract. In order to generalise Walley's theory of lower previsions, which are real-valued maps on bounded random variables, to arbitrary random variables, we introduce extended lower previsions as extended real-valued maps on arbitrary, not necessarily bounded, random variables. We suggest and motivate conditions for avoiding sure loss, coherence and linearity, we construct a natural extension, and we suggest a way to generalise some of the more advanced topological results from the existing theory of lower previsions.

1 Introduction

Walley's theory of lower previsions [8] unifies many of the imprecise probability models in the literature, and from a foundational point of view it seems to be the most satisfactory one. The theory has three important components: (i) *assessment* of a lower prevision in order to represent the available knowledge about a system; (ii) *rationality criteria*, called avoiding sure loss and coherence, which are used to identify conflicts in the assessments and to determine whether they are consistent; and (iii) a *reasoning/inference method*, called natural extension, which tells us how to draw conclusions from, and make decisions based on, the assessments.

A technical problem is that lower previsions are only defined on bounded random variables, whereas in many applications unbounded random variables abound. In particular, the following classes of problems would benefit from a generalisation of the theory of lower previsions: (i) optimisation using an imprecise cost criterion, with an unbounded (e.g., quadratic) cost (Chevé and Congar [1], De Cooman and Troffaes [6]); and (ii) the estimation of unbounded quantities that depend on parameters that are not well known (such as time to failure in reliability theory, Utkin [7]). A number of such generalisations can be found in the literature. Crisma, Gigante and Millossovich [2] have introduced linear previsions for arbitrary real-valued random variables, and these linear previsions may also assume the values $+\infty$ and $-\infty$. Troffaes and De Cooman [5] have constructed an extension for coherent lower previsions defined on bounded random variables to a larger set of random variables, using a limit procedure. But this extension does not assume the values $\pm\infty$, and moreover, the domain of this extension is never the space of all random variables. Since the work of Crisma, Gigante and Millossovich [2] indicates that the domain of linear previsions can be extended to all random variables by including $\pm\infty$ in the range of the prevision, it is now a natural question whether something similar can also be achieved for coherent lower previsions. In this paper, we show that this is indeed possible.

The paper is organised as follows. In Section 2 we define so-called *extended* lower previsions and discuss the types of assessment that give rise to them. Section 3 deals with conditions for avoiding sure loss, coherence, and linearity. We show that all the classical properties of lower previsions are retained with only minor modifications. In Section 4 we construct a natural extension for extended lower previsions, and we use it to define a semi-norm on a special subspace of random variables. Finally, we generalise a topological result from the existing theory of lower previsions that allows us to associate coherent extended lower previsions with sets of dominating extended linear previsions.

Due to limitations of space, we have preferred to stress the underlying ideas rather than to present detailed proofs. Readers interested in the exact details of mathematical reasoning are referred to [4].

2 Assessment through Extended Lower Previsions

2.1 Preliminaries

The set of extended real numbers $R \cup \{-\infty, +\infty\}$ will be denoted by R^*. The operations $+$ and $\cdot$ are defined on R^* as usual, keeping in mind that $0 \cdot (\pm\infty) = 0$. Any expression that cannot be reduced to $-\infty + \infty$ or $+\infty + (-\infty)$ will be called *well defined*. "$\{a \text{ w.d.}; B(a)\}$" means "$\{a; a \text{ well defined}, B(a)\}$", where a denotes an extended real expression and $B(a)$ denotes a Boolean expression that may depend on a.

The set of possible states will be denoted by Ω; one usually thinks of the elements of Ω as possible outcomes of an experiment. A *random variable* is a real-valued map on Ω. The set of random variables on Ω is denoted by $\mathcal{R}(\Omega)$. A bounded random variable is called a *gamble*, and the set of gambles is denoted by $\mathcal{L}(\Omega)$. We stress that random variables only assume values in R. The values of a random variable represent amounts of utility expressed in units of some linear utility scale.

2.2 Definition and Interpretation

A random variable X can be interpreted as an uncertain reward; if $\omega \in \Omega$ turns out to be the true state then the agent receives the amount $X(\omega)$, expressed in units of some linear utility: we say that the agent *receives* X. For any $s \in R$, we say that the agent *buys X for price s* if the agent receives $X - s$. We say that X is *desirable* if the agent is disposed to receive X.

The information the agent has about the outcome of the experiment leads him to accept or reject transactions whose reward or loss depends on this outcome. An extended lower prevision models his uncertainty by looking at a specific type of transactions: buying random variables. In particular, the agent may consider the set ℓ_X of all prices he is disposed to buy X for. If the agent has a disposition to buy X for price s, then he also should have a disposition to buy X for any price less than s. Hence, ℓ_X should take the form of a down-set, which can be characterised almost uniquely by a single extended real number: its supremum.[1] The extended lower prevision of X is then defined as the supremum in R^* of this set: $\underline{P}(X) = \sup \ell_X$.

Definition 1. An *extended lower prevision* $\underline{P}$ on Ω is a (partial) extended real-valued map on $\mathcal{R}(\Omega)$. The domain of this map is denoted by $\operatorname{dom} \underline{P}$.

Thus, $\operatorname{dom} \underline{P}$ is the set of random variables for which the agent assesses supremum buying prices: the agent is willing to buy X for any price strictly less than $\underline{P}(X)$. For example, the map $\inf[\cdot]: \mathcal{R}(\Omega) \to R^*; X \mapsto \inf[X]$ is an extended lower prevision: if X is bounded from below then the agent is willing to pay any price strictly less than the lowest possible reward $\inf[X]$; otherwise the agent is not willing to buy X for any price. We call this map the *vacuous extended lower prevision*.

The conjugate upper prevision $\overline{P}$ of $\underline{P}$ is defined on $-\operatorname{dom} \underline{P}$ by $\overline{P}(X) = -\underline{P}(-X)$ for all $X \in -\operatorname{dom} \underline{P}$, and it can be given the interpretation of an infimum selling price.

3 Rationality Criteria for Extended Lower Previsions

The following is a straightforward generalisation of the axioms of rationality introduced by Walley [8].

Axiom 1 (Axioms of Rationality). Let X and Y be random variables. Any agent whose dispositions conform to the following axioms is called *rational*: if $\sup[X] < 0$ then X is not desirable, if $\inf[X] > 0$ then X is desirable, if X is desirable and $Y \geq X$ then Y is desirable, and finally, if X and Y are desirable then $X + Y$ and λX are desirable for any $\lambda > 0$.

From now on we only consider rational agents. The following lemma tells us how to draw particular conclusions from assessments made by such agents. These will help us to motivate criteria for avoiding sure loss and coherence.

Lemma 1. *Let $\underline{P}$ be an extended lower prevision, representing the assessments of a rational agent. Let $n \in N$, $\lambda_i \geq 0$, $X_i \in \operatorname{dom} \underline{P}$ for every $i \in \{1, \ldots, n\}$ and assume that $\alpha := \sum_{i=1}^{n} \lambda_i \underline{P}(X_i)$ is well defined. If $\alpha = -\infty$ then, for at least one $i \in \{1, \ldots, n\}$, there is no price the agent is willing to buy X_i for. In all other cases, the agent is willing to buy $\sum_{i=1}^{n} \lambda_i X_i$ for any price $s < \alpha$.*

---

[1] The only information that is lost is whether ℓ_X is open or closed (unless the lower prevision is infinite). For simplicity, we therefore always assume that ℓ_X is open. Hence, this model may be a little more conservative than the "true" model.

3.1 Avoiding Sure Loss

For convenience, we shall write $(n; \lambda_i \geq 0; X_i; \underline{P})$ if we mean "$n \in N$, $\lambda_i \geq 0$, $X_i \in \text{dom}\,\underline{P}$ for every $i \in \{1, \ldots, n\}$" and $(n; \lambda_i \geq 0; X_i; \underline{P})_{\text{wd}}$ to mean "$n \in N$, $\lambda_i \geq 0$, $X_i \in \text{dom}\,\underline{P}$ for every $i \in \{1, \ldots, n\}$ and $\sum_{i=1}^{n} \lambda_i \underline{P}(X_i)$ well defined". Note that if in such cases $n = 0$, then $\sum_{i=1}^{n} \lambda_i X_i$ and $\sum_{i=1}^{n} \lambda_i \underline{P}(X_i)$ are zero by definition.

Definition 2. Let $\underline{P}$ be an extended lower prevision. If

$$\sup\left[\sum_{i=1}^{n} \lambda_i X_i\right] \geq \sum_{i=1}^{n} \lambda_i \underline{P}(X_i). \tag{1}$$

for every $(n; \lambda_i \geq 0; X_i; \underline{P})_{\text{wd}}$, then we say that $\underline{P}$ *avoids sure loss*.

Note that our definition coincides with that given by Walley [8] for his lower previsions (in our terminology these are real-valued extended lower previsions defined on sets of gambles). Its motivation is similar to the one given by Walley. Observe what happens if it is not satisfied: assume that there are $(n; \lambda_i \geq 0; X_i; \underline{P})_{\text{wd}}$ such that

$$\alpha := \sum_{i=1}^{n} \lambda_i \underline{P}(X_i) > \sup\left[\sum_{i=1}^{n} \lambda_i X_i\right] =: \gamma.$$

Note that this can only hold if $\alpha \neq -\infty$ and $\gamma \in R$. It implies that we may choose a $\beta \in R$ such that $\alpha > \beta > \gamma \geq \sum_{i=1}^{n} \lambda_i X_i$, which means that if the agent buys $\sum_{i=1}^{n} \lambda_i X_i$ for a price β, he incurs a sure loss of at least $\beta - \gamma$. But by Lemma 1, the rational agent *is* willing to buy $\sum_{i=1}^{n} \lambda_i X_i$ for $\beta < \alpha$ (recall that $\alpha \neq -\infty$).

In contradistinction to the special case considered by Walley, it does *not* suffice in our more general case to consider only integer combinations. As the following counterexample shows, the condition

$$\sup\left[\sum_{i=1}^{n} X_i\right] \geq \sum_{i=1}^{n} \underline{P}(X_i) \tag{2}$$

for every $n \in N$ and $X_i \in \text{dom}\,\underline{P}$, $i \in \{1, \ldots, n\}$ is not equivalent to avoiding sure loss for extended lower previsions in general, but it is equivalent for (real-valued extended) lower previsions on gambles (see Walley [8, Section 2.4.4(a)]).

Example 1. Let $\Omega = R$, and let the identity map on R be denoted by I. Consider the extended lower prevision $\underline{P}$ with domain $\{I, -\pi I\}$, defined by $\underline{P}(I) = 1$ and $\underline{P}(-\pi I) = 2$. Since $n - m\pi \neq 0$ for every $n, m \in N$ not both zero, we find that $\sup_{x \in R}[nI(x) - m\pi I(x)] = +\infty$ for every $n, m \in N$ not both zero. Consequently, we have that condition (2), i.e., $\sup[nI - m\pi I] \geq n + 2m$, holds for every $n, m \in N$ not both zero. If n and m are both zero then the inequality holds trivially. But $\underline{P}$ does not avoid sure loss: take $\lambda_1 = \pi$ and $\lambda_2 = 1$, then $\sup[\lambda_1 I - \lambda_2 \pi I] = 0 < \lambda_1 + 2\lambda_2$.

Many properties of extended lower previsions that avoid sure loss can be generalised from results proven by Walley [8, Section 2.4.7], by adding the requirement that both sides must be well defined for every (in)equality. To give but one example, it can be shown that $\underline{P}(\sum_{i=1}^{n} \lambda_i X_i) \leq \sum_{i=1}^{n} \lambda_i \overline{P}(X_i)$ whenever the right hand side is well defined.

3.2 Coherence

For convenience, we shall write $(n; \lambda_i \geq 0; X_i; \underline{P})'_{\text{wd}}$ to mean "$n \in N$, $\lambda_i \geq 0$, $X_i \in \text{dom}\,\underline{P}$ for every $i \in \{0, \ldots, n\}$ and $\sum_{i=1}^{n} \lambda_i \underline{P}(X_i) - \lambda_0 \underline{P}(X)$ well defined". As before, if in such a case $n = 0$, then $\sum_{i=1}^{n} \lambda_i X_i$ and $\sum_{i=1}^{n} \lambda_i \underline{P}(X_i)$ are zero by definition.

Definition 3. Let $\underline{P}$ be an extended lower prevision. If

$$\sup\left[\sum_{i=1}^{n} \lambda_i X_i - \lambda_0 X_0\right] \geq \sum_{i=1}^{n} \lambda_i \underline{P}(X_i) - \lambda_0 \underline{P}(X_0) \qquad (3)$$

for every $(n; \lambda_i \geq 0; X_i; \underline{P})'_{\text{wd}}$, then we say that $\underline{P}$ is *coherent*.

Obviously coherence implies avoiding sure loss. Walley's [8] definition of coherence for lower previsions is a special case of ours; and its motivation is similar. Observe what happens if the condition is not satisfied: assume that there are $(n; \lambda_i \geq 0; X_i; \underline{P})'_{\text{wd}}$ such that

$$\sum_{i=1}^{n} \lambda_i \underline{P}(X_i) - \lambda_0 \underline{P}(X_0) > \sup\left[\sum_{i=1}^{n} \lambda_i X_i - \lambda_0 X_0\right].$$

We may assume that $\lambda_0 \neq 0$ (we have already given a motivation for the case that $\lambda_0 = 0$ in the previous section), and consequently

$$\alpha := \sum_{i=1}^{n} \frac{\lambda_i}{\lambda_0} \underline{P}(X_i) - \underline{P}(X_0) > \sup\left[\sum_{i=1}^{n} \frac{\lambda_i}{\lambda_0} X_i - X_0\right] =: \beta.$$

Note that this can only hold if $\beta \in R$ and $\alpha \neq -\infty$. Defining $\alpha_1 := \sum_{i=1}^{n} \frac{\lambda_i}{\lambda_0} \underline{P}(X_i)$, we have that $\alpha_1 - \beta > \underline{P}(X_0)$. Observe that also

$$\sum_{i=1}^{n} \frac{\lambda_i}{\lambda_0} X_i - \beta \leq X_0. \qquad (4)$$

Since $\alpha = \alpha_1 - \underline{P}(X_0)$ is well defined and $\alpha \neq -\infty$, it suffices to consider the following cases:

(a) $\underline{P}(X_0) \in R$. The rational agent is disposed to buy the right hand side of (4) for any price strictly less than $\underline{P}(X_0)$. But, we may also infer from the other assessments $\underline{P}(X_1), \ldots, \underline{P}(X_n)$ and Lemma 1 that he is willing to buy the left hand side of (4) for any price strictly less than $\alpha_1 - \beta$. Consequently, he is disposed to buy the right hand side of (4) for any price strictly less than $\alpha_1 - \beta$. But this price $\alpha_1 - \beta$ is strictly larger than the supremum price $\overline{P}(X_0)$ he has specified for X_0, which points to an inconsistency in the assessments.

(b) $\underline{P}(X_0) = -\infty$. Since there is no price the rational agent is disposed to buy the right hand side of (4) for, there is also no price he is disposed to buy the left hand side of (4) for. But from $\alpha_1 > -\infty$ and Lemma 1 we know that the agent is willing to buy $\sum_{i=1}^{n} \frac{\lambda_i}{\lambda_0} X_i$ for any price strictly smaller than α_1, which again points to a contradiction in the assessments.

If $\operatorname{dom} \underline{P}$ is a linear space, then there is a simpler condition.

Proposition 1. *Let $\underline{P}$ be an extended lower prevision and assume that $\operatorname{dom} \underline{P}$ is a linear space. Then $\underline{P}$ is coherent iff the following statements hold for every $X, Y \in \operatorname{dom} \underline{P}$ and every $\lambda \geq 0$.*

(1) $\underline{P}(X) \geq \inf[X]$.
(2) $\underline{P}(\lambda X) = \lambda \underline{P}(X)$.
(3) $\underline{P}(X + Y) \geq \underline{P}(X) + \underline{P}(Y)$ whenever the right hand side is well defined.

The properties of coherence listed in Walley [8, Section 2.6.1] generalise in the same way as the properties of avoiding sure loss. The most important ones are summarised in the following theorem.

Theorem 1. *Let $\underline{P}$ be a coherent extended lower prevision. Let X and Y be random variables. Let μ be a constant random variable. Let λ be a positive, real number. Let X_α be a net of random variables. Then the following statements hold whenever every term and every operation is well defined.*

(i) $\inf[X] \leq \underline{P}(X) \leq \overline{P}(X) \leq \sup[X]$
(ii) $X \leq Y + \mu \implies \underline{P}(X) \leq \underline{P}(Y) + \mu$ and $\overline{P}(X) \leq \overline{P}(Y) + \mu$
(iii) $\underline{P}(X) + \underline{P}(Y) \leq \underline{P}(X+Y) \leq \underline{P}(X) + \overline{P}(Y) \leq \overline{P}(X+Y) \leq \overline{P}(X) + \overline{P}(Y)$
(iv) $\underline{P}(\lambda X) = \lambda \underline{P}(X), \quad \overline{P}(\lambda X) = \lambda \overline{P}(X)$
(v) $|\underline{P}(X) - \underline{P}(Y)| \leq \overline{P}(|X - Y|), \quad |\overline{P}(X) - \overline{P}(Y)| \leq \overline{P}(|X - Y|)$
(vi) $\overline{P}(|X_\alpha - X|) \to 0 \implies \underline{P}(X_\alpha) \to \underline{P}(X)$ and $\overline{P}(X_\alpha) \to \overline{P}(X)$

3.3 Linearity

For convenience we write $(n; \lambda_i \in R; X_i; \underline{P})_{\text{wd}}$ if we mean "$n \in N$, $\lambda_i \in R$, $X_i \in \operatorname{dom} \underline{P}$ for every $i \in \{1, \dots, n\}$ and $\sum_{i=1}^{n} \lambda_i \underline{P}(X_i)$ well defined".

Definition 4. Let $\underline{P}$ be an extended lower prevision. If we have that

$$\sup \left[\sum_{i=1}^{n} \lambda_i X_i \right] \geq \sum_{i=1}^{n} \lambda_i \underline{P}(X_i) \tag{5}$$

for every $(n; \lambda_i \in R; X_i; \underline{P})_{\text{wd}}$ then we say that $\underline{P}$ is *linear*.

This definition is equivalent to Definition 3.1 of Crisma, Gigante and Millossovich [2]. Obviously any linear extended lower prevision is coherent. For any linear extended lower prevision $\underline{P}$ it holds that $\underline{P}(X) = \overline{P}(X)$ whenever both X and $-X$ are in the domain of $\underline{P}$. This establishes that the conjugate upper prevision of a linear extended lower previsions coincides with the extended lower prevision, if its domain is symmetric (dom $\underline{P} = -$ dom $\underline{P}$, this is the case when dom $\underline{P}$ is a linear space, for example). Therefore, if an extended lower prevision $\underline{P}$ is linear and has a symmetric domain, we call it an *extended linear prevision* and denote it by P. It can then be interpreted both as a lower and as an upper prevision.

If dom $\underline{P}$ is a linear space, then there is a simpler condition for linearity. This result has been proven by Crisma, Gigante and Millossovich [2, Theorem 3.2].

Proposition 2. *Let $\underline{P}$ be an extended lower prevision and assume that* dom $\underline{P}$ *is a linear space. Then $\underline{P}$ is linear iff the following statements hold for every $X, Y \in$ dom P and every $\lambda \in R$.*

(1) $\underline{P}(X) \geq \inf[X]$.
(2) $\underline{P}(\lambda X) = \lambda \underline{P}(X)$.
(3) $\underline{P}(X + Y) = \underline{P}(X) + \underline{P}(Y)$ whenever the right hand side is well defined.

In the case of a (real-valued extended) lower prevision on gambles we may drop condition (2) of Proposition 2 (see also [8]). We have not studied in detail whether this condition may be dropped in the more general case as well.

An extended lower prevision $\underline{Q}$ is said to *dominate* an extended lower prevision $\underline{P}$ if dom $\underline{Q} \supseteq$ dom $\underline{P}$ and $\underline{Q} \geq \underline{P}$ point-wise on dom $\underline{P}$. For any extended lower prevision $\underline{P}$, if there is an linear extended lower prevision that dominates $\underline{P}$, then $\underline{P}$ avoids sure loss. A partial converse of this statement will be given in Theorem 2.

4 Natural Extension

4.1 Definition and Properties

Definition 5. Let $\underline{P}$ be an extended lower prevision. The *natural extension* of $\underline{P}$, $\mathfrak{X}_{\underline{P}} : \mathcal{R}(\Omega) \to R^*$, is defined by

$$\mathfrak{X}_{\underline{P}}(Z) = \sup_{(n; \lambda_i \geq 0; X_i; \underline{P})} \left\{ \inf \left[Z - \sum_{i=1}^{n} \lambda_i X_i \right] + \sum_{i=1}^{n} \lambda_i \underline{P}(X_i) \text{ w.d.} \right\}$$

$$= \sup_{(n; \lambda_i \geq 0; X_i; \underline{P})} \left\{ \mu + \sum_{i=1}^{n} \lambda_i \underline{P}(X_i) \text{ w.d.}; \mu \in R, Z \geq \mu + \sum_{i=1}^{n} \lambda_i X_i \right\}.$$

The natural extension of $\overline{P}$ is defined by $\mathfrak{X}_{\overline{P}}(Z) = -\mathfrak{X}_{\underline{P}}(-Z)$.

If $\underline{P}$ is a real-valued extended lower prevision, then the natural extension of $\underline{P}$ is also given by (where $G_{\underline{P}}(X_i)$ denotes $X_i - \underline{P}(X_i)$)

$$\mathcal{X}_{\underline{P}}(Z) = \sup_{(n; \lambda_i \geq 0; X_i; \underline{P})} \left\{ \inf \left[Z - \sum_{i=1}^{n} \lambda_i G_{\underline{P}}(X_i) \right] \right\}$$

$$= \sup_{(n; \lambda_i \geq 0; X_i; \underline{P})} \left\{ \alpha; \alpha \in R, Z - \alpha \geq \sum_{i=1}^{n} \lambda_i G_{\underline{P}}(X_i) \right\}.$$

It coincides with Walley's [8] notion of natural extension in the special case that $\underline{P}$ is a (real-valued extended) lower prevision on gambles.

The following examples show that, even when $\underline{P}$ is real-valued, $\mathcal{X}_{\underline{P}}$ can assume the values $\pm\infty$. This explains why we prefer to include $\pm\infty$ in the range of extended lower previsions when considering unbounded random variables.

Example 2. Let $\underline{P}$ be any coherent, real-valued, extended lower prevision on $\mathcal{L}(\Omega)$. Then it is easily checked that $\mathcal{X}_{\underline{P}}$ assumes the value $-\infty$ at any random variable X that is not bounded from below.

Example 3. Let $\Omega = [0, +\infty)$ and define the increasing sequence of random variables (gambles) X_n by $X_n(x) = x$ if $x < n$, and $X_n(x) = n$ otherwise. Define $\underline{P}$ on $\{X_n; n \in N\}$ by $\underline{P}(X_n) = n$. Use the fact that $G_{\underline{P}}(X_n)(x) = 0$ when $x \geq n$ to show that $\underline{P}$ is coherent. Now let Y be the identity map on $[0, +\infty)$. This random variable is bounded from below but not from above. We find that

$$\mathcal{X}_{\underline{P}}(Y) = \sup \left\{ \inf \left[Y - \sum_{i=1}^{n} \lambda_i G_{\underline{P}}(X_i) \right] ; (n; \lambda_i \geq 0; X_i; \underline{P}) \right\}$$

$$\geq \sup_{n \in N} \inf \left[Y - G_{\underline{P}}(X_n) \right] = \sup N = +\infty,$$

whence $\mathcal{X}_{\underline{P}}(Y) = +\infty$.

One can easily show that $\underline{P}$ avoids sure loss iff $\mathcal{X}_{\underline{P}}$ is a coherent extended lower prevision. A tedious proof shows that all the well-known properties of natural extension for (real-valued extended) lower previsions on gambles (see [8, Section 3.1.2]) generalise as well.

Proposition 3. *Assume that $\underline{P}$ avoids sure loss. Then the following statements hold.*

(i) $\mathcal{X}_{\underline{P}}$ *dominates $\underline{P}$.*

(ii) $\mathcal{X}_{\underline{P}} = \underline{P}$ *on* $\operatorname{dom} \underline{P}$ *iff $\underline{P}$ is coherent.*

(iii) $\mathcal{X}_{\underline{P}}$ *is the smallest coherent extended lower prevision on $\mathcal{R}(\Omega)$ that dominates $\underline{P}$ on* $\operatorname{dom} \underline{P}$.

(iv) *If $\underline{P}$ is coherent then $\mathcal{X}_{\underline{P}}$ is the smallest coherent extension of $\underline{P}$ to $\mathcal{R}(\Omega)$.*

154

4.2 A Semi-Normed Linear Lattice

Definition 6. Let $\underline{P}$ be an extended lower prevision. The $\underline{P}$-*norm* is a map $\|\cdot\|_{\underline{P}} : \mathcal{R}(\Omega) \to R^*$ defined by

$$\|Z\|_{\underline{P}} = \mathcal{X}_{\overline{P}}(|Z|). \tag{6}$$

It should be noted that the terminology $\underline{P}$-"norm" (actually, $\underline{P}$-"semi-norm") is justified if and only if $\underline{P}$ avoids sure loss: if $\underline{P}$ incurs sure loss then $\|\cdot\|_{\underline{P}}$ assumes negative values (e.g., it assumes $-\infty$ at the zero gamble).

Proposition 4. *Let $\underline{P}$ and $\underline{Q}$ be extended lower previsions that avoid sure loss. Assume that $\underline{Q}$ dominates $\underline{P}$. Then the following statements hold for every $X, Y \in \mathcal{R}(\Omega)$, and every $\lambda \in R$.*

(i) $0 \leq \|X\|_{\underline{Q}} \leq \|X\|_{\underline{P}} \leq \sup[|X|]$.
(ii) $\|\lambda X\|_{\underline{P}} = |\lambda| \|X\|_{\underline{P}}$.
(iii) $\|X + Y\|_{\underline{P}} \leq \|X\|_{\underline{P}} + \|Y\|_{\underline{P}}$.

Definition 7. Let $\underline{P}$ be an extended lower prevision that avoids sure loss. A random variable is said to be $\underline{P}$-*norm-finite*, if it belongs to the set

$$\mathcal{L}_{\underline{P}} = \{Z \in \mathcal{R}(\Omega); \|Z\|_{\underline{P}} < +\infty\}, \tag{7}$$

which is called the $\underline{P}$-*space*. If $\underline{P}$ incurs sure loss, then we define $\mathcal{L}_{\underline{P}} = \emptyset$.

Proposition 5. *Let $\underline{P}$ and $\underline{Q}$ be extended lower previsions that avoid sure loss, such that $\underline{Q}$ dominates $\underline{P}$. Then $\mathcal{L}(\Omega) \subseteq \mathcal{L}_{\underline{P}} \subseteq \mathcal{L}_{\underline{Q}}$ and $(\mathcal{L}_{\underline{P}}, \|\cdot\|_{\underline{P}})$ is a semi-normed linear lattice.*

For a (real-valued extended) lower prevision $\underline{P}$ on gambles, where $\operatorname{dom}\underline{P} \subseteq \mathcal{L}(\Omega)$, it turns out that $\mathcal{L}_{\underline{P}} = \mathcal{L}(\Omega)$. We also point out that $\mathcal{L}_{\underline{P}}$ and $\|\cdot\|_{\underline{P}}$ generalise notions previously introduced in the literature for so-called previsible random variables (Troffaes and de Cooman [5]): for a coherent (real-valued extended) lower prevision $\underline{P}$ on gambles, the set $\mathcal{L}_{\underline{P}}^{\times}(\Omega)$ of $\underline{P}$-previsible random variables is included in $\mathcal{L}_{P^\times}$, where $\underline{P}^\times$ is the extension of $\underline{P}$ to the set $\mathcal{L}_{\underline{P}}^{\times}(\Omega)$ of $\underline{P}$-previsible random variables. Moreover, the $\underline{P}$-norm introduced in [5] coincides with the $\underline{P}^\times$-norm on $\mathcal{L}_{\underline{P}}^{\times}(\Omega)$.

In what follows, we shall assume that $\mathcal{L}_{\underline{P}} \supseteq \operatorname{dom}\underline{P}$. Then $\mathcal{M}(\underline{P})$ denotes the set of linear extended lower previsions on $\mathcal{L}_{\underline{P}}$ that dominate $\underline{P}$. It is an important observation that $\mathcal{M}(\underline{P})$ only contains real-valued extended linear previsions; indeed, $|Q(X)| \leq Q(|X|) \leq \mathcal{X}_{\overline{P}}(|X|) < +\infty$ for every $X \in \mathcal{L}_{\underline{P}}$. We can easily show that $\mathcal{M}(\underline{P})$ is a subset of the topological dual $\mathcal{L}_{\underline{P}}^*$ of the semi-normed linear space $\mathcal{L}_{\underline{P}}$. If we equip $\mathcal{L}_{\underline{P}}^*$ with the weak* topology (of point-wise convergence), then $\mathcal{M}(\underline{P})$ is compact with respect to this topology.

The proof of the following theorem relies on the topological structure of $\mathcal{L}_{\underline{P}}$ and the Hahn-Banach separation theorem.

Theorem 2 (Lower Envelope Theorem). *Let $\underline{P}$ be an extended lower prevision and assume that $\mathcal{L}_{\underline{P}} \supseteq \mathrm{dom}\,\underline{P}$. Then $\underline{P}$ avoids sure loss, and there is a (real-valued) extended linear prevision on $\mathrm{dom}\,\underline{P}$ that dominates $\underline{P}$. Moreover, $\mathfrak{X}_{\underline{P}}(X) = \min\{P(X); P \in \mathcal{M}(\underline{P})\}$ for all X in $\mathcal{L}_{\underline{P}}$. Consequently, $\underline{P}$ is coherent iff $\underline{P}(X) = \min\{P(X); P \in \mathcal{M}(\underline{P})\}$ for all X in $\mathrm{dom}\,\underline{P}$.*

Acknowledgements

This paper presents research results of project G.0139.01 of the Fund for Scientific Research, Flanders (Belgium), and of the Belgian Programme on Interuniversity Poles of Attraction initiated by the Belgian state, Prime Minister's Office for Science, Technology and Culture. The scientific responsibility rests with the authors.

References

1. M. Chevé and R. Congar, *Optimal pollution control under imprecise environmental risk and irreversibility*, Risk Decision and Policy **5** (2000), 151–164.
2. Lucio Crisma, Patrizia Gigante, and Pietro Millossovich, *A notion of coherent prevision for arbitrary random quantities*, Journal of the Italian Statistical Society **6** (1997), no. 3, 233–243.
3. G. de Cooman, F. G. Cozman, S. Moral, and P. Walley (eds.), *ISIPTA '99 – proceedings of the first international symposium on imprecise probabilities and their applications*, Ghent, Imprecise Probabilities Project, 1999.
4. Gert de Cooman and Matthias C. M. Troffaes, *Lower previsions for unbounded random variables*, Tech. report, Universiteit Gent, Onderzoeksgroep SYSTeMS, 2002, In progress.
5. Matthias C. M. Troffaes and Gert de Cooman, *Extension of coherent lower previsions to unbounded random variables*, accepted for IPMU 2002 (The 9th International Conference on Information Processing and Management of Uncertainty in Knowledge-Based Systems, 1–5 July 2002, Annecy, France).
6. Matthias C. M. Troffaes and Gert de Cooman, *Dynamical programming for deterministic discrete-time systems with uncertain gain*, Submitted, 2002.
7. Lev V. Utkin and Sergey V. Gurov, *Imprecise reliability models for the general lifetime distribution classes*, In De Cooman et al. [3], pp. 333–342.
8. Peter Walley, *Statistical reasoning with imprecise probabilities*, Chapman and Hall, London, 1991.

A Hierarchical Uncertainty Model under Essentially Incomplete Information

Lev V. Utkin

Munich University, Institute of Statistics, Ludwigstr. 33, 80539, Munich Germany

Abstract. A hierarchical uncertainty model for combining the different judgements is studied in the paper. The model is general enough for many applications. The presented approach for dealing with such model allows us to combine the available heterogeneous information in the following ways: computing new probability bounds for some predefined interval of previsions, computing an "average" interval of first-order previsions, and updating the second-order probabilities after observing new event.

1 Introduction

There are judgements taking into account a degree of belief to assessments, for example, "Mean time to failure of a component is between 3 and 5 hours with the probability greater than 0.95". They can be operated by the *second-order uncertainty models* (*hierarchical uncertainty models*) on which much attention have been focused due to their quite commonality. De Cooman *et al.* [10,1,3] proposed that *fuzzy probabilities* can be interpreted as a special type of hierarchical uncertainty model. Gardenfors and Sahlin [2] introduced the notion of an "*epistemic reliability*" function for modelling the second-order probabilities. Nau [7] described beliefs by lower and upper probabilities qualified by numerical *confidence weights* and studied their properties. The study of some tasks related to the homogeneous second-order previsions has been illustrated by Kozine and Utkin in [5].

In this paper, we study a hierarchical uncertainty model for combining the different evidences where the second-order probabilities can be regarded as confidence weights and the first-order uncertainty is modelled by the lower and upper previsions of different continuous gambles [9,11,12]. This model is more general for many applications than the model considered in [5]. The presented approach for dealing with such model allows us to combine the available heterogeneous information in the following ways: computing new probability bounds (weights) for some predefined interval, computing an "average" interval of first-order previsions, and updating the second-order probabilities (weights) after observing new event. This approach is self-adapted to special situations. An experts does not need to think about a type of an averaging operator. The natural extension is automatically doing that.

2 Preliminary Definitions

Suppose there is a continuous random variable x defined on the sample space Ω and information about this variable is represented as a set of m interval-valued expectations of functions $f_1(x), ..., f_m(x)$. Denote these lower and upper expectations $\underline{a}_i = \underline{M}f_i$ and $\overline{a}_i = \overline{M}f_i$, $i = 1, ..., m$. In terms of the theory of imprecise probabilities the corresponding functions $f_i(x)$ and interval-valued expectations $\underline{M}f_i$ and $\overline{M}f_i$, $i = 1, ..., m$, are called gambles and lower and upper previsions, respectively. The previsions $\underline{M}f_i$ and $\overline{M}f_i$ can be regarded as the bounds for an unknown precise prevision Mf_i which is called a linear prevision. For computing new previsions $\overline{M}g$ and $\underline{M}g$ of a gamble $g(x)$ from the available information, the natural extension can be used in the following form:

$$\underline{M}g = \inf_{\mathcal{P}} \int_{\Omega} g(x)\rho(x)\mathrm{d}x, \overline{M}g = \sup_{\mathcal{P}} \int_{\Omega} g(x)\rho(x)\mathrm{d}x, \tag{1}$$

subject to

$$\int_{\Omega} \rho(x)\mathrm{d}x = 1, \ \rho(x) \geq 0, \ \underline{a}_i \leq \int_{\Omega} f_i(x)\rho(x)\mathrm{d}x \leq \overline{a}_i, \ i \leq m. \tag{2}$$

Here the infimum and supremum are taken over the set $\mathcal{P}$ of all possible probability density functions $\{\rho(x)\}$ satisfying conditions (2).

Problems (1)-(2) are linear and the dual optimization problems can be written as [4,6,9,12]:

$$\overline{M}g = \inf_{c_0,c_i,d_i} \left(c_0 + \sum_{i=1}^{m} (c_i\overline{a}_i - d_i\underline{a}_i) \right), \ \underline{M}g = -\overline{M}(-g), \tag{3}$$

subject to $c_i, d_i \in \mathbf{R}^+$, $c_0 \in \mathbf{R}$, $i = 1, ..., n$, and $c_0 + \sum_{i=1}^{m} (c_i - d_i) f_i(x) \geq g(x)$ for all $x \in \Omega$.

Natural extension is a general mathematical procedure for calculating new previsions from initial judgements. It produces a coherent overall model from a certain collection of imprecise probability judgements and may be seen as the basic constructive step in interval-valued statistical reasoning.

3 The Problem Statement

Suppose that we have a set of weighted expert judgements related to some measures of the system behaviour Mf_i, $i = 1, ..., m$, i.e. there are lower and upper previsions $\underline{a}_i$, $\overline{a}_i$. Suppose that each of m experts is characterized by a subjective probability γ_i or interval of probabilities $[\underline{\gamma}_i, \overline{\gamma}_i]$, $i = 1, ..., m$. Generally, the judgements can be written as follows:

$$\Pr\{\underline{a}_i \leq Mf_i(x) \leq \overline{a}_i\} \in [\underline{\gamma}_i, \overline{\gamma}_i], \ i \leq m. \tag{4}$$

Here the set $\{\underline{a}_i, \overline{a}_i\}$ contains the first-order previsions, the set $\{\underline{\gamma}_i, \overline{\gamma}_i\}$ contains the second-order probabilities and $Mf_i(x) = \int_{\mathbf{R}_+} f_i(x)\rho(x)\mathrm{d}x$. Our aim is to produce new judgement which can be regarded as a combination of available judgements. In other words, the following tasks should be solved:

1. *Computing the probability bounds $[\underline{\gamma}, \overline{\gamma}]$ for some new interval $A = [\underline{a}, \overline{a}]$ of new linear previsions $Mg(x)$.*
2. *Computing an "average" interval $[a_*, a^*]$ of new previsions $Mg(x)$.*
3. *Updating probabilities $[\underline{\gamma}_i, \overline{\gamma}_i]$ after observing an event B.*

In order to give the reader the essence of the subject analyzed and make all the formulas more readable, we will mainly consider the natural extension only for the upper bound. Furthermore, throughout the paper the obvious constraints for densities ρ to optimization problems such that $\rho(x) \geq 0$, $\int_{\mathbf{R}_+} \rho(x)\mathrm{d}x = 1$ will not be written.

4 Computing the Probability Bounds (Task 1)

Suppose that the set of linear previsions Mf_i, $i = 1, ..., m$, Mg is an outcome set. Then we have the set of lower $\underline{\gamma}_i$ and upper $\overline{\gamma}_i$ probabilities of events $A_i = \{\underline{a}_i \leq Mf_i(x) \leq \overline{a}_i\}$. In this case the linear previsions $Mf_i(x)$ and Mg can be regarded as continuous random variables denoted z_i and z having desities ψ_i and ψ, respectively. At that the variable z is some function of variables $z_1, ..., z_m$ whose implicit form is unknown. If all gambles are identical, i.e. for all i there holds $f_i(x) = g(x)$, then $z_i = z$ for all i and the stated tasks are reduced [5] to (1)-(3). However, the main problem is the difference of all gambles f_i and g. By regarding the linear previsions as random variables, we can not define a functional relationship between them except some simplest special cases. At the same time, we can not regard the corresponding variables as independent different ones because they are joined through the common density ρ. Therefore, the following approach is proposed.

Let J be a set of indices and $J \subseteq N = \{1, 2, ..., m\}$. Denote the following sets of constraints:

$$\mathcal{A}_J = \{A_i, \; i \in J\} = \{\underline{a}_i \leq Mf_i \leq \overline{a}_i, \; i \in J\},$$
$$\mathcal{A}_0 = \{A\} = \{\underline{a} \leq Mg \leq \overline{a}\}, \; \mathcal{A}_J^c = \{A_i^c, \; i \in J\}.$$

$$A_j^c = \left(\inf f_j \leq Mf_j \leq \underline{a}_j\right) \cup \left(\overline{a}_j \leq Mf_j \leq \sup f_j\right). \tag{5}$$

Note that probabilities of events A_i, $i \in J$, can be represented as previsions of gambles being the indicator functions

$$I_{A_i}(z_i) = I_{A_i}(Mf_i) = I_{A_i}\left(\int_{\mathbf{R}_+} f_i(x)\rho(x)\mathrm{d}x\right).$$

Suppose that $\Psi(z_1, ..., z_m)$ is a joint density of random variables $z_1, ..., z_m$. Then the upper probability $\overline{\gamma} = \overline{M} I_A(Mg) = \overline{M} I_A(z)$ can be obtained from the following optimization problem:

$$\overline{\gamma} = \sup_{\mathcal{R}} \int_{\mathbf{R}_+^n} I_A(z) \Psi(z_1, ..., z_m) \mathrm{d}z_1 \cdots \mathrm{d}z_m,$$

subject to $\underline{\gamma}_i \le \int_{\mathbf{R}_+^n} I_{A_i}(z) \Psi(z_1, ..., z_m) \mathrm{d}z_1 \cdots \mathrm{d}z_m \le \overline{\gamma}_i, \ i \le m.$

Here $\mathcal{R}$ is the set of all possible joint densities $\{\Psi(z_1, ..., z_m)\}$; $I_{A_i}(z)$ is the indicator function such that $I_{A_i}(z) = 1$ if $z \in A_i$ and $I_{A_i}(z) = 0$ if $z \notin A_i$. The corresponding dual optimization problem is of the form:

$$\overline{\gamma} = \inf_{c_0, c_i, d_i} \left\{ c_0 + \sum_{i=1}^{m} \left(c_i \overline{\gamma}_i - d_i \underline{\gamma}_i \right) \right\}, \tag{6}$$

subject to $c_i, d_i \in \mathbf{R}^+$, $c_0 \in \mathbf{R}$, $i = 1, ..., m$,

$$c_0 + \sum_{i=1}^{m} (c_i - d_i) I_{A_i}(z_i) \ge I_A(z). \tag{7}$$

Constraints (7) can be rewritten as

$$c_0 + \sum_{i=1}^{m} (c_i - d_i) I_{A_i} \left(\int_{\mathbf{R}_+} f_i(x) \rho(x) \mathrm{d}x \right) \ge I_A \left(\int_{\mathbf{R}_+} g(x) \rho(x) \mathrm{d}x \right), \ \rho \in \mathcal{P}.$$

Here $\mathcal{P}$ is the set of all probability density functions $\{\rho(x)\}$. Let us consider constraints to the above problem in detail. In order to compute the indicator functions, it is necessary to substitute the different functions ρ from $\mathcal{P}$ and calculate corresponding integrals. Obviously, this task can not be practically solved. Therefore, we propose another way to do it. Let $\mathcal{P}_i$ be a set of densities ρ satisfying the i-th constraint $\underline{a}_i \le M f_i \le \overline{a}_i$ and $\mathcal{P}_0$ be a set of densities ρ satisfying the constraint $\underline{a} \le Mg \le \overline{a}$. We call the set $\mathcal{A}_J$ consistent if there is at least one density ρ satisfying all constraints whose indices belong to J, i.e. $\bigcap_{i \in J} \mathcal{P}_i \ne \emptyset$. Let $\mathcal{C}$ be a set of all consistent sets $\mathcal{A}_J$. Now we can see that if the set $\mathcal{A}_J \cup \mathcal{A}_{N \setminus J}^c$ is consistent, then $I_{A_i}(M f_i) = 1$ if $i \in J$ and $I_{A_i}(M f_i) = 0$ if $i \in N \setminus J$. This implies that

$$c_0 + \sum_{i \in J} (c_i - d_i) \ge I_A \left(\int_{\mathbf{R}_+} g(x) \rho(x) \mathrm{d}x \right).$$

Moreover, if the set $\mathcal{A}_{N \setminus J}^c \cup \mathcal{A}_J \cup \mathcal{A}_0$ is consistent, then there holds $c_0 + \sum_{i \in J} (c_i - d_i) \ge 1$, otherwise $c_0 + \sum_{i \in J} (c_i - d_i) \ge 0$. In other words, if the set $\mathcal{A}_{N \setminus J}^c \cup \mathcal{A}_J$ is consistent, then there exists at least one density ρ such that all linear previsions $M f_i$, $i \in J$, are in intervals $[\underline{a}_i, \overline{a}_i]$ and their indicator functions equal to 1, all linear previsions Mf_j, $j \in N \backslash J$, do not belong to intervals $[\underline{a}_j, \overline{a}_j]$ and their indicator functions equal to 0. If the extended set $\mathcal{A}^c_{N \backslash J} \cup \mathcal{A}_J \cup \mathcal{A}_0$ is consistent, then there exists at least one density ρ satisfying the additional constraint $\mathcal{A}_0$. It should be noted that the union of two events in (5) means that for consistency of $\mathcal{A}^c_j \cup \mathcal{A}_J$ at least one of these events has to give the consistency of $\mathcal{A}^c_j \cup \mathcal{A}_J$. So, to simplify constraints (7) it is necessary to look over all consistent sets $\mathcal{A}^c_{N \backslash J} \cup \mathcal{A}_J$. Then constraints (7) can be rewritten $\forall J \subseteq N$ as follows:

$$c_0 + \sum_{i \in J} (c_i - d_i) \geq \begin{cases} 1, \ \mathcal{A}^c_{N \backslash J} \cup \mathcal{A}_J \cup \mathcal{A}_0 \in C \\ 0, \ \mathcal{A}^c_{N \backslash J} \cup \mathcal{A}_J \cup \mathcal{A}_0 \notin C \end{cases}. \qquad (8)$$

If $\mathcal{A}^c_{N \backslash J} \cup \mathcal{A}_J$ is inconsistent, then the inequality $c_0 + \sum_{i \in J} (c_i - d_i) \geq 0(1)$ is excluded from the list of all constraints.

Now the question arises: how to determine the consistency of sets $\mathcal{A}^c_{N \backslash J} \cup \mathcal{A}_J$ or $\mathcal{A}^c_{N \backslash J} \cup \mathcal{A}_J \cup \mathcal{A}_0$. Condition $\bigcap_{i \in J} \mathcal{P}_i \neq \emptyset$ is valid if an optimization problem with constraints $\mathcal{A}_J$ has any solution. At that the objective function can be arbitrary. In other words, it is necessary to solve the following optimization problem:

$$\inf_{\rho} \left(\sup_{\rho} \right) \int_{\mathbf{R}_+} \varphi(x)\rho(x)\mathrm{d}x,$$

subject to $\underline{a}_i \leq \int_{\mathbf{R}_+} f_i(x)\rho(x)\mathrm{d}x \leq \overline{a}_i$, $i \in J$. Here φ is any function. It can be seen that the above problem can be regarded as the natural extension of first-order previsions $\{\underline{a}_i, \overline{a}_i, i \in J\}$ on one of previsions $\underline{M}\varphi$ or $\overline{M}\varphi$. Since the consistency of constraints does not depend on the function φ, then it should be chosen as simpler as possible. The similar reasoning can be made for lower previsions. In this case

$$\underline{\gamma} = \sup_{c_0, c_i, d_i} \left(c_0 + \sum_{i=1}^{m} \left(c_i \underline{\gamma}_i - d_i \overline{\gamma}_i \right) \right), \qquad (9)$$

subject to $c_i, d_i \in \mathbf{R}^+$, $c_0 \in \mathbf{R}$, $i \in J$, $\forall J \subseteq N$

$$c_0 + \sum_{i \in J} (c_i - d_i) \leq \begin{cases} 1, \ \mathcal{A}^c_{N \backslash J} \cup \mathcal{A}_J \cup \mathcal{A}^c_0 \notin C \\ 0, \ \mathcal{A}^c_{N \backslash J} \cup \mathcal{A}_J \cup \mathcal{A}^c_0 \in C \end{cases}. \qquad (10)$$

So, we write a general algorithm for computing $\underline{\gamma}$ and $\overline{\gamma}$:

Step 1. By considering all possible binary vectors $(y_1, ..., y_m)$, $y_i \in \{0, 1\}$, the sets $\mathcal{A}^c_{N \backslash J} \cup \mathcal{A}_J$ of constraints are formed, where $i \in J$ if $y_i = 1$ and $i \in N \backslash J$ if $y_i = 0$.

Step 2. Choosing a set $\mathcal{A}^c_{N \backslash J} \cup \mathcal{A}_J$ from the list obtained at Step 1.

Step 3. If $\mathcal{A}^c_{N \backslash J} \cup \mathcal{A}_J \in C$, then (8) is used for computing $\overline{\gamma}$ and (10) is used for computing $\underline{\gamma}$. If $\mathcal{A}^c_{N \backslash J} \cup \mathcal{A}_J \notin C$, then go to Step 2.

Step 4. From systems of constraints obtained at Step 3 and objective functions (6), (9), the probabilities $\overline{\gamma}$ and $\underline{\gamma}$ are computed as solutions to corresponding optimization problems.

5 Computing an "Average" Interval (Task 2)

The second task of computing an "average" interval $[a_*, a^*] = [\underline{M}Mg, \overline{M}Mg]$ of the linear prevision Mg can be solved as follows. Let us rewrite problem (6)-(7) as

$$a^* = \inf_{c_0, c_i, d_i} \left\{ c_0 + \sum_{i=1}^{m} \left(c_i \overline{\gamma}_i - d_i \underline{\gamma}_i \right) \right\}, \tag{11}$$

subject to $c_i, d_i \in \mathbf{R}^+$, $c_0 \in \mathbf{R}$, $i = 1, ..., m$, $\forall \rho \in \mathcal{P}$

$$c_0 + \sum_{i=1}^{m} (c_i - d_i) I_{A_i}(Mf_i) \geq Mg, \tag{12}$$

Denote $M^{(k)} f_i = \int_{\mathbf{R}_+} f_i(x) \rho^{(k)}(x) \mathrm{d}x$, $k = 1, 2$. Let $\rho^{(1)}$ and $\rho^{(2)}$ be some densities satisfying constraints $\mathcal{A}^c_{N \backslash J} \cup \mathcal{A}_J$ and $M^{(2)} g \geq M^{(1)} g$. Then the constraint $c_0 + \sum_{i \in J} (c_i - d_i) I_{A_i}(M^{(1)} f_i) \geq M^{(1)} g$ follows from the constraint $c_0 + \sum_{i \in J} (c_i - d_i) I_{A_i}(M^{(2)} f_i) \geq M^{(2)} g$ and can be removed. This implies that (12) is equivalent to

$$c_0 + \sum_{i \in J} (c_i - d_i) I_{A_i}(Mf_i) \geq \sup_{\mathcal{R}_J} Mg.$$

Here the supremum is taken over the set $\mathcal{R}_J$ of all possible density functions $\{\rho(x)\}$ satisfying the set of consistent constraints $\mathcal{A}^c_{N \backslash J} \cup \mathcal{A}_J$. Hence the supremum of Mg is replaced by the upper prevision $\overline{M}_J g$ under consistent constraints $\mathcal{A}^c_{N \backslash J} \cup \mathcal{A}_J$. As a result, we obtain constraints

$$c_0 + \sum_{i \in J} (c_i - d_i) \geq \sup_{\mathcal{R}_J} \int_{\mathbf{R}_+} g(x) \rho(x) \mathrm{d}x, \quad \forall J \subseteq N. \tag{13}$$

In the same way the lower value a_* can be computed.

6 Updating Second-Order Probabilities (Task 3)

Suppose that there is the set of expert judgements (4) and we observe an event $B(x) = I_{[\underline{b}, \overline{b}]}(x)$, for example, "failure occurs between 10 and 12 hours". Now we can update beliefs $\underline{\gamma}_i$ and $\overline{\gamma}_i$ after observing the event B, i.e. construct posterior lower and upper probabilities $\underline{\eta}_k$ and $\overline{\eta}_k$ from the prior lower and upper previsions $\underline{\gamma}_k$ and $\overline{\gamma}_k$ and statistical data B such that

$$\Pr\left\{ \underline{a}_k \leq M(f_k | B) \leq \overline{a}_k \right\} \in [\underline{\eta}_k, \overline{\eta}_k].$$

The linear prevision $M(f_k|B)$ can be represented as follows [6,8]:

$$M(f_k|B) = \frac{Mf_kB}{\Pr(B)} = \frac{\int_{\mathbf{R}_+} f_k(x)B(x)\rho(x)\mathrm{d}x}{\int_{\mathbf{R}_+} B(x)\rho(x)\mathrm{d}x}.$$

After observing the event B, we obtain new previsions $\underline{M}(f_k|B)$ and $\overline{M}(f_k|B)$. This implies that the probability that the linear prevision $M(f_k|B)$ is in the previous interval $[\underline{a}_k, \overline{a}_k]$ is updated and its bounds are $\underline{\eta}_k$ and $\overline{\eta}_k$. Then the natural extension for computing the posterior upper probability $\overline{\eta}_k = \overline{MI}_{A_k}(M(f_k|B)$ is of the form:

$$\overline{\eta}_k = \inf_{c_0, c_i, d_i} \left(c_0 + \sum_{i=1}^{m} \left(c_i \overline{\gamma}_i - d_i \underline{\gamma}_i \right) \right), \tag{14}$$

subject to $c_i, d_i \in \mathbf{R}^+$, $c_0 \in \mathbf{R}$, $i = 1, ..., m$, $\forall \rho \in \mathcal{P}$

$$c_0 + \sum_{i=1}^{m} (c_i - d_i) I_{A_i} (Mf_i) \geq I_A (M(f_k|B)). \tag{15}$$

Here $\mathcal{P}$ is a set of all possible density functions $\{\rho(x)\}$. The following solution is similar to task 1.

Acknowledgements

The work was supported by the Alexander von Humboldt Foundation (Germany). I am very grateful to Prof. Kurt Weichselberger, Dr. Anton Wallner, Dr. Thomas Augustin (Munich University, Germany), and Dr. Igor Kozine (Risoe National Laboratory, Denmark) for their very valuable remarks and comments.

References

1. de Cooman G. (1998) Possibilistic previsions. In: EDK (Ed.) Proceedings of IPMU'98, Volume 1, Paris, 2–9
2. Gärdenfors P., N.-E. Sahlin N.-E. (1982) Unreliable probabilities, risk taking, and decision making. Synthese 53:361–386
3. Gilbert L., de Cooman G., Kerre E. (2000) Practical implementation of possibilistic probability mass functions. In: Proceedings of Fifth Workshop on Uncertainty Processing (WUPES 2000), Jindvrichouv Hradec, Czech Republic, 90–101
4. Gurov S.V., Utkin L.V. (1999) Reliability of Systems under Incomplete Information. Saint Petersburg, Lubavich Publ, in Russian
5. Kozine I.O., Utkin L.V. (2001) Constructing coherent interval statistical models from unreliable judgements. In: Zio E., Demichela M,, Piccini N. (Eds.) Proceedings of the European Conference on Safety and Reliability ESREL2001, Volume 1, Torino, Italy, 173–180

6. Kuznetsov V. P. (1991) Interval Statistical Models. Moscow, Radio and Communication, in Russian

7. Nau R. F. (1992) Indeterminate probabilities on finite sets. The Annals of Statistics 20:1737–1767

8. Utkin L.V., Kozine I.O. (2000) Conditional previsions in imprecise reliability. In: Ruan H. A. D., D'Hondt P. (Eds.) Intelligent Techniques and Soft Computing in Nuclear Science and Engineering, Bruges, Belgium, 72–79

9. Walley P. (1991) Statistical Reasoning with Imprecise Probabilities. London, Chapman and Hall

10. Walley P. (1997) Statistical inferences based on a second-order possibility distribution. International Journal of General Systems 9:337–383

11. Weichselberger K. (2000) The theory of interval-probability as a unifying concept for uncertainty. International Journal of Approximate Reasoning 24:149–170

12. Weichselberger K. (2001) Elementare Grundbegriffe einer allgemeineren Wahrscheinlichkeitsrechnung, Volume I Intervallwahrscheinlichkeit als umfassendes Konzept. Heidelberg, Physica

Imprecise Calculation with the Qualitative Information about Probability Distributions

Lev V. Utkin

Munich University, Institute of Statistics, Ludwigstr. 33, 80539, Munich Germany

Abstract. One of the advantages of imprecise probabilities is the possibility to carry out computations over the set of all possible probability distributions that, in turn, leads to rather imprecise results. To reduce the imprecision, it is proposed to use additional qualitative information about kurtosis, skewness, variance, and unimodality. It is shown how this information can be involved into the natural extension, which is the main tool for dealing with imprecise data in the framework of imprecise probabilities. Numerical examples illustrate the proposed approach.

1 Introduction

For combining and processing the partial information about some system of random variables and to make maximal use of the available information without additional assumptions about probability distributions or independency of random variables, the theory of imprecise probabilities (also called the theory of lower previsions [6], the theory of interval statistical models [3], the theory of the interval probabilities [7,8]) may be successfully applied. A general framework for the theory is provided by upper and lower previsions. They can model a very wide variety of kinds of uncertainty, partial information, and ignorance. The rules used in this theory are based on a general procedure called natural extension and can be applied to various measures. In fact, the natural extension is a linear optimization problem for computing new interval-valued previsions from the available set of lower and upper previsions. It has been shown by P. Walley and G. de Cooman that the possibility measures and belief functions can be regarded as special cases of the imprecise probabilities.

When applying the natural extension, it is declared that the lower and upper previsions are sought over the set of all possible probability distributions, i.e. the widest class, which is really appealing. It should be noted that, on the one hand, this is one of the advantages of imprecise probabilities because some additional and often unjustified assumptions about a type of probability distributions are avoided and the risk of obtaining the incorrect results is reduced. From the other hand, the natural extension results in many applications too imprecise intervals of computed characteristics that restricts the wide usage of the imprecise probabilities. One of the ways for reducing the obtained imprecision is to use the additional qualitative information about the probability distributions which may be very often available and is related to special applications. For example, it is known that the most distributions considered in reliability analysis are unimodal and it is difficult to expect that a lifetime distribution is multimodal. We may know that, for example, the variance of a random variable is less than the expectation squared (information about the variance) or the left tail of a possible distribution is heavier than its right tail (negative values for the skewness). Sometimes, it is known that a random variable has typically a "flat" density function, which is rather constant near zero, and very small for larger values of the variable (negative kurtosis). How to take into account this additional information and incorporate it into the natural extension?

The problem here is that such characteristics as the variance, skewness, kurtosis can not be represented as previsions because they can not be regarded as expectations of some functions. The same can be said about unimodality. Therefore, the paper presents the methods for involving the above peculiarities of probability distributions into imprecise calculations. At that, for involving the unimodality, it is shown how to use Khintchine's condition in the natural extension. By using the information about the variance, kurtosis, skewness, the natural extension becomes a parametric linear optimization problem. The various numerical examples illustrate the impact of additional information on results of imprecise calculations.

2   Kurtosis, Skewness and Variance

Kurtosis is a measure of whether the data are peaked or flat relative to a normal distribution. That is, data sets with a high kurtosis tend to have a distinct peak near the mean, decline rather rapidly, and have heavy tails. Data sets with low kurtosis tend to have a flat top near the mean rather than a sharp peak. The kurtosis of $x \in \Omega$ is classically defined by $kurt(x) = E\{x^4\} - 3\left(E\{x^2\}\right)^2$. Obviously, $kurt(x)$ can not be represented as a prevision, because it contains the "non-linear" term $\left(E\{x^2\}\right)^2$. This implies that it is necessary to find a way in order to avoid this non-linearity.

Suppose that there is an additional judgement $kurt(x) \geq 0$. Then for computing new previsions of a gamble g on the basis the available information in the form of lower $\underline{a}_i$ and upper $\overline{a}_i$ previsions of the gamble φ_i, $i = 1, ..., n$, the natural extension can be written as the following optimization problems:

$$\underline{M}(g) = \min_{\mathcal{P}} \int_\Omega g(x)\rho(x)\mathrm{d}x, \ \overline{M}(g) = \max_{\mathcal{P}} \int_\Omega g(x)\rho(x)\mathrm{d}x, \qquad (1)$$

subject to

$$\int_\Omega \rho(x)\mathrm{d}x = 1, \ \underline{a}_i \leq \int_\Omega \varphi_i(x)\rho(x)\mathrm{d}x \leq \overline{a}_i, \ i \leq n, \ kurt(x) \geq 0. \qquad (2)$$

Here the minimum and maximum are taken over the set $\mathcal{P}$ of all possible probability density functions $\{\rho(x)\}$ satisfying conditions (2).

166

The last constraint can be rewritten as follows:

$$\int_\Omega x^4 \rho(x)\mathrm{d}x - 3\left(\int_\Omega x^2\rho(x)\mathrm{d}x\right)^2 \geq 0.$$

Now we have the non-linear optimization problem. However, it can be reduced to a parametric linear programming problem. Denote $\int_\Omega x^2\rho(x)\mathrm{d}x = h$. Then the constraint $kurt(x) \geq 0$ can be represented as two constraints

$$\int_\Omega x^2\rho(x)\mathrm{d}x - h = 0, \quad \int_\Omega x^4\rho(x)\mathrm{d}x - 3h^2 \geq 0.$$

Here h is the parameter in the parametric linear programming problem. Finally, we obtain

$$\underline{M}(g) = \min_{\mathcal{P},h}\int_\Omega g(x)\rho(x)\mathrm{d}x, \quad \overline{M}(g) = \max_{\mathcal{P},h}\int_\Omega g(x)\rho(x)\mathrm{d}x,$$

subject to

$$\int_\Omega \rho(x)\mathrm{d}x = 1, \ \rho(x) \geq 0, \ \underline{a}_i \leq \int_\Omega \varphi_i(x)\rho(x)\mathrm{d}x \leq \overline{a}_i, \ i \leq n,$$

$$\int_\Omega (x^2 - h)\rho(x)\mathrm{d}x = 0, \quad \int_\Omega (x^4 - 3h^2)\rho(x)\mathrm{d}x \geq 0.$$

The dual optimization problem for computing the lower prevision is of the form:

$$\underline{M}(g) = \sup_{c,c_i,d_i}\left(c + \sum_{i=1}^n (c_i\underline{a}_i - d_i\overline{a}_i) + (v_1 - w_1)h + 3v_2h^2\right),$$

subject to $c_i \in \mathbf{R}_+$, $d_i \in \mathbf{R}_+$, $v_i \in \mathbf{R}_+$, $w_i \in \mathbf{R}_+$, $c \in \mathbf{R}$ and $\forall x \in \Omega$,

$$c + \sum_{i=1}^n (c_i - d_i)\varphi_i(x_i) + (v_1 - w_1)x^2 + (v_2 - w_2)x^4 \leq g(x).$$

So, the task of computing the lower and upper previsions of the function g is reduced to the solution of a number of linear optimization problems with different values of the parameter h whose results are lower and upper previsions depending on h, i.e. $\underline{M}_h(g)$ and $\overline{M}_h(g)$. The final result is determined as follows: $\underline{M}(g) = \min_h \underline{M}_h(g)$, $\overline{M}(g) = \max_h \overline{M}_h(g)$.

Skewness is a measure of the lack of symmetry. The skewness for a normal distribution is zero, and any symmetric data should have a skewness near zero. Negative values for the skewness indicate data that are skewed left and positive values for the skewness indicate data that are skewed right.

The skewness of x is defined to be

$$skew(x) = \frac{E\{(x - E(x))^3\}}{[E\{(x - E(x))^2\}]^{3/2}}.$$

Suppose the additional judgement is $skew(x) \geq 0$. It can be rewritten as $E(x^3) - 3E(x^2)E(x) + 2(E(x))^3 \geq 0$. Denote $\int_\Omega x\rho(x)\mathrm{d}x = h$. By using the same reasoning, the constraint $skew(x) \geq 0$ can be written as two constraints

$$\int_\Omega x\rho(x)\mathrm{d}x = h, \quad \int_\Omega x^3\rho(x)\mathrm{d}x - 3h\int_\Omega x^2\rho(x)\mathrm{d}x + 2h^3 \geq 0.$$

There are judgements concerning the variance of a random variable, for example, the variance is less than the expectation squared. The variance of x is defined to be $var(x) = E\{x^2\} - (E\{x\})^2$. Suppose that there is the additional judgement $var(x) \leq (E\{x\})^2$. Hence $\int_\Omega x^2\rho(x)\mathrm{d}x - 2\left(\int_\Omega x\rho(x)\mathrm{d}x\right)^2 \leq 0$. Denote $\int_\Omega x\rho(x)\mathrm{d}x = h$. Then there hold $\int_\Omega x\rho(x)\mathrm{d}x = h$ and $\int_\Omega x^2\rho(x)\mathrm{d}x - 2h^2 \leq 0$.

The following computations with the information about skewness and variance are similar.

Example 1. Suppose that the available information is represented as follows: lower and upper probabilities of failure before 4 hours are 0.6 and 0.7, lower and upper probabilities of failure after 5 hours are 0.1 and 0.2. By using the information about kurtosis, we obtain bounds for the probability of failure before 6 hours: 0.7 and 0.9. By using the information about skewness, we obtain bounds: 0.6 and 1. Without this additional information bounds for the same probability are 0.6 and 1. Now suppose that the mean time to failure is between 5 and 6 hours (additional information). By using the information about kurtosis, we obtain bounds: 0.75 and 0.9. By using the information about skewness, we obtain bounds: 0.8 and 0.9. Without the additional information bounds for the same probability are 0.75 and 1.

It can be seen from Example 1 that the information about kurtosis and skewness may reduce the intervals of probabilities. At the same time, there are cases when the additional qualitative information does not change the available imprecision.

3 Unimodal Distributions

Let us consider another type of the additional qualitative information when it is known that the distribution of the considered random variable is unimodal. A univariate distribution is said to be unimodal with mode r if its cumulative probability distribution function is convex on the interval $(-\infty, r)$ and concave on the interval $(r, +\infty)$. Denote the set of all unimodal distributions or densities $\mathcal{U}$. Then the minimum and maximum in the natural extension must be taken over the set $\mathcal{U}$. According to [2,1], the probability distribution F belong to $\mathcal{U}$ if it can be represented as $F(t) = \int_0^1 G(t/x)\mathrm{d}x$ (Khintchine's condition). Here G is an arbitrary distribution function. Denote the corresponding densities as $\rho(t) = \mathrm{d}F(t)/\mathrm{d}t$, $\omega(t) = \mathrm{d}G(t)/\mathrm{d}t$. Then

$$\rho(t) = \frac{\mathrm{d}\int_0^1 G(t/x)\mathrm{d}x}{\mathrm{d}t} = \int_0^1 \frac{1}{x}\omega(t/x)\mathrm{d}x. \tag{3}$$

It should be noted that Khintchine's condition allows us to restrict the set of possible distributions by unimodal ones having the mode value at point 0. Here we consider a more general case when r is a possible mode value, $r \geq 0$ and $a \leq r \leq b$. Then (3) can be rewritten as $\rho_r(t) = \int_0^1 \frac{1}{x}\omega((t-r)/x)\mathrm{d}x$. Suppose that $\Omega = [a, b]$, $a \leq 0 \leq b$. After substituting ρ_r into optimization problem (1)-(2), for example, into the objective function, we obtain

$$M(g) = \int_{a+r}^{b+r} g(t)\rho_r(t)\mathrm{d}t = \int_a^b g(t+r) \int_0^1 \frac{1}{x}\omega(t/x)\mathrm{d}x\mathrm{d}t.$$

Hence we obtain the objective function with the density $\omega(z)$ and new gambles

$$\widetilde{g}_r(z) = z^{-1} \int_a^b g(t+r) \left[I_{[0,b]}(t)I_{[t,b]}(z) - I_{[a,0]}(t)I_{[a,t]}(z) \right] \mathrm{d}t,$$

which can be written in another form:

$$\widetilde{g}_r(z) = z^{-1} \begin{cases} -\int_z^0 g(t+r)\mathrm{d}t, & z < 0 \\ \int_0^z g(t+r)\mathrm{d}t, & z \geq 0 \end{cases}.$$

Finally, we have the following optimization problem with variables $\omega(z) \in \mathcal{U}$:

$$\underline{M}(g) = \min_{\mathcal{U}} \int_\Omega \widetilde{g}_r(z)\omega(z)\mathrm{d}z, \quad \overline{M}(g) = \max_{\mathcal{U}} \int_\Omega \widetilde{g}_r(z)\omega(z)\mathrm{d}z, \tag{4}$$

subject to

$$\int_\Omega \omega(z)\mathrm{d}z = 1, \quad \omega(z) \geq 0, \quad \underline{a}_i \leq \int_\Omega \widetilde{\varphi}_{ri}(z)\omega(z)\mathrm{d}z \leq \overline{a}_i, \quad i \leq n, \tag{5}$$

where

$$\widetilde{\varphi}_{ri}(z) = z^{-1} \begin{cases} -\int_z^0 \varphi_i(t+r)\mathrm{d}t, & z < 0 \\ \int_0^z \varphi_i(t+r)\mathrm{d}t, & z \geq 0 \end{cases}.$$

Consider the most useful gambles. If $g(z) = z^k$, $k = 1, 2, ...$, then

$$\widetilde{g}_r(z) = \frac{(z+r)^{k+1} - r^{k+1}}{z(k+1)}.$$

If $g(z) = I_{[c,d]}(z)$, $0 \leq c \leq d$, then

$$\widetilde{g}_r(z) = z^{-1} \left[\min(d, \max(c, z + r)) - \max(c, \min(d, r)) \right].$$

It should be noted that the obtained linear optimization problem is parametric with the parameter r. So, the problem of involving the unimodality condition in the natural extension is reduced to some changes of gambles in the objective function and in constraints.

Example 2. Consider data from Example 1. By using the information about unimodality, we obtain bounds for the probability of failure before 6 hours: 0.8 and 1. At that the value of r corresponding to the lower bound is 4 and the value of r corresponding to the upper bound is 5.

Acknowledgements

The work was supported by the Alexander von Humboldt Foundation (Germany). I am very grateful to Dr. Igor Kozine (Risoe National Laboratory, Denmark) for his very valuable remarks and comments.

References

1. Feller W. (1971) An Introduction to Probability Theory and its Applications, volume 2, 2nd edition, Wiley, New York
2. Khintchine A.Y. (1938) On unimodal distributions. Izv. Nauchno-Isled. Inst. Mat. Mech. 2:1–7
3. Kuznetsov V. P. (1991) Interval Statistical Models. Moscow, Radio and Communication, in Russian
4. Shapiro A., Kleywegt A. (2000) Robust analysis of stochastic problems. Technical report, School of Industrial and Systems Engineering, Georgia Institute of Technology
5. Utkin L.V., Kozine I.O. (2001) Different faces of the natural extension. In: de Comman G., Fine T.L., Seidenfeld T. (Eds.) Imprecise Probabilities and Their Applications. Proc. of the 2nd Int. Symposium ISIPTA'01, Ithaca, USA, Shaker Publishing, 316–323
6. Walley P. (1991) Statistical Reasoning with Imprecise Probabilities. London, Chapman and Hall
7. Weichselberger K. (2000) The theory of interval-probability as a unifying concept for uncertainty. International Journal of Approximate Reasoning 24:149–170
8. Weichselberger K. (2001) Elementare Grundbegriffe einer allgemeineren Wahrscheinlichkeitsrechnung, Volume I Intervallwahrscheinlichkeit als umfassendes Konzept. Heidelberg, Physica

Approximation of Belief Functions by Minimizing Euclidean Distances

Thomas Weiler and Ulrich Bodenhofer

Software Competence Center Hagenberg
A-4232 Hagenberg, Austria
e-mail: {thomas.weiler,ulrich.bodenhofer}@scch.at

Abstract. This paper addresses the approximation of belief functions by minimizing the Euclidean distance to a given belief function in the set of probability functions. The special case of Dempster-Shafer belief functions is considered in particular detail. It turns out that, in this case, an explicit solution by means of a projective transformation can be given. Furthermore, we also consider more general concepts of belief. We state that the approximation by means of minimizing the Euclidean distance, unlike other methods that are restricted to Dempster-Shafer belief, works as well. However, the projective transformation formula cannot necessarily be applied in these more general settings.

Key words: belief functions, Dempster-Shafer theory, evidence reasoning, probability theory, theory of evidence, uncertain reasoning.

1 Introduction

The relation between belief and probability plays an important role in the theory of uncertain reasoning and its applications. Belief, e.g. as in Dempster-Shafer theory, can be viewed as an extension of probability. An advantage of more general concepts of belief is that they relate to incomplete information. Probability theory, however, provides a well-established decision making theory [1,2]. In order to be able to apply probabilistic decision principles also in the presence of more general concepts of belief, therefore, different ways to transform belief functions into probability functions have been developed. So far, Dempster-Shafer belief functions have been studied most extensively in this context [5–9].

This paper is concerned with the problem how a given belief function can be approximated by minimizing the Euclidean distance in the set of probability functions, i.e. we consider the following minimization problem.

Optimization Problem (OP) For a given belief function $D : \overline{SL} \to [0,1]$ minimize the objective function

$$\sqrt{\sum_{\overline{\theta} \in \overline{SL}} \left(D(\overline{\theta}) - P(\overline{\theta}) \right)^2}$$

with respect to a probability function $P : \overline{SL} \to [0,1]$, where $\overline{SL}$ denotes the Lindenbaum algebra (i.e. the Boolean algebra of well-formed formulae, where two formulae are considered as equal if all their evaluations coincide) of a finite propositional language L with n propositional variables $p_1, \ldots, p_n$.

In this paper, we first consider (OP) in relation with Dempster-Shafer belief functions. We show that (OP) can be solved explicitly by means of a projection transformation. Finally, we discuss more general concepts of belief. However, we demonstrate that the projective transformation does not necessarily give the correct result if we go beyond Dempster-Shafer belief functions.

2 Belief Functions

Definition 1. On the Lindenbaum algebra $\overline{SL}$, we define the following ordering: For all $\overline{\theta}, \overline{\psi} \in \overline{SL}$,

$$\overline{\theta} \leq \overline{\psi} \text{ if and only if } \overline{\theta} \models \overline{\psi}.$$

A formula $\overline{\alpha} \in \overline{SL}$ is called an *atom* of $\overline{SL}$ if and only if for each propositional sentence $\overline{\theta} \in \overline{SL}$ either $\overline{\alpha} \leq \overline{\theta}$ or $\overline{\alpha} \leq \neg\overline{\theta}$ holds.

Note that atoms uniquely correspond to conjunctions of the form

$$[\neg]p_1 \wedge [\neg]p_2 \wedge \cdots \wedge [\neg]p_n,$$

where the brackets indicate that each propositional variable may be prefixed with a negation or not. Therefore, $J = 2^n$ different atoms exist for a language L with n propositional variables.

Definition 2. A mapping $V : \overline{SL} \to \{0,1\}$ is called a *valuation* if and only if there exists an atom $\overline{\alpha} \in \overline{At}$ such that for all $\overline{\theta} \in \overline{SL}$,

$$V(\overline{\theta}) = \begin{cases} 1 \text{ for } \overline{\theta} \geq \overline{\alpha} \\ 0 \text{ otherwise} \end{cases}$$

where $\overline{At}$ stands for the set of atoms of $\overline{SL}$. Moreover, we denote the valuation induced by an atom $\overline{\alpha}$ with $V_{\overline{\alpha}}$.

As easy to see, a valuation is a function that assigns a truth value to each formula in $\overline{SL}$ with the particular property that the truth values assigned to the propositional variables uniquely determine the truth value of any compound formula.

Definition 3. A mapping $P : \overline{SL} \to [0,1]$ is called a *probability function* if and only if there exists a mapping $m_P : \overline{At} \to [0,1]$ satisfying

(P1) $\sum_{\overline{\alpha} \in \overline{At}} m_P(\overline{\alpha}) = 1$,
(P2) $P(\overline{\theta}) = \sum_{\overline{\alpha} \in \overline{At}} m_P(\overline{\alpha}) \cdot V_{\overline{\alpha}}(\overline{\theta})$ for all $\overline{\theta} \in \overline{SL}$.

In order to treat Dempster-Shafer belief functions in a similar way, we generalize the concept of valuations.

Definition 4. A mapping $V' : \overline{SL} \rightarrow \{0,1\}$ is called an *information function* if and only if there exists a $\overline{\theta} \in \overline{SL} \setminus \{\mathbf{0}\}$ such that for all $\overline{\psi} \in \overline{SL}$,

$$V'(\overline{\psi}) = \begin{cases} 1 \text{ for } \overline{\psi} \geq \overline{\theta} \\ 0 \text{ otherwise} \end{cases}$$

where $\mathbf{0}$ denotes the equivalence class of contradictions and $V'_{\overline{\theta}}$ stands for the information function which is generated by $\overline{\theta}$.

The crucial difference between a valuation and an information function is that an information function does not need to be generated by an atom. Information functions are closely related to the simple support functions defined by Shafer [10]. By using information functions, we are able to extend the definition of probability to Dempster-Shafer belief.

Definition 5. A mapping $D : \overline{SL} \rightarrow [0,1]$ is called a *Dempster-Shafer belief function* if and only if there exists a mapping $m_D : \overline{SL} \setminus \{\mathbf{0}\} \rightarrow [0,1]$ which satisfies

(DS1) $\sum_{\overline{\theta} > 0} m_D(\overline{\theta}) = 1$,
(DS2) $D(\overline{\theta}) = \sum_{\overline{\psi} > 0} m_D(\overline{\psi}) \cdot V'_{\overline{\psi}}(\overline{\theta})$ for all $\overline{\theta} \in \overline{SL}$.

Dempster-Shafer belief functions are, therefore, convex combinations of information functions (equivalent definitions can be found in the literature[3]). Now let us come to the most general case.

Definition 6. An arbitrary mapping $Bel : \overline{SL} \rightarrow [0,1]$ is called a *general belief function*. We denote the set of all general belief functions (with respect to a given finite propositional language) with GB.

In the following section, we define a vector space structure on the above set of belief functions. In this vector space, the optimization problem (OP) turns out to be equivalent to projecting a vector into a subspace, at least if we restrict to Dempster-Shafer belief functions.

3   The Vector Space of Belief Functions

Definition 7. The *vector space of functions from* $\overline{SL}$ *to* $\mathbb{R}$ is defined as

$$\mathbf{B} = (B, \oplus, \ominus, \odot, O, \|.\|, \langle ., . \rangle)$$

where B is the set of all functions from $\overline{SL}$ to $\mathbb{R}$. For all $\overline{\theta} \in \overline{SL}$, $Bel_1, Bel_2 \in B$ and $\lambda \in \mathbb{R}$, the operations $\oplus$, $\ominus$ and $\odot$ are defined by

$$(Bel_1 \oplus Bel_2)(\overline{\theta}) = Bel_1(\overline{\theta}) + Bel_2(\overline{\theta})$$
$$(Bel_1 \ominus Bel_2)(\overline{\theta}) = Bel_1(\overline{\theta}) - Bel_2(\overline{\theta})$$
$$(\lambda \odot Bel_1)(\overline{\theta}) = \lambda \cdot Bel_1(\overline{\theta})$$

O denotes the zero function $O(\overline{\theta}) = 0$. The inner product is defined by

$$\langle Bel_1, Bel_2 \rangle = \sum_{\overline{\theta} \in \overline{SL}} Bel_1(\overline{\theta}) \cdot Bel_2(\overline{\theta}).$$

The norm is uniquely given by the inner product in the usual way:

$$\|Bel_1\| = \sqrt{\langle Bel_1, Bel_1 \rangle}.$$

The precise interpretation of probability and Dempster-Shafer belief functions in $\mathbf{B}$ becomes clear by recalling the definitions of the convex and affine hull.

Definition 8. Let $\mathbf{V}$ be a vector space and $\{V_1, \ldots, V_l\}$ be a set of vectors in $\mathbf{V}$. Then

$$C(V_1, \ldots, V_l) = \{(m_1 \odot V_1) \oplus \cdots \oplus (m_l \odot V_l) \mid m_1, \ldots, m_l \geq 0 \text{ and } \sum_{i=1}^{l} m_i = 1\}$$

is called the *convex hull* of $\{V_1, \ldots, V_l\}$ and

$$A(V_1, \ldots, V_l) = \{(m_1 \odot V_1) \oplus \cdots \oplus (m_l \odot V_l) \mid \sum_{i=1}^{l} m_i = 1\}$$

defines the *affine hull* of $\{V_1, \ldots, V_l\}$.

From the definitions above, we see that the set of probability functions is obviously the convex hull of all valuations, and that the set of Dempster-Shafer belief functions is the convex hull of all information functions. The optimization problem (OP) formulated in the framework of $\mathbf{B}$ is equivalently given as: for a given belief function $D \in GB$, minimize the objective function $\|D \ominus P\|$ with respect to

$$P \in C(V_{\overline{\alpha_1}}, \ldots, V_{\overline{\alpha_J}}).$$

We proceed in the following way: we define an operator P^* on the set of Dempster-Shafer belief functions and show that $P_D^*(D) \in A(V_{\overline{\alpha_1}}, \ldots, V_{\overline{\alpha_J}})$ and, moreover, that $P_D^*(D)$ solves the modified optimization problem given next.

Simplified Optimization Problem (SOP) For a given belief function $D \in GB$, minimize the objective function $\|D \ominus P\|$ with respect to

$$P \in A(V_{\overline{\alpha_1}}, \ldots, V_{\overline{\alpha_J}}).$$

The advantage of considering $A(V_{\overline{\alpha_1}}, \ldots, V_{\overline{\alpha_J}})$ instead of $C(V_{\overline{\alpha_1}}, \ldots, V_{\overline{\alpha_J}})$ is that every affine hull is an affine subspace. Minimizing $\|D \ominus P\|$ in a subspace of $\mathbf{B}$ means that we look for the projection of D onto the subspace. A projection has the property of the projection vector standing perpendicular on the given subspace, i.e. the inner product of the projection vector and any vector of the subspace is zero [11]. The same holds true for affine subspaces if we consider that linear subspace which is parallel to the affine subspace.

Finally, we show that even $P_D^*(D) \in C(V_{\overline{\alpha_1}}, \ldots, V_{\overline{\alpha_J}})$ holds, which proves that $P_D^*(D)$ indeed solves (OP).

4 The Transformation Formula for Dempster-Shafer Belief Functions

Assume throughout this section that $D \in GB$ is a Dempster-Shafer belief function with

$$D = \sum_{\overline{\theta} > 0} m_D(\overline{\theta}) \cdot V_{\overline{\theta}}'.$$

Definition 9. The formula

$$P_D^*(D) = \sum_{\overline{\alpha} \in \overline{At}} m_{P_D^*(D)}(\overline{\alpha}) \cdot V_{\overline{\alpha}}$$

defines the *projective transformation* of $D : \overline{SL} \rightarrow [0,1]$, such that for all $\overline{\alpha} \in \overline{At}$,

$$m_{P^*(D)}(\overline{\alpha}) = \sum_{\overline{\theta} \geq \overline{\alpha}} 2^{1-|S_{\overline{\theta}}|} \cdot m_D(\overline{\theta}) - \sum_{\overline{\theta} > 0} |S_{\overline{\theta}}| \cdot 2^{1-|S_{\overline{\theta}}|-n} \cdot m_D(\overline{\theta}) + J^{-1},$$

where $S_{\overline{\theta}} = \{\overline{\alpha} \in \overline{At} \mid \overline{\alpha} \leq \overline{\theta}\}$, i.e. the set of atoms from which $\overline{\theta}$ can be inferred. Consequently, $|S_{\overline{\theta}}|$ denotes the cardinality of this set.

It can be shown that $P_D^*(D)$ maps D into $A(V_{\overline{\alpha_1}}, \ldots, V_{\overline{\alpha_J}})$.

Lemma 1. $P_D^*(D) \in A(V_{\overline{\alpha_1}}, \ldots, V_{\overline{\alpha_J}})$.

In order to prove that $P_D^*(D)$ solves (SOP) it is to show that the projection is perpendicular onto the plane of probability functions.

Lemma 2. $\langle V_{\overline{\alpha_i}} \ominus V_{\overline{\alpha_1}}, P_D^*(D) \ominus D \rangle = 0$ *holds for all* $i \in \{2, \ldots, J\}$.

By applying basic functional analysis it follows the required result [11].

Theorem 1. $P_D^*(D)$ *solves (SOP)*.

An alternative way to prove Theorem 1 can be based on using Lagrange multipliers [9]. In order to show that $P_D^*(D)$ solves (OP), it remains to be proved that $P_D^*(D)$ is indeed contained in $C(V_{\overline{\alpha_1}}, \ldots, V_{\overline{\alpha_J}})$, i.e. that all belief values are positive.

Theorem 2. $P_D^*(D) \in C(V_{\overline{\alpha_1}}, \ldots, V_{\overline{\alpha_J}})$ *and, therefore, solves the optimization problem (OP)*.

5 Generalized Projective Transformation

Alternatively to a relative frequency interpretation, it is popular to define a belief function by an expert's subjective opinion that an event might take place [3]. The motivation for this is that it is often not possible to get hold of some appropriate data to construct a belief function. Nevertheless, in more complex situations, it would be very hypothetical to assume that belief of a human being exactly matches a Dempster-Shafer belief function or a probability function. Even if an expert intends to represent her/his belief in terms of probability, in most non-trivial cases it would be far beyond her/his mental capacity to meet precisely all requirements. Nevertheless, probability functions have the advantage that they can be applied to a rational decision making process.

In fact, it has been shown by using the so-called Dutch book argument that when applying belief functions different to probability, a rational decision making process can lead to completely irrational decisions [12]. The Dutch-Book argument works when the decision maker incorrectly assumes that her/his belief is similar to probability and uses a decision making principle (e.g. maximizing expected value) that has been developed for probabilistic belief. The special case of the Dutch book argument applied to Dempster-Shafer belief is also discussed in the literature [13].

In order to address a more intuitive concept of belief and decision making, our intention is as follows. We continue with the optimization problems (OP) and (SOP) admitting arbitrary belief functions $Bel \in GB$. For this purpose, we proceed analogously to Section 4.

Definition 10. For $Bel \in GB$, the formula

$$P_{GB}^*(Bel) = \sum_{\overline{\alpha} \in \overline{At}} m_{P_{GB}^*(Bel)}(\overline{\alpha}) \cdot V_{\overline{\alpha}}$$

defines the *projective transformation* of Bel, where, for all $\overline{\alpha} \in \overline{At}$,

$$m_{P_{GB}^*(Bel)}(\overline{\alpha}) = \frac{1}{2^{J+n-2}} \left(J \sum_{\overline{\theta} \geq \overline{\alpha}} Bel(\overline{\theta}) - \sum_{\overline{\theta} \in \overline{SL}} |S_{\overline{\theta}}| \, Bel(\overline{\theta}) + 2^{J-2} \right).$$

Analogously to Lemma 1, it can be shown that $P_{GB}^*(Bel)$ is a member of the affine hull of valuations and, analogously to Theorem 1, we can prove that $P_{GB}^*(Bel)$ solves (SOP).

Lemma 3. $P_{GB}^*(Bel) \in A(V_{\overline{\alpha_1}}, \dots, V_{\overline{\alpha_J}})$.

Next, it is to show that $P_{GB}^*(Bel)$ solves the (SOP). As previously mentioned, the (SOP) is meant to be re-defined in such a way that the Dempster-Shafer belief functions $D : \overline{SL} \to [0,1]$ is replaced by a belief function $Bel : \overline{SL} \to [0,1] \in GB$.

Theorem 3. $P_{GB}^*(Bel)$ *solves (SOP).*

Essentially, the projection function defined for general belief functions is exactly an extension of the projection functions for Dempster-Shafer belief. This is due to the fact that both are unique solutions of the (SOP).

Corollary 1. *Let* $D : \overline{SL} \to [0,1]$ *be a Dempster-Shafer belief function then* $P_D^*(D) = P_{GB}^*(D)$.

Unfortunately, as the following example demonstrates, it turns out that the generalized projective transformation $P_{GB}^*(Bel)$ does not necessarily solve the optimization task (OP).

Example 1. For a two-variable propositional language $L = \{p_1, p_2\}$, let $Bel : \overline{SL} \to [0,1]$ be given by

$$Bel(\overline{\theta}) = \begin{cases} 0 \text{ for } \overline{\theta} \geq \overline{\alpha}_1 \\ 1 \text{ otherwise,} \end{cases}$$

where $\overline{\alpha}_1 = \overline{p_1 \wedge p_2}$. For this belief function the projective transformation returns a negative value for $\overline{\alpha}_1$.

6 Concluding Remarks

In this paper, we explicitly found the probability function which minimizes the Euclidean distance to a given Dempster-Shafer belief function. This result was accomplished by re-formulating the optimization problem within the framework of the linear space of functions from $\overline{SL}$ to $\mathbb{R}$ and using methods from linear algebra. But it is not obvious how a geometric motivated transformation method for Dempster-Shafer belief can be justified.

The geometrical interpretation of distance seems to be more natural when the concept of belief is extended. For such a more general belief function, we solved the simplified optimization problem (SOP). However, it turned out the projective transformation does not always work in such settings.

Acknowledgements

The authors gratefully acknowledge partial support of the K*plus* Competence Center Program which is funded by the Austrian Government, the Province of Upper Austria and the Chamber of Commerce of Upper Austria. Many thanks also to Jeff Paris and Alena Vencovska for mathematical support and encouragement.

References

1. H. Laux, *Entscheidungstheorie*, fourth edition, (Springer, Heidelberg, 1997).
2. S. French and D. R. Insua, *Statistical Decision Theory*, first edition, (Arnold, London, 2000).
3. J. B. Paris, *The Uncertain Reasoner's Companion*, fourth edition, (Cambridge University Press, Cambridge, 1994).
4. I. Pratt, *Artificial Intelligence*, first edition, (MacMillan Press, Houndmills, 1994).
5. M. Bauer, "Approximation algorithms and decision making in the Dempster-Shafer theory of evidence — an empirical study", *Internat. J. Approx. Reason.* **17** (1997) 217–237.
6. P. Smets, "Constructing the pignistic probability function in a context of uncertainty", in *Uncertainty in Artificial Intelligence V*, eds. M. Herion, R. D. Shachter, L. N. Kanal, and J. Lemmer, (North-Holland, Amsterdam, 1990), pp. 29–39.
7. B. Tessem, "Approximations for efficient computation in the theory of evidence", *Artificial Intelligence* **61** (1993) 315–329.
8. F. Voorbraak, "A computationally efficient approximation of Dempster-Shafer theory", *Int. J. Man-Mach. Stud.* **30** (1989) 525–536.
9. T. Weiler, "Updating Towards Probability Belief", Master's thesis, University of Manchester, 1996.
10. G. Shafer, *A Mathematical Theory of Evidence*, (Princeton University Press, Princeton, NJ, 1976).
11. R. F. Curtain and A. J. Pritchard, *Functional Analysis in Modern Applied Mathematics*, (Academic Press, London, 1977).
12. B. de Finetti, *Theory of Probability*, first edition, volume I, (John Wiley & Sons, New York, 1974).
13. P. Smets, "No Dutch book can be built against the TBM even though update is not obtained by Bayes rule of conditioning", in *Proc. Workshop on Probabilistic Expert Systems*, eds. R. Scozzafava, SIS, 1993.

Variants of Defining the Cardinalities of Fuzzy Sets

Maciej Wygralak

Faculty of Mathematics and Computer Science
Adam Mickiewicz University
Matejki 48/49, 60-769 Poznań, Poland
Email:wygralak@math.amu.edu.pl

Abstract. Cardinality, one of the most basic characteristics of a fuzzy set, is a notion having many applications. One of them is elementary probability theory of imprecise events. Fuzzy and nonfuzzy approaches to probabilities of events like "a ball drawn at random from an urn containing balls of various sizes is large" do require an appropriate notion of the cardinality of a fuzzy set, e.g. of the fuzzy set of large balls in an urn. Contemporary fuzzy set cardinality theory offers a variety of options, including the use of triangular norms. This paper presents their overview encompassing scalar approaches as well as approaches in which cardinalities of fuzzy sets are themselves fuzzy sets of usual cardinals.

1   Introduction

Dealing with the notion of the cardinality of a fuzzy set has strong motivations. Indeed, cardinality belongs to the most fundamental mathematical characteristics of a fuzzy set. Besides this theoretical motivation, one should mention multiple applications, e.g. to computing with words, communication with data bases, modeling the meaning of imprecise quantifiers and, last but not least, to (elementary) probability theory of imprecise events (see [1], [9–11]). In each case, we mean the problem of satisfactory answers to queries of the form "How many x's are p?" or, say, "Are there more x's which are p than x's which are q?"; p, q - generally imprecise properties. They are cardinal queries, i.e. queries about cardinalities of fuzzy sets or comparisons of such cardinalities. In particular, the question of probabilities of events like "a ball drawn at random from an urn containing balls of various sizes is large" leads to the question what is the cardinality of the fuzzy set of large balls (see e.g. [9–11]). This paper presents an overview of constructive approaches to cardinalities of fuzzy sets with triangular norm-based operations. Our attention will be focused on finite fuzzy sets, i.e. on fuzzy sets with finite supports, which play the central role from the viewpoint of applications. From now on, if not emphasized otherwise, the phrase "fuzzy set" will mean "finite fuzzy set". The family of all (finite) fuzzy sets in a universe M will be denoted by FFS. FCS symbolizes the family of all finite crisp sets in M.

There are two main general approaches to the cardinality $|A|$ of a fuzzy set $A \in FFS$:

(a) $|A|$ = a convex fuzzy set in $\mathbb{N} = \{0, 1, 2, \dots\}$,
(b) $|A|$ = a single nonnegative integer or real number.

The "fuzzy" approach in (a), offering the fuzzy perception of cardinality, gives us the most complete and adequate cardinal information about A at the price of relatively high complexity. Variants of that approach for fuzzy sets with the standard *min* and *max* operations are discussed in [1,10]. In Section 2 and Subsection 3.2 of this paper, we like to present a few cardinality concepts oriented to fuzzy sets with triangular norms (see [7,8]).

The approach (b) is scalar, nonfuzzy. The scalar optics of cardinality is simple and convenient in every respect and, therefore, it is favoured by many practitioners despite of its disadvantages. A review of basic scalar cardinalities and references to them can be found e.g. in [1]. In Section 3, we present a general axiomatic approach to scalar cardinalities ([4–6,8]).

Let us recall some notions and facts from the theory of triangular norms which will be useful in the main part of this paper. Details and further references are given in [2,3].

A binary operation $t : [0,1] \times [0,1] \to [0,1]$ is called a *triangular norm (t-norm)* if t is commutative, associative, nondecreasing in the first and, hence, in each argument, and has 1 as neutral element. If $s : [0,1] \times [0,1] \to [0,1]$ does satisfy the conditions of commutativity, associativity and nondecreasingness, but has 0 as neutral element, it is said to be a *triangular conorm (t-conorm)*. Triangular norms together with triangular conorms will be called *triangular operations (t-operations)*. If $a\,s\,b = 1 - (1-a)\,t\,(1-b)$ for each a and b, we say that s and t are *associated*, and we write $s = t^*$. Simplest instances of a t-norm and the associated t-conorm are $\wedge = min$ and $\vee = max$.

We say that a continuous t-norm t (t-conorm s, respectively) is *Archimedean* if $a\,t\,a < a\,(a\,s\,a > a$, respectively) for each $a \in (0,1)$. An Archimedean t-operation is called *strict* if it is a strictly increasing function on $(0,1) \times (0,1)$. Strictly increasing and, thus, strict t-norms do not have zero divisors ($a, b > 0 \Rightarrow a\,t\,b > 0$). Nonstrict Archimedean t-norms do have. The following examples of t-norms will be useful in the further discussion:

$$a\,t_a\,b = ab, \qquad\qquad (algebraic\ t\text{-}norm)$$
$$a\,t_L\,b = 0 \vee (a+b-1), \qquad\qquad (Łukasiewicz\ t\text{-}norm)$$
$$a\,t_{Y,p}\,b = 0 \vee [1 - ((1-a)^p + (1-b)^p)^{1/p}], \quad p > 0, \qquad (Yager\ t\text{-}norms)$$
$$a\,t_{S,p}\,b = [0 \vee (a^p + b^p - 1)]^{1/p}, \quad p > 0, \qquad (Schweizer\ t\text{-}norms)$$
$$a\,t_{F,\lambda}\,b = \log_\lambda \left(1 + \frac{(\lambda^a-1)(\lambda^b-1)}{\lambda-1}\right), \quad 1 \neq \lambda > 0, \qquad (Frank\ t\text{-}norms)$$
$$a\,t_{W,\lambda}\,b = 0 \vee \frac{a+b-1+\lambda ab}{1+\lambda}, \quad \lambda > -1. \qquad (Weber\ t\text{-}norms)$$

The t-conorms associated with these t-norms will be denoted and called in a similar way, e.g. $s_L = t_L^*$ (*Łukasiewicz t-conorm*) and $s_{S,p} = t_{S,p}^*$ (*Schweizer t-conorm*). Exceptional limit properties of Frank t-operations allow to put $t_{F,0} = \wedge$, $t_{F,1} = t_a$, $t_{F,\infty} = t_L$, $s_{F,0} = \vee$, $s_{F,1} = s_a$ and $s_{F,\infty} = s_L$. The extended families $(t_{F,\lambda})_{\lambda\in[0,\infty]}$ and $(s_{F,\lambda})_{\lambda\in[0,\infty]}$ are called *Frank families* of t-operations. Archimedean t-operations do have a useful characterization given by Ling.

Theorem 1. *(a) t is an Archimedean t-norm iff there exists a strictly decreasing and continuous function $g : [0,1] \to [0,\infty]$ such that $g(1) = 0$ and $a\,t\,b = g^{-1}(g(0) \wedge (g(a) + g(b)))$ for each $a, b \in [0,1]$. Moreover, t is strict iff $g(0) = \infty$.*
(b) s is an Archimedean t-conorm iff there exists a strictly increasing and continuous function $h : [0,1] \to [0,\infty]$ such that $h(0) = 0$ and $a\,s\,b = h^{-1}(h(1) \wedge (h(a) + h(b)))$ for each $a, b \in [0,1]$. s is strict iff $h(1) = \infty$.

The function g in the thesis (a) is called a *generator of* t; h from (b) is said to be a *generator of* s. One says that g (h, respectively) is normed if $g(0) = 1$ ($h(1) = 1$, respectively).

Each nonincreasing function $\nu : [0, 1] \to [0, 1]$ with $\nu(0) = 1$ and $\nu(1) = 0$ will be called a *negation*. The negation ν_* such that $\nu_*(a) = 0$ for $a > 0$ is thus the smallest possible negation, whereas ν^* with $\nu^*(a) = 1$ for $a < 1$ is the largest possible one. Strictly decreasing and continuous negations are called *strict negations*. Involutive strict negations ($\nu(\nu(a)) = a$) are said to be *strong*. A typical example of a strong negation is the *Łukasiewicz negation* ν_L with $\nu_L(a) = 1 - a$. Each t-norm t and t-conorm s do generate negations ν_t and ν_s defined as

$$\nu_t(a) = \bigvee\{c \in [0, 1] : a \, t \, c = 0\}, \quad \nu_s(a) = \bigwedge\{c \in [0, 1] : a \, s \, c = 1\}.$$

These negations are strong whenever t and s are Archimedean and nonstrict. Moreover, the binary operations t° and s°, respectively, such that $a \, t^\circ b = \nu_t(\nu_t(a) \, t \, \nu_t(b))$ and $a \, s^\circ b = \nu_s(\nu_s(a) \, s \, \nu_s(b))$ are then a nonstrict Archimedean t-conorm and a nonstrict Archimedean t-norm, respectively. t and t° are called *complementary t-operations*. Instances of pairs of complementary t-operations are $(t_{S,p}, s_{Y,p})$ and $(t_{Y,p}, s_{S,p})$.

Triangular operations and negations are suitable tools for defining operations on arbitrary fuzzy sets. We shall use the *sum* $A \cup_s B$ of A and B induced by a t-conorm s with $(A \cup_s B)(x) = A(x) \, s \, B(x)$, the *intersection* $A \cap_t B$ and the *cartesian* product $A \times_t B$ induced by a t-norm t with $(A \cap_t B)(x) = A(x) \, t \, B(x)$ and $(A \times_t B)(x, y) = A(x) \, t \, B(y)$, and the *complement* A^ν of A induced by a negation ν with $A^\nu(x) = \nu(A(x))$. $\cup = \cup_\vee$, $\cap = \cap_\wedge$, $\times = \times_\wedge$, and $' = {}^\nu$ with $\nu = \nu_L$ are the standard operations. Inclusions and equalities of fuzzy sets will be understood in the standard way via pointwise $\leq$ and $=$, no matter which t-operations are used. The *t-cut set* and the *sharp t-cut set* of A, respectively, will be denoted by A_t and A^t, respectively. So, $A_t = \{x \in M : A(x) \geq t\}$ and $A^t = \{x \in M : A(x) > t\}$.

2 Cardinalities of fuzzy sets as fuzzy sets of nonnegative integers

We like to present three groups of approaches in which the cardinality of a fuzzy set is a fuzzy set of nonnegative integers (of finite cardinals, in other words). Throughout, $A \in FFS$, $n = |supp(A)|$ and $m = |core(A)|$. Let $[A]_k = \bigvee\{t : |A_t| \geq k\}$ for $k \in \mathbb{N}$. $[A]_k$ with $0 < k \leq n$ is the kth element in the nonincreasingly ordered sequence of all positive values $A(x)$, including their possible repetitions. So, $[A]_k = 1$ if $k \leq m$, and $[A]_k = 0$ for $k > n$.

2.1 Generalized FGCounts

For a t-norm t, let $FGCount_t : FFS \to [0, 1]^\mathbb{N}$ be defined as follows (see [7,8]):

$$FGCount_t(A)(k) = [A]_1 \, t \, [A]_2 \, t \, \ldots \, t \, [A]_k.$$

$FGCount_\wedge(A)$ becomes the well-known $FGCount(A)$ with $FGCount(A)(k) = [A]_k$ ([10]). $FGCount_t(A)$ is its appropriate generalization to fuzzy sets with t-norms. In the language of many-valued sentential calculus, $FGCount_t(A)(k)$ is the truth degree of the sentence saying that A contains (at least) i elements for each $i \leq k$; the quantification "for each" is here interpreted via t. If $A \in FCS$, then $FGCount_t(A) = 1_{\{0,1,\ldots,n\}}$. Our further discussion of generalized FGCounts will be restricted to the case of an Archimedean t, including $t = \wedge$. Let

$$e(A) = \bigvee\{k \in \mathbb{N} : [A]_1 \ t \ [A]_2 \ t \ \ldots \ t \ [A]_k > 0\}.$$

The equipotency relation $\sim_t$ guaranteeing that $FGCount_t(A) = FGCount_t(B)$ iff $A \sim_t B$ is of the form

$$A \sim_t B \Leftrightarrow e(A) = e(B) \ \& \ \forall k \leq e(A) : [A]_k = [B]_k.$$

If t is strict or $t = \wedge$, this collapses to

$$A \sim_t B \Leftrightarrow \forall k \in \mathbb{N} : [A]_k = [B]_k$$
$$\Leftrightarrow \forall t \in (0,1] : |A_t| = |B_t|$$
$$\Leftrightarrow \forall t \in [0,1) : |A^t| = |B^t|,$$

i.e. the dependence on t vanishes. A partial ordering of generalized FGCounts and basic arithmetical operations on them are defined in the classical manner:

$$FGCount_t(A) \leq FGCount_t(B) \Leftrightarrow \exists B^* \subset B : A \sim_t B^*,$$

$$FGCount_t(A) + FGCount_t(B) = FGCount_t(A \cup B) \text{ whenever } A \cap B = 1_\emptyset,$$

$$FGCount_t(A) \cdot FGCount_t(B) = FGCount_t(A \times B).$$

The relation $\leq$ collapses to $\subset$ only if $t = \wedge$. The addition $+$ can be equivalently realized via the triangular norm-based extension principle $(P + Q)(k) = \bigvee\{P(i)tQ(j) : i+j = k\}$. For the multiplication, this can be done only if $\wedge$ is used as t-norm. Let α, β and δ denote generalized FGCounts of fuzzy sets; as previously, t is Archimedean or equal to $\wedge$. Then

$\alpha(\beta + \gamma) = \alpha\beta + \alpha\gamma,$	(*distributivity*)
$(\alpha \leq \beta \ \& \ \gamma \leq \delta) \Rightarrow (\alpha + \gamma \leq \beta + \delta \ \& \ \alpha\gamma \leq \beta\delta),$	(*monotonicity*)
$\alpha < \gamma \nRightarrow \exists\beta : \alpha + \beta = \gamma.$	(*lack of compensation*)

The failure of the compensation property is one of the most important differences between the arithmetic of ordinary cardinals and the arithmetic of generalized FG-Counts. The valuation property $FGCount_t(A) + FGCount_t(B) = FGCount_t(A \cap_t B) + FGCount_t(A \cup_s B)$ is generally satisfied iff $t = \wedge$ and $s = \vee$. Some further classical properties are lost if one uses nonstrict Archimedean t-norms. Instances are the *cancellation laws*

$$\alpha + \beta = \alpha + \gamma \Rightarrow \beta = \gamma \text{ and } \alpha\beta = \alpha\gamma \Rightarrow \beta = \gamma \ (\alpha \geq |1_{\{x\}}|)$$

which hold true only if t is strict or equal to $\wedge$.

2.2 Generalized FLCounts

Let

$$FLCount_{t,\nu}(A)(k) = \nu([A]_{k+1}) \, t \, \nu([A]_{k+2}) \, t \, \ldots \, t \, \nu([A]_n)$$

with a t-norm t and a negation ν. This time $FLCount_{t,\nu}(A)(k)$ is a degree to which A contains at most i elements for each $i \geq k$. $FLCount_{t,\nu}(A)(k) = 1$ if $k \geq n$. Putting $t = \wedge$ and $\nu = \nu_L$, $FLCount_{t,\nu}(A)$ collapses to the classical $FLCount(A)$ from [10].

2.3 Generalized FECounts

$$FECount_{t,\nu}(A)(k) = [A]_1 \, t \, \ldots \, t \, [A]_k \, t \, \nu([A]_{k+1}) \, t \, \ldots \, t \, \nu([A]_n)$$

defines the generalized FECount of $A \in FFS$, the intersection of its generalized FGCount and generalized FLCount induced by t. It is always convex. In particular, $t = \wedge$ and $\nu = \nu_L$ do lead to the usual FECount of A ([10]). Putting $t = \wedge$ and $\nu = \nu^*$, $FGCount_{t,\nu}(A)$ becomes the cardinality of A due to Dubois ([1]). Inequalities between and arithmetical operations on generalized FECounts can be introduced in the classical way used for generalized FGCounts. However, a restriction to strict t-norms, including $t = \wedge$, is then required. A detailed study of generalized FGCounts, FLCounts and FECounts is placed in [7,8].

3 Scalar cardinalities of fuzzy sets

3.1 Generalized sigma counts

Definition 1. A function $\sigma : FFS \rightarrow [0, \infty)$ is called a *scalar cardinality* if the following axioms are satisfied for each $a, b \in [0, 1]$, $x, y \in M$ and $A, B \in FFS$:

(A1) $\sigma(1/x) = 1$,
(A2) $a \leq b \Rightarrow \sigma(a/x) \leq \sigma(b/y)$,
(A3) $A \cap B = 1_\emptyset \Rightarrow \sigma(A \cup B) = \sigma(A) + \sigma(B)$.

If the postulates (A1)-(A3) are satisfied by a function σ, one says that $\sigma(A)$ is a *scalar cardinality of* A. The following properties are simple consequences of the axioms:

$\sigma(A) = |supp(A)|$ if $A \in FCS$,   (*coincidence*)
$\sigma(A) \leq \sigma(B)$ if $A \subset B$,   (*monotonicity*)
$|core(A)| \leq \sigma(A) \leq |supp(A)|$.   (*boundedness*)

Theorem 2. $\sigma : FFS \rightarrow [0, \infty)$ *is a scalar cardinality iff there exists a nondecreasing function* $f : [0, 1] \rightarrow [0, 1]$ *with* $f(0) = 0$ *and* $f(1) = 1$ *such that*

$$\sigma(A) = \sum_{x \in supp(A)} f(A(x)) \quad \text{for each } A.$$

So, the scalar cardinalities from Definition 1 can be called *generalized sigma counts* because they are in essence natural generalizations of the well-known concept of the sigma count $sc_A = \sum_{x \in supp(A)} A(x)$ of A. Each function f satisfying the conditions of Theorem 2 is said to be a *cardinality pattern* as it does express our understanding of the (scalar) cardinality of a singleton. Let us present a few basic examples of cardinality patterns and the resulting scalar cardinalities of a fuzzy set $A \in FFS$.

(i) $f_{1,t}(a) = (1$ if $a \geq t$, else 0) with $t \in (0, 1]$. Then $\sigma(A) = |A_t|$. $f_{1,1}$ is the smallest possible cardinality pattern, and it leads to the smallest scalar cardinality $core(A)$ of A.

(ii) $f_{2,t}(a) = (1$ if $a > t$, else 0) with $t \in [0, 1)$. Now $\sigma(A) = |A^t|$. The largest possible cardinality pattern $f_{2,0}$ generates the largest scalar cardinality $|supp(A)|$. $|A_t|$ and $|A^t|$ are integer scalar cardinalities which, without any preassumptions, can be applied to infinite fuzzy sets. In the next examples, scalar cardinalities are generally noninteger numbers.

(iii) $f_{3,p}(a) = a^p$ with $p > 0$. Then $\sigma(A) = \sum_{x \in supp(A)} (A(x))^p$, i.e. $\sigma(A) = sc_A$ for $p = 1$.

(iv) $f_{4,p}(a) = (2^{p-1} a^p$ if $a \leq 0.5$, else $1 - 2^{p-1}(1-a)^p)$ with $p > 0$. So, $f_{4,2}$ is the classical contrast enhancement function and $\sigma(A) = sc_A$ for $p = 1$.

(v) By definition, normed generators of nonstrict Archimedean t-conorms are cardinality patterns, e.g. the function $f_{5,\lambda}(a) = ln(1+\lambda a)/ln(1+\lambda)$ as normed generator of $(t_{W,\lambda})^\circ$.

Theorem 3. *(a) The valuation property*

$$\forall A, B \in FFS: \ \sigma(A \cap_t B) + \sigma(A \cup_s B) = \sigma(A) + \sigma(B)$$

holds true for a t-norm t, a t-conorm s, and a scalar cardinality σ based on a cardinality pattern f iff f, t and s are such that

$$\forall a, b \in [0, 1] : f(a\ t\ b) + f(a\ s\ b) = f(a) + f(b).$$

(b) The cartesian product rule

$$\forall A, B \in FFS : \sigma(A \times_t B) = \sigma(A) \cdot \sigma(B)$$

is satisfied iff

$$\forall a, b \in [0, 1] : f(a\ t\ b) = f(a) \cdot f(b).$$

(c) Assume M is finite. The complementarity rule

$$\forall A \in FFS : \sigma(A) + \sigma(A^v) = |M|$$

is fulfilled iff

$$\forall a \in [0, 1] : f(a) + f(\nu(a)) = 1.$$

The following are thus examples of triples (f, t, s) satisfying the valuation property:
(a) $(f, \wedge, \vee)$ with any cardinality pattern f,
(b) $(id, t_{F,\lambda}, s_{F,\lambda})$ with $\lambda \in [0, \infty]$,
(c) (h, s°, s), where s is a nonstrict Archimedean t-conorm with normed generator h, e.g. $(f_{3,p}, t_{S,p}, s_{Y,p})$ with $p > 0$ and $(f_{5,\lambda}, t_{W,\lambda}, (t_{W,\lambda})^\circ)$ with $\lambda > -1$.

184

By Theorem 3(b),

(d) if $t = \wedge$, then the cartesian product rule holds true iff $f = f_{1,t}$ or $f = f_{2,t}$,

(e) if t is strict with a generator g, then $f = e^{-g}$ and t do satisfy the cartesian product rule (e.g. $f = id$ and $t = t_a$),

(f) if $t = t_a$, the criterion in Theorem 3(b) collapses to the Cauchy functional equation whose unique continuous solutions are the cardinality patterns $f = f_{3,p}$ with $p > 0$.

Finally, Theorem 3(c) leads to the following examples of pairs (f, ν) satisfying the complementarity rule:

(g) if $\nu = \nu_L$, then the complementarity rule holds true whenever $(0.5, 0.5)$ is the symmetry point of the diagram of f (e.g. $f_{4,p}$),

(h) if s is a nonstrict Archimedean t-conorm with normed generator h, then (h, ν_s) fulfils the complementarity rule; an example is $h = f_{3,p}$ and $\nu_s(a) = (1 - a^p)^{1/p}$ for $s = s_{Y,p}$.

A unique nontrivial quadruple (f, t, s, ν) leading to the simultaneous fulfilment of the valuation property and the cartesian product and complementarity rules seems to be (id, t_a, s_a, ν_L).

3.2 Triangular norm-based generalized sigma counts

There exists an obvious connection between the sigma count and the FGCount of A, namely $sc_A = \sum_{k=1}^{n} FGCount(A)(k)$. So, sc_A can be viewed as a "summary" of $FGCount(A)$. This suggests three other possible variants of a generalization of sc_A:

$$sc_{A,t} = \sum_{k=1}^{n} [A]_1 \ t \ [A]_2 \ t \ \ldots \ t \ [A]_k, \quad sc_{A,f,t} = \sum_{k=1}^{n} f([A]_1 \ t \ [A]_2 \ t \ \ldots \ t \ [A]_k),$$

$$sc_{A,t,f} = \sum_{k=1}^{n} f([A]_1) \ t \ f([A]_2) \ t \ \ldots \ t \ f([A]_k)$$

with a cardinality pattern f. Conversely, the concept of the FGCount can be generalized in the following three ways:

$$FGCount_f(A)(k) = f([A]_k),$$

$$FGCount_{f,t}(A)(k) = f([A]_1 \ t \ [A]_2 \ t \ \ldots \ t \ [A]_k),$$

$$FGCount_{t,f}(A)(k) = f([A]_1) \ t \ f([A]_2) \ t \ \ldots \ t \ f([A]_k).$$

References

1. Dubois, D., Prade, H., Fuzzy cardinality and the modeling of imprecise quantification, Fuzzy Sets and Systems 16 (1985) 199-230.
2. Gottwald, S., A Treatise on Many-Valued Logics, Research Studies Press, Baldock, Hertfordshire 2001.

3. Klement, E. P., Mesiar, R., Pap, E., Triangular Norms, Kluwer, Dordrecht Boston London 2000.

4. Wygralak, M., Triangular operations, negations, and scalar cardinality of a fuzzy set, in: Zadeh, L. A., Kacprzyk, J. (Eds.), Computing with Words in Information/Intelligent Systems, Vol. 1 - Foundations, Physica-Verlag, Heidelberg New York (1999) 326-341.

5. Wygralak, M., An axiomatic approach to scalar cardinalities of fuzzy sets, Fuzzy Sets and Systems 110 (2000) 175-179.

6. Wygralak, M., A generalizing look at sigma counts of fuzzy sets, in: Bouchon-Meunier, B., Yager, R. R., Zadeh, L. A. (Eds.), Uncertainty in Intelligent and Information Systems, World Scientific, Singapore New Jersey London New York (2000) 34-45.

7. Wygralak, M., Fuzzy sets with triangular norms and their cardinality theory, Fuzzy Sets and Systems 124 (2001) 1-24.

8. Wygralak, M., Cardinalities of Fuzzy Sets (book in preparation).

9. Zadeh, L. A., Fuzzy probabilities and their role in decision analysis, in: Proc. IFAC Symp. on Theory and Applications of Digital Control, New Dehli (1982) 15-23.

10. Zadeh, L. A., A computational approach to fuzzy quantifiers in natural languages, Comput. and Math. with Appl. 9 (1983) 149-184.

11. Zadeh, L. A., From computing with numbers to computing with words - From manipulation of measurements to manipulation of perceptions, IEEE Trans. on Circuits and Systems - I: Fundamental Theory and Appl. 45 (1999) 105-119.

Soft Methods in Statistics:
Fuzzy Stochastic Models

Probabilistic Reasoning in Fuzzy Rule-Based Systems

Jan van den Berg, Uzay Kaymak, and Willem-Max van den Bergh

Faculty of Economics, Erasmus University Rotterdam
Room H9-19, P. O. Box 1738, 3000 DR, Rotterdam, the Netherlands
jvandenberg@few.eur.nl, u.kaymak@ieee.org, vandenbergh@few.eur.nl

Abstract. We concentrate on Takagi–Sugeno (TS) probabilistic fuzzy systems where interpretability of fuzzy systems is combined with the statistical properties of probabilistic systems. After having sketched the general architecture of TS probabilistic fuzzy systems, we present an appropriate mathematical framework and introduce two probabilistic fuzzy reasoning schemes which have a different interpretation but, eventually, yield the same input-output mapping. We illustrate our theoretical considerations by presenting some simulation results concerning a financial time series analysis.
Keywords. Probabilistic fuzzy systems, fuzzy reasoning, fuzzy rule base, fuzzy event.

1 Introduction

Fuzzy systems (FSs) are widely applied in fields like classification, decision support, process simulation, and control [1,2]. Original applications of FSs have concentrated on their design from expert knowledge [3]. In the past decade, however, data-driven techniques for designing FSs have gained much attention, partly due to the availability of large amounts of data from modern sensory, measurement and computer systems. One important advantage of fuzzy inference systems is their linguistic interpretability (when designed appropriately), whereby the results from the data-driven approach can be combined with or compared to the knowledge available from experts. Various methods have been developed to design fuzzy inference systems that are interpretable (transparent) and numerically sufficiently accurate [4].

The focus for the design of transparent FSs has been on the identification of rules that describe system behavior and of membership functions that represent the linguistic values used in the fuzzy rules. Hence, the focus has been on the modelling of linguistic vagueness of classes without clear boundaries. However, another type of uncertainty, namely probabilistic uncertainty, is often also present. Design of FSs has dealt with this type of uncertainty only implicitly, and the statistical properties of the available data have often been neglected. Conversely, statistical learning techniques exploit the statistical properties of the data [5], but they ignore the fuzziness and the linguistic vagueness. The emphasis is then on the numerical accuracy for learning, but the transparency of the system is not considered explicitly. Since fuzziness and probability model different types of uncertainty, a paradigm for combining both types uncertainty can provide "the best of the two worlds". In this paper, we argue that *probabilistic fuzzy systems* (PFSs) are such a paradigm, leading to a generalization of deterministic rule-based fuzzy systems. PFSs combine *interpretability* of fuzzy systems with the *statistical properties* of probabilistic systems.

For the scope of this paper, we concentrate on zero-order Takagi–Sugeno FSs. The core of most FSs is a *fuzzy rule-base* consisting of a set of IF-THEN rules, together with a *fuzzy inference mechanism* for reasoning. Rules in a zero-order TS fuzzy system have the general form

$$\text{Rule } R_q: \text{If } x_1 \text{ is } A_{q1} \text{ and } \dots \text{ and } x_M \text{ is } A_{qM}$$
$$\text{then } y = y_q, \quad q = 1, 2, \dots, Q. \tag{1}$$

Here, $\mathbf{x} = (x_1, x_2, \dots, x_M) \in X$ is an M-dimensional input vector, each A_{qi} is an antecedent linguistic value defined by a fuzzy membership function $\mu_{qi}(x_i)$, y is consequent rule variable having the crisp output value y_q. Note that some $\mu_{qi}(x_i)$ may be (and in practice will be) the same for different q. In short, we write (1) as

$$\text{Rule } R_q: \text{If } \mathbf{x} \text{ is } A_q \text{ then } y = y_q, \quad q = 1, 2, \dots, Q. \tag{2}$$

Using the standard TS reasoning mechanism, the value of consequent variable y is calculated as a weighted sum of rule contributions y_q according to

$$y = \sum_{q=1}^{Q} \beta_q y_q = \frac{\sum_{q=1}^{Q} A'_q y_q}{\sum_{q=1}^{Q} A'_q}, \tag{3}$$

where the weight $\beta_q = A'_q / \sum_{q=1}^{Q} A'_q$ with A'_q being the degree of activation of the q-th rule, which can be defined by

$$A'_q(\mathbf{x}) = \prod_{i=1}^{M} \mu_{qi}(x_i) = \mu_q(\mathbf{x}). \tag{4}$$

It is clear that the above-mentioned TS zero-order FSs implement a deterministic mapping $X \to Y$ based on a deterministic fuzzy reasoning mechanism. The goal of this paper is to generalize this type of TS systems to zero-order TS PFSs having a probabilistic fuzzy rule base and having a probabilistic fuzzy reasoning mechanism.

The rest of this paper is structured as follows. In section 2, we introduce the general architecture of zero-order TS PFSs and present some key results from a mathematical framework for calculating probabilities on fuzzy events. In section 3, we generalize the deterministic reasoning mechanism (3) to two probabilistic fuzzy reasoning procedures and present the similarities and differences between them. In section 4, we present some simulation results concerning a GARCH-type of time series and we finalize this paper by summarizing what we have achieved so far.

2   Architecture

For our probabilistic fuzzy framework, we generalize the set of deterministic fuzzy rules (2) to a set of probabilistic fuzzy rules

$$\text{Rule } R_q: \text{ If } \mathbf{x} \text{ is } A_q \text{ then}$$
$$\underline{y} = y_{q1} \text{ with } \Pr(y_1|A_q) \text{ and}$$
$$\underline{y} = y_{q2} \text{ with } \Pr(y_2|A_q) \text{ and } \dots \text{ and}$$
$$\underline{y} = y_{qN} \text{ with } \Pr(\mathbf{y}_N|A_q). \tag{5}$$

Our interpretation is as follows. Given the occurrence of the antecedent 'fuzzy event' [6] A_q, the value of the stochastic consequent variable $\underline{y}$ equals one of the values $y_{q1}, y_{q2}, \dots, y_{qN}$. The selection of this consequent value is done proportionally to the conditional probabilities $\Pr(y_{q1}|A_q), \dots, \Pr(y_{qN}|A_q)$, with $\forall j : \Pr(y_{qj}|A_q) = \Pr(\underline{y} = y_{qj}|\mathbf{x} \text{ is } A_q)$.

For the scope of this paper, we will use fuzzy rules (5) where the consequent values y_{qj} are the same for all rules or, mathematically expressed, we assume that

$$\forall j, q, q' : y_{qj} = y_{q'j} = y_j. \tag{6}$$

In addition to the probabilistic fuzzy rules (5), we need a probabilistic fuzzy inference mechanism. Before introducing two probabilistic fuzzy inference schemes however, we first present some key results, to be used below, from an appropriate mathematical (probabilistic fuzzy) framework. Such a framework, actually an elaboration of the probability theory on fuzzy sets as introduced by [7], was introduced in [6]. Using mathematical statistics, probabilities on fuzzy events can be assessed. For example, given a set of S samples $\mathbf{x}_s, (s = 1, \dots, S)$ in a 'well-defined' [6] sample space X with fuzzy events $A_1, A_2, \dots, A_c, \dots$, then

$$\Pr(A_c) \approx \tilde{f}_{A_c} = \frac{f_{A_c}}{S} = \frac{1}{S} \sum_{\mathbf{x}_s} \mu_{A_c}(\mathbf{x}_s) = \hat{\mu}_{A_c}, \tag{7}$$

where $\tilde{f}_{A_c}$ denotes the relative frequency and f_{A_c} the absolute frequency of the fuzzy sample values $\mu_{A_c}(\mathbf{x}_s)$ for fuzzy class A_c. In addition, conditional probabilities on fuzzy sets can be assessed according to

$$\Pr(A_c|A_b) = \frac{\Pr(A_c \cap A_b)}{\Pr(A_b)} \approx \frac{\sum_{\mathbf{x}_s} \mu_{A_b}(\mathbf{x}_s)\mu_{A_c}(\mathbf{x}_s)}{\sum_{\mathbf{x}_s} \mu_{A_b}(\mathbf{x}_s)}. \tag{8}$$

Below, we will meet expressions like $\Pr(y_j|A_q)$ describing the probability of a crisp event $\underline{y} = y_j$, given the occurrence of fuzzy event A_q. Having a training set of data pairs $(\mathbf{x}_s, y_s), s = 1, \ldots, S$, such a conditional probability can be calculated by means of an adapted version of (8):

$$\Pr(y_j | A_q) \approx \frac{\sum_{(\mathbf{x}_s, y_s)} \chi_j(y_s) \mu_{A_q}(\mathbf{x}_s)}{\sum_{\mathbf{x}_s} \mu_{A_q}(\mathbf{x}_s)}, \tag{9}$$

with $\chi_j(y)$ defined as

$$\chi_j(y) \begin{cases} = 1 & \text{if } y = y_j \\ = 0 & \text{if } y \neq y_j. \end{cases} \tag{10}$$

3 Probabilistic fuzzy reasoning

3.1 Probabilistic fuzzy reasoning I

Using this scheme, we start to estimate conditional probabilities $\Pr(y_j | \mathbf{x})$ for arbitrary $\mathbf{x}$ and next calculate the regression hyperplane y on $\mathbf{x}$. Inspired by (3), we may estimate the conditional probabilities $\Pr(y_j | \mathbf{x})$ as of a weighted sum of conditional probabilities $\Pr(y_j | A_q)$ according to

$$\Pr(y_j | \mathbf{x}) = \sum_{q=1}^{Q} \phi_q \Pr(y_j | A_q) = \frac{\sum_{q=1}^{Q} \mu_q(\mathbf{x}) \Pr(y_j | A_q)}{\sum_{q=1}^{Q} \mu_q(\mathbf{x})}, \tag{11}$$

with $\phi_q = \mu_q(\mathbf{x}) / \sum_{q=1}^{Q} \mu_q(\mathbf{x})$. However, in this way we do not take the probability density $f(\mathbf{x})$ into account, i.e., the fact that certain values of $\mathbf{x}$ are more frequent than other ones. We can repair this omission by generalizing (11) to

$$\Pr(y_j | \mathbf{x}) = \sum_{q=1}^{Q} \phi_q \Pr(y_j | A_q) = \frac{\sum_{q=1}^{Q} \Pr(A_q) \mu_q(\mathbf{x}) \Pr(y_j | A_q)}{\sum_{q=1}^{Q} \Pr(A_q) \mu_q(\mathbf{x})}, \tag{12}$$

with $\phi_q = \Pr(A_q) \mu_q(\mathbf{x}) / \sum_{q=1}^{Q} \Pr(A_q) \mu_q(\mathbf{x})$. The regression hyperplane of y on X defined [5] as the location of the mathematical expectations $E(y | \mathbf{x})$, can now be calculated according to

$$y = E(\underline{y} | \mathbf{x}) = \sum_{j=1}^{N} y_j \Pr(y_j | \mathbf{x}). \tag{13}$$

3.2 Probabilistic fuzzy reasoning II

Here, we start calculating the expectations $E(\underline{y} | A_q), q = 1, 2, \ldots, Q$, according to

$$E(\underline{y} | A_q) = \sum_{j=1}^{N} y_j \Pr(y_j | A_q). \tag{14}$$

Inspired by (3), we may now estimate y (as function of $\mathbf{x}$) by the weighted sum of these expectations[1]:

$$y = \sum_{q=1}^{Q} \phi_q E(\underline{y}|A_q) = \frac{\sum_{q=1}^{Q} \mu_q(\mathbf{x}) E(\underline{y}|A_q)}{\sum_{q=1}^{Q} \mu_q(\mathbf{x})}, \qquad (15)$$

with $\phi_q = \mu_q(\mathbf{x})/\sum_{q=1}^{Q} \mu_q(\mathbf{x})$. Again however, we do not take the probability density $f(\mathbf{x})$ into account in this way. We can repair this omission by generalizing (15) to

$$y = \sum_{q=1}^{Q} \phi_q E(\underline{y}|A_q) = \frac{\sum_{q=1}^{Q} \Pr(A_q) \mu_q(\mathbf{x}) E(\underline{y}|A_q)}{\sum_{q=1}^{Q} \Pr(A_q) \mu_q(\mathbf{x})}, \qquad (16)$$

with $\phi_q = \Pr(A_q) \mu_q(\mathbf{x})/\sum_{q=1}^{Q} \Pr(A_q) \mu_q(\mathbf{x})$.

Theorem 1. *Equations (13) and (16) describe the same hyperplane.*

Proof. The proof is straightforward. Substitution of (12) in (13) yields an expression that is also found by substitution of (14) in (16).

4 Simulation

We introduce an artificially constructed GARCH time series training set and show how zero-order TS PFSs can be used to obtain a nice intuitive description of the underlying data generating process.

4.1 Experimental setting: GARCH modelling

GARCH (Generalized Auto Regressive Conditional Hetero-skedasticity) models [8] are often used in financial literature to describe the volatility behavior of asset return series. Being able to infer something about the volatility of tomorrow from the volatility as per today has important implications for the valuation of many financial contracts, more particularly *contingent claims*. Typically the value of such contract depends on the probability that the price, $\underline{S}$, of some underlying asset attains a pre-specified level. We define the asset return $\underline{u}(t)$ at time t as the instantaneous relative price change: $\forall t : \underline{u}(t) = \ln\left(\underline{S}(t)/\underline{S}(t-1)\right)$. Then $\underline{\sigma}(t)$ is the volatility of the return $\underline{u}(t)$, i.e. the standard deviation over a given previous period. This *local* volatility $\underline{\sigma}(t)$ is assumed to move around the constant *global* volatility $\bar{\sigma}$.

The GARCH process used in the simulation is defined as follows:

1. Each return $\underline{u}(t)$ is drawn from a normal distribution with a constant mean μ and with a standard deviation equal to the local volatility $\underline{\sigma}(t-1)$: $\underline{u}(t) \sim N(\mu, \underline{\sigma}(t-1))$.

---

[1] This step can be considered as an interpolation step.

2. Each period the local volatility estimate is updated according to $\underline{\sigma}^2(t) = \gamma\overline{\sigma}^2 + \alpha\underline{u}^2(t) + \beta\underline{\sigma}^2(t-1)$.

3. The parameter values used are in line with those found empirically in stock return series: $\overline{\sigma} = 0.03$, $\gamma = 0.02$, $\alpha = 0.2$ and $\beta = 0.78$. The series is initiated with $\sigma_0 = \overline{\sigma}$.

In figure 1 we show simulation results for 1000 consecutive samples. The

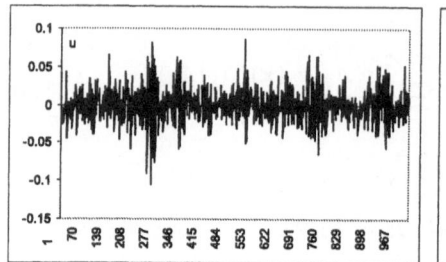 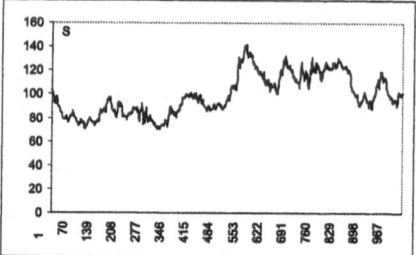

Fig. 1. (left) Return path **(right)** Price path from a simulated GARCH process

return series in the left graph exhibit volatility clusters that are typical for the process. The right graph shows the price development that, starting with $S_0 = 100$, is calculated from the instantaneous return as $S(t) = S(t-1)\, e^{u(t)}$.

4.2 Simulation results

In the left panel of figure 2 we have scattered the antecedent space with values $\underline{u}(t-1)$ on the x-axis against the consequent space with $\underline{u}(t)$ on the y-axis. In the probabilistic fuzzy rule base, we consider 3 antecedent linguis-

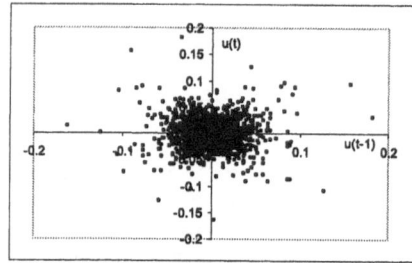 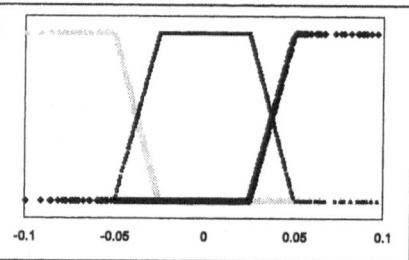

Fig. 2. (left) Scatter of $\underline{u}(t-1)$ against $\underline{u}(t)$, **(right)** memberships value for $\underline{u}(t-1)$

tic values A_q, defined by a fuzzy membership functions $\mu_{A_q}(u), q = 1, 2, 3$, (see the right panel of figure 2) . The corresponding linguistic values "Low" resp. "Average", "High" describe return values in linguistic terms. Using equation (7), we have estimated the corresponding probabilities yielding $\Pr(\underline{u}(t-1)$ is "Low") $= 0.0594$, $\Pr(\underline{u}(t-1)$ is "Average") $= 0.8722$, and $\Pr(\underline{u}(t-1)$ is "High") $= 0.0684$.

We examine 5 crisp consequent prototype values $u_1 = -0.050, u_2 = -0.025, u_3 = 0.000, u_4 = 0.025, u_5 = 0.050$, describing future returns. These values are named "very low" resp. "low", "average", "high", "very high". Each return value from the time series is classified according to the nearest prototype value using the Euclidian norm. By simply counting all u-values and determining the relative score, we can make an estimate of the (unconditional) probability distribution of $\Pr(u_j) = \Pr(\underline{u}(t) = u_j), j = 1, 2, \ldots, 5$. The results of these calculations are shown in the (emphasized) row of table 1 labelled 'All'. Using equation (9), we can also calculate $\Pr(u_j|A_q), j = 1, 2, 3, 4, 5; q = 1, 2, 3$. It concerns probabilities like "the probability that the future return is high given that the current return is Low". All these conditional probabilities are also summarized in table 1. Analyzing these results,

Future return	very low (-0.05)	low (-0.025)	average (0)	high (0.025)	very high (0.05)	Prob
Current return						
All	*0.0550*	*0.2265*	*0.4435*	*0.2140*	*0.0610*	*1.0000*
Low	0.1271	0.2084	0.2954	0.2302	0.1390	0.0594
Average	0.0437	0.2293	0.4666	0.2136	0.0468	0.8722
High	0.1374	0.2077	0.2808	0.2063	0.1679	0.0684

Table 1. *Unconditional and conditional probabilities* $\Pr(u_j)$ *and* $\Pr(u_j|A_q)$

it becomes clear that for both low and high current returns, the probability for very high and very low future returns is higher than the overall probability. To a lesser content this is also true for high and low future returns. We may also attach linguistic values to the magnitude of the difference between the conditional probability and the overall probability, for example: More than 5 percent is "much higher" or "much lower" and more than 2 percent is "higher" or "lower". The above results can thus be summarized as:

If *current* return is Low or High, then the probability of a high or low *future* return is higher and the probability of a very high or very low *future* return is much higher.

This looks like a pretty good intuitive description of the GARCH process used in the simulation.

Finally, we show two additional results. First, we have plotted the regression line of $u(t)$ on $u(t-1)$ (estimated according to equation (13)) in the right panel of figure 3. As expected for this problem, we found that $u(t) \approx 0$. In the left panel of the same figure, we show the difference between the conditional probabilities $\Pr(u_j|u(t-1))$ and the unconditional probability $\Pr(u_j)$, for $j = 1, 2, 3, 4, 5$. If current returns are Average, we observe that all con-

196

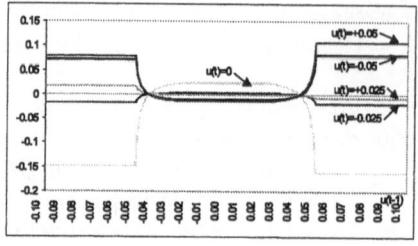 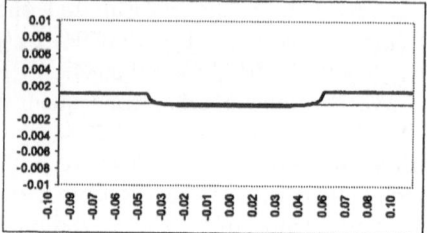

Fig. 3. (left) $\Pr(u_j|u(t-1)) - \Pr(u_j)$, (**right**) regression line of $u(t)$ on $u(t-1)$

ditional probabilities are almost equal to the unconditional one. However, if current returns are Low or High, we observe differences in the probability distribution of the future returns $u(t)$: average future returns are less dominating here while lower and higher future returns are more probable.

5 Conclusions

In this paper we have introduced two probabilistic fuzzy reasoning schemes. Using the first scheme as introduced in section 3.1, we can estimate a stochastic mapping $X \to Y$ conform equation (12). This is illustrated in section 4.2. If desired, the stochastic fuzzy mapping can be made deterministic using a regression approach (equation (13)). We have shown in section 3.2 that the resulting deterministic mapping can also be found using another probabilistic fuzzy reasoning scheme. Besides averaging, we apply interpolation here.

References

1. G. J. Klir and B. Yuan, *Fuzzy Sets and Fuzzy Logic: theory and applications*, Prentice Hall, Upper Saddle River, 1995.
2. Mohammad Jamshidi, André Titli, Lotfi Zadeh, and Serge Boverie, Eds., *Applications of Fuzzy Logic*, Prentice Hall, New Jersey, 1997.
3. E. H. Mamdani, "Application of fuzzy logic to approximate reasoning using linguistic systems," *IEEE Transactions on Computers*, vol. 26, no. 12, pp. 1182–1191, 1977.
4. Serge Guillaume, "Designing fuzzy inference systems from data: an interpretability-oriented review," *IEEE Transactions on Fuzzy Systems*, vol. 9, no. 3, pp. 426–443, June 2001.
5. V. Kecman, *Learning and Soft Computing*, MIT Press, Cambridge, MA, 2001.
6. Jan van den Berg, Willem Max van den Bergh, and Uzay Kaymak, "Probabilistic and statistical fuzzy set foundations of competitive exception learning," in *Proceedings of the Tenth IEEE International Conference on Fuzzy Systems*, Melbourne, Australia, Dec. 2001, vol. 2, pp. 1035–1038.
7. L. A. Zadeh, "Probability measures and fuzzy events," in *Fuzzy Sets and Applications, Selected Papers by L.A. Zadeth*, R.R. Yager, S. Ovchinnikov, R.M Tong, and H.T. Nguyen, Eds., pp. 45–51. John Wiley and Sons, USA, 1987.
8. J. C. Hull, *Options, Futures, & Other Derivatives*, Prentice Hall, Upper Saddle River, 4-th edition, 2000.

Acceptance Sampling Plans by Variables for Vague Data

Przemysław Grzegorzewski[1,2]

[1] Systems Research Institute, Polish Academy of Sciences,
   Newelska 6, 01-447 Warsaw, Poland
[2] Faculty of Math. and Inf. Sci., Warsaw University of Technology,
   Plac Politechniki 1, 00-661 Warsaw, Poland

Abstract. All classical sampling plans were constructed for exact data. However, sometimes we are not able to obtain such data but we deal with imprecise or even linguistic data. Therefore, a method for designing acceptance sampling plans by variables for vague data is considered.
Key Words: acceptance sampling plans by variables, hypotheses testing, vague data, fuzzy numbers.

1   Introduction

Acceptance sampling is a major field of statistical quality control. Its main goal is to assess whether goods (or services) are fulfilling certain requirements that relate to their fitness of use. If during the inspection the produced items are only classified into two disjoint categories – either "good" or "bad" – then we have, so called, acceptance sampling by attributes. However, if the characteristic of interest is a continuous variable, like weight, length, diameter, life time, etc., then we deal with acceptance sampling by variables (or acceptance control for measurements).

The statistical tools of acceptance sampling are called sampling plans and are equivalent to some statistical tests. Classical acceptance sampling plans have been studied by many researchers. They are thoroughly elaborated e.g. in Schilling (1982). Despite of the fact that both statisticians and practitioners have been considered sampling plans for many years, all the time new questions arrises (see Grzegorzewski and Hryniewicz, 2000). For example, before designing a sampling plan a producer and consumer have to specify statistical requirements of the plan concerning their risks and quality levels. Since various economic and technological factors must be taken into account while defining these parameters it makes difficult for producers and consumers to uniquely specify these factors and it happens sometimes that a desired plan requires needlessly large samples if their standpoints are too rigid. Therefore, to relax this rigidity, single sampling plans by attributes with relaxed requirements were discussed by Ohta and Ichihashi (1988), Kanagawa and Ohta (1990), Tamaki, Kanagawa and Ohta (1991) and Grzegorzewski (1998, 2001b). Grzegorzewski (2000b, 2002) also considered sampling plans by variables with fuzzy requirements.

All classical sampling plans were constructed for exact data. However, sometimes we are not able to obtain exact numerical data but we deal with imprecise or even linguistic data. To use classical tools in such situations we should compress these vague observations to exact data, but by doing this we often loose too much information. Thus it seems reasonable to use fuzzy sets for modelling vague or linguistic data and then to design sampling plans for these fuzzy data. Sampling plans by attributes for vague data were considered by Hryniewicz (1992, 1994). In the present paper we propose a method for designing acceptance sampling plans by variables for imprecise data.

2   Sampling plans by variables

Let X denote a quality characteristic under study of a product item (length, diameter, weight, pressure strength, etc.). An item is called nonconforming if the measurement x of X lies below a certain lower specification limit LSL or above an upper specification limit USL or if it lies outside a tolerance interval, i.e. a specified closed interval $[LSL, USL]$. Which of these cases is appropriate depends on the specific situation. In practice we generally assume that X is normally distributed, i.e. $X \sim N(m, \sigma^2)$.

To perform a single sampling plan by variables one draws a random sample of n items $X_1, \ldots, X_n$ from the lot of size N, takes measurements $X_1 = x_1, \ldots, X_n = x_n$ and compute the realization of a test statistic $T(X_1, \ldots, X_n)$. If $T(x_1, \ldots, x_n)$ falls in a specified acceptance region then the whole lot is accepted, otherwise, it is rejected. From the mathematical point of view, applying a single-sampling plan by variables is equivalent to performing a hypothesis test for the location, especially for the mean. It is seen easily that a sampling plan with a given lower specification limit is equivalent to the hypotheses testing problem

$$H : m \geq m_0 \quad \text{against} \quad K : m < m_0, \tag{1}$$

a sampling plan with a given upper specification limit is equivalent to

$$H : m \leq m_0 \quad \text{against} \quad K : m > m_0, \tag{2}$$

while the sampling plan with a double specification limits is equivalent to the hypotheses testing problem

$$H : m = m_0 \quad \text{against} \quad K : m \neq m_0. \tag{3}$$

However, a specific feature of these sampling plans is that the particular parameter m_0 is not explicitly specified.

The test statistic and corresponding acceptance and rejection regions can be specified in many ways. Under our assumptions the most natural test statistic is the sample mean $T(X_1, \ldots, X_n) = \overline{X} = \frac{1}{n} \sum_{i=1}^{n} X_i$. Since observations are normally distributed, i.e. $X_i \sim N(m, \sigma^2)$, the sample mean is also normally distributed and $E\overline{X}_n = m$. In real quality control we have a sampling without replacement and then $Var(\overline{X}) = \frac{\sigma^2}{n}\frac{N-n}{N-1}$. However, in practice, the sample size is very small compared to the lot size (say, $\frac{n}{N} \leq 0.1$) and we can assume that the approximate distribution of the sample mean is $\overline{X} \sim N\left(m, \frac{\sigma^2}{n}\right)$.

A large realization of $\overline{X}$ is interpreted as evidence of a large process mean m. Therefore in the case of plan with a lower specification limit we will find that the fraction nonconforming in the lot is small and, consequently, we will accept the lot. Thus we have a following decision criteria for a sampling plan with the lower specification limit LSL

$$\begin{aligned} &\text{if} \quad \overline{X} \geq LSL + k\sigma \quad \text{then accept the lot} \\ &\text{if} \quad \overline{X} < LSL + k\sigma \quad \text{then reject the lot,} \end{aligned} \qquad (4)$$

where $k > 0$ is called the acceptance factor.

Similarly, the decision criteria for a sampling plan with the upper specification limit USL are as follows

$$\begin{aligned} &\text{if} \quad \overline{X} \leq USL - k\sigma \quad \text{then accept the lot} \\ &\text{if} \quad \overline{X} > USL - k\sigma \quad \text{then reject the lot.} \end{aligned} \qquad (5)$$

while for a sampling plan with the double specification limits we have

$$\begin{aligned} &\text{if} \quad LSL + k'\sigma \leq \overline{X} \leq USL - k'\sigma \quad &&\text{then accept the lot} \\ &\text{if} \quad \overline{X} < LSL + k'\sigma \quad \text{or} \quad \overline{X} > USL - k'\sigma \quad &&\text{then reject the lot.} \end{aligned} \qquad (6)$$

If the true population variance σ^2 is unknown one has to use sample data to estimate it and then to substitute σ in (4)–(6) by a sample standard deviation $\hat{\sigma} = S = \sqrt{\frac{1}{n-1}\sum_{i=1}^{n}(X_i - \overline{X})^2}$.

It is seen that an acceptance sampling plan by variables is completely described by an ordered pair (n, k), i.e. by the number of items one has draw from the lot and the acceptance factor.

It should be mentioned that acceptance sampling does not reduce to single-sampling only. There are also double-sampling plans, sequential-sampling plans, variables sampling schemes and other techniques. For more details we refer the reader to Montgomery (1991), Mittag and Rinne (1993), Schilling (1982), etc. However, further on by a sampling plan (or a plan, for short) we will understand a single acceptance sampling plan by variables.

The common way to determine a sampling plan is to specify four numbers: the quality levels – the acceptable quality level (AQL) and rejectable quality level (RQL), where $AQL < RQL$, with corresponding risks – the producer's risk (δ) and consumer's risk (β), where $\beta < 1 - \delta$. Let us now briefly discuss the interpretation of the parameters used in the plan determination. Suppose the quality level is measured by the fraction nonconforming in the lot. Then the *acceptable quality level (AQL)* represents the poorest level of quality for the producer's process that the consumer would consider to be acceptable as a process average (compare ISO 2859). The consumer often tries to design the sampling plan so that it gives a high probability $1-\delta$ of acceptance at the AQL. Thus the probability of rejecting the lot having good quality is equal to δ and is called the *producer's risk*. It is worth noting that the AQL is a property of the producer's process and is not a property of the sampling plan. It is also not a target value for the producer's process but it is just a simple standard against which to judge the lots. However the consumer also needs the protection against lots of poor quality. Thus he establishes the poorest level of quality that he is willing to accept, called the *rejectable quality level* (RQL). The probability of accepting the lot having such poor quality is called the *consumer's risk* β. Note that RQL is also not a property of the plan but a level of quality specified by the consumer to protect himself.

Now the plan parameters n and k are the solutions of

$$\begin{cases} P(\text{ acceptance } | \ AQL) = 1 - \delta \\ P(\text{ acceptance } | \ RQL) = \beta. \end{cases} \tag{7}$$

Usually (7) cannot be realized since the sample size n must be integer valued and a desired plan does not exist. Therefore instead of (7) we express the requirements in the following way

$$\begin{cases} P(\text{ acceptance } | \ AQL) \geq 1 - \delta \\ P(\text{ acceptance } | \ RQL) \leq \beta, \end{cases} \tag{8}$$

which ensures that we are on the safe side.

It can be shown that for one-sided sampling plans, i.e. plans with given either lower specification limit LSL or upper specification limit USL, with known variance σ^2, we have

$$n = \left(\frac{u_{1-\beta} + u_{1-\delta}}{u_{1-AQL} - u_{1-RQL}} \right)^2, \tag{9}$$

$$k = \frac{u_{1-\beta} \ u_{1-AQL} + u_{1-\delta} \ u_{1-RQL}}{u_{1-\beta} + u_{1-\delta}}, \tag{10}$$

where u_γ denotes the quantile of order γ from the standard normal distribution. Of course, the sample size n is always rounded to the next bigger integer. Since (8) could have many solutions, our optimal sampling plan is one with the smallest sample size among all plans satisfying (8). If the variance σ^2 is unknown, we first compute k form (10) and then we get

$$n = \left(1 + \frac{k^2}{2} \right) \left(\frac{u_{1-\beta} + u_{1-\delta}}{u_{1-AQL} - u_{1-RQL}} \right)^2. \tag{11}$$

Direct formulas for n and k for plans with known or unknown process variance can be found in Montgomery (1991), Mittag and Rinne (1993), Schilling (1982), etc.

Further on, without loss of generality, we restrict our considerations to single specification limit problem since the construction of plans for given double specification limits is approached in the same way as for plans single specification limits.

3 Vague data

It may happen that a sample used for making decision on the lot consists of observations that are not necessarily crisp but may be vague as well. In order to describe the vagueness of data we use the notion of a fuzzy number, introduced by Dubois and Prade (1978). We say that a fuzzy subset A of the real line $\mathcal{R}$, with the membership function $\mu_A : \mathcal{R} \to [0,1]$, is a fuzzy number if and only if A is normal (i.e. there exists an element x_0 such that $\mu_A(x_0) = 1$), A is fuzzy convex (i.e. $\mu_A(\lambda x_1 + (1-\lambda)x_2) \geq \mu_A(x_1) \wedge \mu_A(x_2)$, $\forall x_1, x_2 \in \mathcal{R}, \forall \lambda \in [0,1]$), μ_A is upper semicontinuous and supp$A$ is bounded.

A useful notion for dealing with a fuzzy number is a set of its α−cuts. The α−cut of a fuzzy number A is a nonfuzzy set defined as

$$A_\alpha = \{x \in \mathcal{R} : \mu_A(x) \geq \alpha\}. \tag{12}$$

A family $\{A_\alpha : \alpha \in (0,1]\}$ is a set representation of the fuzzy number A. According to the definition of a fuzzy number it is easily seen that every α−cut of a fuzzy number is a closed interval. Hence we have $A_\alpha = [A_\alpha^L, A_\alpha^U]$, where

$$\begin{aligned} A_\alpha^L &= \inf\{x \in \mathcal{R} : \mu_A(x) \geq \alpha\}, \\ A_\alpha^U &= \sup\{x \in \mathcal{R} : \mu_A(x) \geq \alpha\}. \end{aligned} \tag{13}$$

A space of all fuzzy numbers will be denoted by $\mathcal{FN}(\mathcal{R})$.

A notion of fuzzy random variable was introduced by Kwakernaak (1978, 1979). Other definitions of fuzzy random variables are due to Kruse (1982) or to Puri and Ralescu (1986). Our definition is similar to those of Kwakernaak and Kruse. Suppose that a random experiment is described as usual by a probability space $(\Omega, \mathcal{A}, \mathcal{P})$, where Ω is a set of all possible outcomes of the experiment, $\mathcal{A}$ is a σ−algebra of subsets of Ω (the set of all possible events) and P is a probability measure. Then mapping $X : \Omega \to \mathcal{FN}(\mathcal{R})$ is called a fuzzy random variable if it satisfies the following properties:
(a) $\{X_\alpha(\omega) : \alpha \in [0,1]\}$ is a set representation of $X(\omega)$ for all $\omega \in \Omega$,
(b) for each $\alpha \in [0,1]$ both $X_\alpha^L = X_\alpha^L(\omega) = \inf X_\alpha(\omega)$ and $X_\alpha^U = X_\alpha^U(\omega) = \sup X_\alpha(\omega)$, are usual real-valued random variables on $(\Omega, \mathcal{A}, \mathcal{P})$.

Thus a fuzzy random variable X is considered as a perception of an unknown usual random variable $V : \Omega \to \mathcal{R}$, called an *original* of X (if only vague data are available, it is of course impossible to show which of the possible originals is the true one). Similarly n−dimensional fuzzy random sample $X_1, \ldots, X_n$ may be treated as a fuzzy perception of the usual random sample $V_1, \ldots, V_n$ (where $V_1, \ldots, V_n$ are independent and identically distributed crisp random variables). For more information we refer the reader to Kruse and Meyer (1987).

4 Sampling plans for fuzzy data

Now without loss of generality we will consider a sampling plan with given lower specification limit LSL in situation with known variance σ^2. As it was mentioned above, such a plan is equivalent to the well-known test for the one-sided hypothesis testing problem $H : m \geq m_0$ against $K : m < m_0$. If the true mean m_0 were known, we have a following test $\phi : \mathcal{R}^n \to \{0, 1\}$

$$\phi(X_1, \ldots, X_n) = \begin{cases} 0, & \text{if } \overline{X} \geq m_0 - u_{1-\delta/2}\frac{\sigma}{\sqrt{n}}, \\ 1, & \text{if } \overline{X} < m_0 - u_{1-\delta/2}\frac{\sigma}{\sqrt{n}}. \end{cases} \tag{14}$$

However, since instead of m_0 we have only the lower specification limit LSL, we get a following test $\phi' : \mathcal{R}^n \to \{0, 1\}$

$$\phi'(X_1, \ldots, X_n) = \begin{cases} 0, & \text{if } \overline{X} \geq LSL + k\sigma, \\ 1, & \text{if } \overline{X} < LSL + k\sigma, \end{cases} \tag{15}$$

which agrees with (4). It is obvious that we can rewrite (15) in a slightly different way

$$\phi'(X_1, \ldots, X_n) = \begin{cases} 0, & \text{if } LSL \leq \overline{X} - k\sigma, \\ 1, & \text{if } LSL > \overline{X} - k\sigma, \end{cases} \tag{16}$$

which would be our starting point for the forthcoming generalization.

Now suppose our data are no longer crisp but they are rather vague and they are described by fuzzy numbers. Let $X_1, \ldots, X_n$ denote a fuzzy random sample. Grzegorzewski (2000a) has shown how to construct a statistical test for such data.

Let $\mathcal{F}(\{0, 1\})$ denote a family of fuzzy subsets of $\{0, 1\}$. In our case of the one-sided hypothesis testing problem discussed above we get a fuzzy test $\varphi : (\mathcal{FN}(\mathcal{R}))^n \to \mathcal{F}(\{0, 1\})$ with the following α−cuts

$$\varphi_\alpha(X_1, \ldots, X_n) = \begin{cases} \{0\}, & \text{if } m_0 \in (\Pi_\alpha \setminus (\neg\Pi)_\alpha), \\ \{1\}, & \text{if } m_0 \in ((\neg\Pi)_\alpha \setminus \Pi_\alpha), \\ \{0, 1\}, & \text{if } m_0 \in (\Pi_\alpha \cap (\neg\Pi)_\alpha), \\ \emptyset, & \text{if } m_0 \notin (\Pi_\alpha \cup (\neg\Pi)_\alpha), \end{cases} \tag{17}$$

where

$$\Pi_\alpha = \left(-\infty, \frac{1}{n}\sum_{i=1}^{n}(X_i)_\alpha^U + u_{1-\delta}\frac{\sigma}{\sqrt{n}}\right], \tag{18}$$

and where $(X_i)_\alpha^U$ stands for the upper bound of the α−cut corresponding to X_i. After simple calculations we get

$$\varphi(X_1, \ldots, X_n) = \begin{cases} 1/0 + 0/1, & \text{if } m_0 \in \Pi_{\alpha=1}, \\ 0/0 + 1/1, & \text{if } m_0 \notin \Pi_{\alpha=0}, \\ \mu_\Pi(m_0)/0 + (1 - \mu_\Pi(m_0))/1, & \text{otherwise.} \end{cases} \tag{19}$$

In situation with crisp data we accept hypothesis H if the test statistic belongs to the acceptance region and reject H otherwise. In a fuzzy situation a fuzzy test leads to fuzzy decisions. We may get $1/0+0/1$ which indicates that we should accept H, or $0/0+1/1$ which means that H should be rejected, but we may also get $\mu_0/0+(1-\mu_0)/1$, where $\mu_0 \in (0,1)$, which can be interpreted as a degree of conviction that we should accept (μ_0) or reject $(1-\mu_0)$ the hypothesis H, respectively.

Now, utilizing the equivalence between test ϕ (14) and ϕ' (16) one get a fuzzy test φ' equivalent to φ

$$\varphi'_\alpha(X_1,\ldots,X_n) = \begin{cases} \{0\}, & \text{if } LSL \in (\Delta_\alpha \setminus (\neg\Delta)_\alpha), \\ \{1\}, & \text{if } LSL \in ((\neg\Delta)_\alpha \setminus \Delta_\alpha), \\ \{0,1\}, & \text{if } LSL \in (\Delta_\alpha \cap (\neg\Delta)_\alpha), \\ \emptyset, & \text{if } LSL \notin (\Delta_\alpha \cup (\neg\Delta)_\alpha), \end{cases} \tag{20}$$

where

$$\Delta'_\alpha = \left(-\infty, \frac{1}{n}\sum_{i=1}^{n}(X_i)^U_\alpha - k\sigma\right]. \tag{21}$$

And again, after some calculations, we get

$$\varphi'(X_1,\ldots,X_n) = \begin{cases} 1/0 + 0/1, & \text{if } LSL \leq \frac{1}{n}\sum_{i=1}^{n}(X_i)^U_{\alpha=1} - k\sigma, \\ 0/0 + 1/1, & \text{if } LSL > \frac{1}{n}\sum_{i=1}^{n}(X_i)^U_{\alpha=0} - k\sigma, \\ \mu_0/0 + (1-\mu_0)/1, & \text{otherwise,} \end{cases} \tag{22}$$

where μ_0 is given by the inverse function of the upper bound of α−cut of the fuzzy number (21), i.e. $\mu_0 = ((\Delta')^U_\alpha(LSL))^{-1}$.

Since ϕ' (16) is actually a single sampling plan by variables for crisp data, therefore φ' (22) is a sampling plan by variables for fuzzy data. And this sampling plan for fuzzy data is itself also fuzzy. It is no longer univocal as the plan (4) for crisp data, which often leads to acceptance or rejection of the lot. Now, in particular, $\varphi' = 1/0 + 0/1$ indicates that the lot under study should be accepted, while $\varphi' = 0/0 + 1/1$ means that given lot should be rejected. However, output of the form $\varphi' = \mu_0/0 + (1-\mu_0)/1$, where $\mu_0 \in (0,1)$, corresponds to more ambiguous situation that cannot be interpreted strictly as acceptance or rejection. In such a case we get only a suggestion of the type: "rather accept" or "rather reject", depending on whether μ_0 is close to 1 or close to 0, respectively. Moreover, μ_0 and $(1-\mu_0)$ might be interpreted as a degree of conviction that we should accept or reject the lot, respectively.

In the same manner we can obtain a sampling plan with a given lower specification limit LSL in situation when the true variance σ^2 is unknown.

The corresponding plan is equivalent to the following test

$$
\varphi''(X_1,\ldots,X_n) = \begin{cases} 1/0 + 0/1, & \text{if } LSL \leq \frac{1}{n}\sum_{i=1}^{n}(X_i)_{\alpha=1}^{U} - kS_{\alpha=1}^{U}, \\ 0/0 + 1/1, & \text{if } LSL > \frac{1}{n}\sum_{i=1}^{n}(X_i)_{\alpha=0}^{U} - kS_{\alpha=0}^{U}, \\ \mu_0/0 + (1-\mu_0)/1, & \text{otherwise,} \end{cases}
$$

(23)

where μ_0 is given by the inverse function of the upper bound of $\alpha-$cut of the fuzzy number Δ'' given by

$$
\Delta''_\alpha = \left(-\infty, \frac{1}{n}\sum_{i=1}^{n}(X_i)_\alpha^U - kS_\alpha^U\right],
$$

(24)

and where $S_\alpha^U = \sqrt{\frac{1}{n-1}\sum_{i=1}^{n}((X_i)_\alpha^U - (\overline{X})_\alpha^U)^2}$ is the upper bound of $\alpha-$cut of the sample standard deviation (see Höppner (1994), Höppner and Wolf (1995)). Hence $\mu_0 = ((\Delta'')_\alpha^U(LSL))^{-1}$.

Similarly, a plan with a given upper specification limit USL and known variance σ^2 is given by

$$
\varphi'''(X_1,\ldots,X_n) = \begin{cases} 1/0 + 0/1, & \text{if } USL \geq \frac{1}{n}\sum_{i=1}^{n}(X_i)_{\alpha=1}^{L} + k\sigma, \\ 0/0 + 1/1, & \text{if } USL < \frac{1}{n}\sum_{i=1}^{n}(X_i)_{\alpha=0}^{L} + k\sigma, \\ \mu_0/0 + (1-\mu_0)/1, & \text{otherwise,} \end{cases}
$$

(25)

where μ_0 is given by the inverse function of the upper bound of $\alpha-$cut of the fuzzy number Δ''' given by

$$
\Delta'''_\alpha = \left[\frac{1}{n}\sum_{i=1}^{n}(X_i)_\alpha^L + k\sigma, +\infty\right),
$$

(26)

i.e. $\mu_0 = ((\Delta''')_\alpha^L(USL))^{-1}$. If σ^2 is unknown than the desired plan is given by

$$
\varphi''''(X_1,\ldots,X_n) = \begin{cases} 1/0 + 0/1, & \text{if } USL \geq \frac{1}{n}\sum_{i=1}^{n}(X_i)_{\alpha=1}^{L} + kS_{\alpha=1}^{U}, \\ 0/0 + 1/1, & \text{if } USL < \frac{1}{n}\sum_{i=1}^{n}(X_i)_{\alpha=0}^{L} + kS_{\alpha=0}^{U}, \\ \mu_0/0 + (1-\mu_0)/1, & \text{otherwise,} \end{cases}
$$

(27)

where μ_0 is given by the inverse function of the upper bound of $\alpha-$cut of the fuzzy number Δ'''' given by

$$
\Delta''''_\alpha = \left[\frac{1}{n}\sum_{i=1}^{n}(X_i)_\alpha^L + kS_\alpha^U, +\infty\right),
$$

(28)

i.e. $\mu_0 = ((\Delta'''')_\alpha^L(USL))^{-1}$.

Plans with double specification limits are "mixtures" of the corresponding plans with lower and upper limits, i.e. (22) and (25) or (23) and (27).

5 Conclusions

In the present paper we have proposed a method for designing single acceptance sampling plans by variables for fuzzy data. These plans are well defined since if all the data are crisp they reduce to classical plans by variables. As it was shown, in the presence of vague data sampling plans do not necessarily lead to univocal decisions: to accept or to reject a given lot. In general, a user obtains a coefficient which might be interpreted as a degree of conviction that he should accept or reject the lot, respectively. Thus our fuzzy sampling plan leads to fuzzy decisions. Of course, if a crisp conclusion is required, a defuzzification method have to be used (the problem of defuzzification of fuzzy tests was considered by Grzegorzewski, 2001a).

As it was mentioned above, acceptance sampling doesn't reduce to single-sampling only. However, our method for designing sampling plans may be also applied for determining more complicated acceptance sampling plans by variables.

References

1. Dubois D., Prade H. (1978), Operations on fuzzy numbers, Int. J. Syst. Sci. **9**, 613–626.
2. Grzegorzewski P. (1998), A Soft Design of Acceptance Sampling Plans by Attributes, In: Proceedings of the VIth International Workshop on Intelligent Statistical Quality Control, Würzburg, September 14–16, pp. 29–38.
3. Grzegorzewski P. (2000a), Testing Statistical Hypotheses with Vague Data. Fuzzy Sets and Systems **112**, 501–510.
4. Grzegorzewski P. (2000b), A Soft Design of Acceptance Sampling Plans by Variables, Proceedings of the Eight International Conference "Information Processing and Management of Uncertainty in Knowledge-based Systems IPMU'2000", Madrid, July 3-7, pp. 208-214.
5. Grzegorzewski P. (2001a), Fuzzy Tests - Defuzzification and Randomization, Fuzzy Sets and Systems, 118 (2001), 437-446.
6. Grzegorzewski P. (2001b), Acceptance Sampling Plans by Attributes with Fuzzy Risks and Quality Levels, In: Frontiers in Frontiers in Statistical Quality Control, vol. 6, Eds. Wilrich P.Th., Lenz H.J., Springer, Heidelberg, pp. 36–46.
7. Grzegorzewski P. (2002), A Soft Design of Acceptance Sampling Plans by Variables, In: Technologies for Contructing Intelligent Systems, Eds. Bouchon-Meunier B., Gutierrez-Rios J., Magdalena L. and Yager R.R., Springer, Vol.2, pp. 275–286.
8. Grzegorzewski P., Hryniewicz O. (2000), Soft methods in statistical quality control, Control and Cybernetics, 29, 119–140.

9. Höppner J. (1994), Statistiche Prozeßkontrolle mit Fuzzy-Daten. Ph.D. Dissertation, Ulm University.
10. Höppner J., Wolff H. (1995), The Design of a Fuzzy-Shewhart Control Chart. Research Report, Würzburg University.
11. Hryniewicz O. (1992), Statistical Acceptance Sampling with Uncertain Information from a Sample and Fuzzy Quality Criteria. Working Paper of SRI PAS, Warsaw, (in Polish).
12. Hryniewicz O. (1994), Statistical Decisions with Imprecise Data and Requirements. In: Systems Analysis and Decisions Support in Economics and Technology, Proceedings of the 9th Polish-Italian and 6th Polish-Finnish Conference, Eds. R. Kulikowski, K. Szkatuła, J. Kacprzyk, Omnitech Press, 135–143.
13. ISO 2859, International Standard, Sampling Procedures for Inspection by Attributes.
14. Kanagawa A., Ohta H. (1990), A Design for Single Sampling Attribute Plan Based on Fuzzy Sets Theory, Fuzzy Sets and Systems, **37**, 173–181.
15. Kruse R. (1982), The Strong Law of Large Numbers for Fuzzy Random Variables. Inform. Sci. **28**, 233–241.
16. Kruse R., Meyer K. D. (1987), Statistics with Vague Data. D. Riedel Publishing Company.
17. Kwakernaak H. (1978), Fuzzy Random Variables, Part I: Definitions and Theorems. Inform. Sci. **15**, 1–15.
18. Kwakernaak H. (1979), Fuzzy Random Variables, Part II: Algorithms and Examples for the Discrete Case. Inform. Sci. **17**, 253–278.
19. Mittag H.J., Rinne H. (1993), Statistical Methods of Quality Assurance, Chapman and Hall.
20. Montgomery D.C. (1991), Introduction to Statistical Quality Control, Wiley, New York.
21. Ohta H., Ichihashi H. (1988), Determination of Single-Sampling Attribute Plans Based on Membership Functions, Int. J. Prod. Res. **26**, 1477–1485.
22. Puri M. L., Ralescu D. A. (1986), Fuzzy Random Variables. J. Math. Anal. Appl. **114**, 409–422.
23. Schilling E.G. (1982), Acceptance Sampling in Quality Control, Dekker, New York.
24. Tamaki F., Kanagawa A., Ohta H. (1991), A Fuzzy Design of Sampling Inspection Plans by Attributes, Japanese Journal of Fuzzy Theory and Systems, **3**, 315–327.

Possibilistic Approach to the Bayes Statistical Decisions

Olgierd Hryniewicz

Systems Research Institute, Newelska 6, 01-447 Warsaw, Poland
hryniewi@ibspan.waw.pl

Abstract. In the paper we consider the problem of Bayes verification of statistical hypotheses. We consider different cases where statistical data, assumed loss functions and considered hypotheses may be described vaguely. The formulae for the calculation of expected fuzzy risks are given. The tools of the possibility theory such as Possibility of Dominance (PD) and Necessity of Strict Dominance (NSD) indices are proposed for final decision making.

1 Introduction

Testing statistical hypotheses is one of the most important parts of statistical inference. On the other hand it can be regarded as a part of the decision theory. In the decision theory we assume that decisions (actions belonging to a certain action space) should depend upon a certain state which is uncontrollable and unknown for a decision maker. We usually assume that unknown states are generated by random mechanisms. However, all we could know about these mechanism is their description in terms of the probability distribution P_θ that belongs to a family of distributions $\mathcal{P} = \{P_\theta : \theta \in \Theta\}$ indexed by a parameter θ (one or multidimensional). In such a case a state space is often understood as equivalent to the parameter space Θ. If we knew the true value of θ we would be able to take a correct decision. The choice of an appropriate decision depends upon a value of a certain utility function that has to be defined on the product of the action space and the state space. If we had known the unknown state we would have been able to choose the most preferred action looking for the action with the highest value of the assigned utility. In practice, we define the expected reward (or the loss) associated with the given action for the given state $\theta \in \Theta$, and then we define the utility $u \in U$ that 'measures' the preference the decision maker assigns to that reward (loss).

In the Bayesian setting of the decision theory we assume that there exists the prior information about the true state, and that this information is expressed in terms of the probability distribution $\pi(\theta)$ defined on the parameter space Θ. By doing this we identify each action with probability distribution on a set of possible utilities $\mathcal{U}$. According to the Bayesian decisions paradigm we choose the action with the highest value of the *expected* utility, where expectation is calculated with respect to the probability distribution defined on $\mathcal{U}$.

When a decision maker has an opportunity to observe a random variable (or a random vector) X that is related to the state θ, such an observation provides him with additional information which may be helpful in making proper decisions. In such a case the decision problem is called the statistical decision problem. Comprehensive presentation of the Bayesian decision theory is presented in a classical textbook of Raiffa and Schleifer [21], and the Bayesian approach to statistical decision problems may be found in DeGroot [6].

In the statistical decision theory we deal with many quantities which may be vague and imprecise. First, our observation may be imprecise, described in linguistic terms. In such a case we deal with imprecise (fuzzy) statistical data. Many books and papers have been written on the generalization of classical statistical methods for the analysis of fuzzy data. Classical problems of statistical decisions have been discussed, e.g., in the paper of Grzegorzewski and Hryniewicz [15]. First results presenting the Bayesian decision analysis for imprecise data were given in papers by Casals, Gil and Gil [3], [4], and Gil [11]. In these papers the authors described fuzzy observations using the notion of the fuzzy information system by Zadeh [26] and Tanaka, Okuda and Asai [22]. Other approach has been proposed by Viertl [24]. Further results concerning the decisions based on fuzzy statistical information have been published by Casals [2] and Gil and Lopez-Diaz [12]. Imprecise information about the parameters of the prior distribution have been considered in Hryniewicz [16] and Frühwirth-Schnatter [10]. In the statistical decision theory we may also face practical problems when verified hypotheses are imprecise. This problem was considered by Delgado, Verdegay and Vila [7] and Casals [2]. Finally, the loss function (or the utility function) may be expressed in a fuzzy way as in [12]. Recent results on Bayes fuzzy hypotheses testing have been presented by Taheri and Behboodian [23] who proposed another approach using the posterior odds ratio as the criterion for decision making.

The crucial problem of the fuzzy approach to the Bayes statistical decision analysis is to compare fuzzy risks related to considered decisions. This problem arises from the fact that fuzzy numbers that describe fuzzy risks are not naturally ordered. Thus, the decisions depend upon the method used for such an ordering. In this paper we propose to use the Necessity of Strict Dominance Index introduced by Dubois and Prade [9]. We claim that in specific situations this approach is preferable to the others.

The paper is organized in four sections. In the second section we present methods for the calculation of fuzzy risks in the presence of fuzzy data, fuzzy prior information and fuzzy statistical hypotheses. This result may be easily extended to the case of the fuzzy utility function. The section describes the results obtained by different authors together with some generalizations. In the third section we propose a possibilistic method for choosing the optimal decision. Finally, in the fourth section we present the discussion of the proposed solution and the areas requiring further investigations.

2   Calculation of the Bayes risks

2.1   Bayes risk in crisp environment

In the Bayesian approach to statistical decisions we take into consideration potential losses and rewards associated with each considered decision. Let $\theta \in \Theta$ be a parameter describing an element of the state space, and $\delta \in \Delta$ be a decision (action) from a space of possible (admissible) decisions. Usually we define a loss function $L(\theta, \delta)$ which assigns a certain loss (or reward) to the decision δ if the true state is described by θ. In a more general setting, when we face losses (rewards) of a different character we introduce a more general notion of utility $u \in U$ which describes a decision maker's level of preference for the decision δ if the true state, is described by θ.

Assume now that the decision maker knows the conditional probability density function $f(\mathbf{x}|\theta) = f(x_1, x_2, ..., x_n|\theta)$ describing the observed values of a random sample $(X_1, X_2, ..., X_n)$. Moreover, we assume that the decision maker has some prior information about possible values of θ. This information, according to the Bayes decision theory is represented by the prior probability distribution $\pi(\theta)$. This information is merged with the information contained in the prior probability distribution $\pi(\theta)$. The updated information about the true state is calculated using the Bayes theorem, and expressed in the form of the posterior probability distribution

$$g(\theta|\mathbf{x}) = \frac{f(\mathbf{x}|\theta)\,\pi(\theta)}{\int\limits_{\Theta} f(\mathbf{x}|\theta)\,d\pi(\theta)}, \tag{1}$$

where $\mathbf{x} = (x_1, x_2, ..., x_n)$. Further analysis is performed in exactly the same way with the posterior probability distribution $g(\theta|\mathbf{x})$ replacing the prior probability distribution $\pi(\theta)$.

Let $\delta(\mathbf{x}) = \delta(x_1, x_2, ..., x_n)$ be a decision function which is used for choosing an appropriate decision for given sample values $x_1, x_2, ..., x_n$. The risk function, interpreted as an expected loss incurred by the decision δ, is calculated as

$$\rho(\delta) = \int\limits_{\Theta}\int\limits_{X} L(\theta, \delta(x_1, x_2, ..., x_n))\,f(x_1, x_2, ..., x_n|\theta)\,\pi(\theta)\,d\mathbf{x}d\theta. \tag{2}$$

Let Δ be the space of possible decision functions. Function δ^* that fulfils the following condition

$$\rho\left(\delta^*\right) = \inf_{\delta \in \Delta} \rho\left(\delta\right) \tag{3}$$

we call the Bayes decision function, and the corresponding risk $\rho\left(\delta^*\right)$ we call the Bayes risk. Statistical decisions with the risk equal to the Bayes risk are called optimal. In this paper we restrict ourselves to a particular problem of the Bayes decisions, namely to the Bayes test of statistical hypothesis $H_0 : \theta \in \Theta_0$ against the alternative $H_1 : \theta \in \Theta_1$, where Θ_0 and Θ_0 are the subsets of the state space Θ such that $\Theta_0 \cap \Theta_1 = \emptyset$. Let us define two functions:

$$H_0\left(\theta\right) = \begin{cases} 1, \, \theta \in \Theta_0 \\ 0, \, \theta \in \Theta_1 \end{cases} \tag{4}$$

and

$$H_1\left(\theta\right) = \begin{cases} 0, \, \theta \in \Theta_0 \\ 1, \, \theta \in \Theta_1 \end{cases} \tag{5}$$

Now, let us define loss functions:

$$L\left(\theta, a_0\right) = a\left(\theta\right)\left[1 - H_0\left(\theta\right)\right] \tag{6}$$

that describes the loss related to the acceptance of H_0, and

$$L\left(\theta, a_1\right) = b\left(\theta\right)\left[1 - H_1\left(\theta\right)\right] \tag{7}$$

that describes the loss related to the acceptance of H_1. Functions $a\left(\theta\right)$ and $b\left(\theta\right)$ are two arbitrary nonnegative functions. In such a case we may consider only two risks: the risk of accepting H_1 when H_0 is true given by

$$R_1 = \int_{\Theta_1} L\left(\theta, a_1\right) g\left(\theta|\mathbf{x}\right) d\theta. \tag{8}$$

and the risk of accepting H_0 when H_1 is true given by

$$R_0 = \int_{\Theta_0} L\left(\theta, a_0\right) g\left(\theta|\mathbf{x}\right) d\theta. \tag{9}$$

In general, we have to calculate the risk R_h, $h = 0, 1$ given by

$$R_h = \int_{\Theta_h} L\left(\theta, a_h\right) g\left(\theta|\mathbf{x}\right) d\theta \tag{10}$$

In the following subsections we present methods for the computation of such a risk in different cases representing situations when different parts of the decision model are described in an imprecise way.

2.2 Bayes risk for fuzzy statistical data and fuzzy prior information

Let us consider situation when available statistical data are vague and are described by fuzzy random variables. The notion of a fuzzy random variable has been defined by many authors. First definition of the fuzzy random variable has been proposed by Kwakernaak [19] and described in details in Kruse [17] and Kruse and Meyer [18]. According to this definition the fuzzy random variable $\widetilde{X}$ may be considered as a fuzzy (vague) perception of an unknown usual random variable $X : \Omega \to \mathcal{R}$, called an original of $\widetilde{X}$. Puri and Ralescu [20] gave another definition of the fuzzy random variable that has been used in the context of Bayes decision problems by some authors, e.g. by Gil and Lopez-Diaz [12]. In this paper we assume that all fuzzy sets are normal, and in such a case both definitions of the fuzzy variable are equivalent (see Zhong and Zhou [27]). Therefore, all the presented results do not depend upon the assumed definition.

In the presence of fuzzy statistical data the posterior distribution of the state variable θ can be obtained by the application of Zadeh's extension principle to (1). Let $\widetilde{x}_i^\alpha = ((\widetilde{x}_i^\alpha)_L, (\widetilde{x}_i^\alpha)_U)$, $i = 1, ..., n$ be the α-cuts of the fuzzy observations $\widetilde{x}_1, \widetilde{x}_2, ..., \widetilde{x}_n$. Following Frühwirth-Schnatter [10] we denote by $C(\widetilde{\mathbf{x}})_\alpha$ the α-cut of the fuzzy sample which is equal to the Cartesian product of the α-cuts $\widetilde{x}_i^\alpha$, $i = 1, ..., n$. The fuzzy posterior distribution $\widetilde{g}(\theta|\widetilde{\mathbf{x}})$ is, according to [25], given by a-contours

$$g_\alpha^L(\theta) = \min_{\mathbf{x} \in C(\widetilde{\mathbf{x}})_\alpha} \frac{f(\mathbf{x}|\theta)\,\pi(\theta)}{n(\mathbf{x})}, \tag{11}$$

$$g_\alpha^U(\theta) = \max_{\mathbf{x} \in C(\widetilde{\mathbf{x}})_\alpha} \frac{f(\mathbf{x}|\theta)\,\pi(\theta)}{n(\mathbf{x})}, \tag{12}$$

where $n(\mathbf{x})$ is a normalizing constant equal to the denominator of the right hand side of (1). Now, we can compute the fuzzy risks $\widetilde{R}_h$, $h = 0, 1$ using the general methodology for integrating fuzzy functions presented in [8].

Let us denote by $C\left(\widetilde{\mathbf{R}}_h\right)_\alpha = \left(\widetilde{\mathbf{R}}_h^{\alpha,L}, \widetilde{\mathbf{R}}_h^{\alpha,U}\right)$ the α-cut of the fuzzy risk $\widetilde{\mathbf{R}}_h$. The lower and upper bounds are calculated from the following formulae:

$$\widetilde{\mathbf{R}}_h^{\alpha,L} = \int_{\Theta_h} L(\theta, a_h)\, g_\alpha^L(\theta|\mathbf{x})\, d\theta \tag{13}$$

and

$$\widetilde{\mathbf{R}}_h^{\alpha,U} = \int_{\Theta_h} L(\theta, a_h)\, g_\alpha^U(\theta|\mathbf{x})\, d\theta. \tag{14}$$

Thus, we can calculate the membership functions of $\widetilde{\mathbf{R}}_0$ and $\widetilde{\mathbf{R}}_1$, respectively.

Frühwirth-Schnatter [10] proposed a further generalization of the fuzzy risk model by allowing some imprecision in the description of prior information. Probability density function $\pi(\theta)$ that describes the prior knowledge about the values of the state variable θ is usually specified as a certain function $\pi(\theta, \eta)$ of a vector of known parameters η. It seems to be reasonable, however, to assume a vague description of η by the fuzzy vector $\tilde{\eta}$.

Let us denote the fuzzy prior distribution by $\tilde{\pi}(\theta) = \pi(\theta, \tilde{\eta})$, and the α-cut of the fuzzy vector $\tilde{\eta}$ by $C(\tilde{\eta})_\alpha$. In the presence of vagueness in the description of prior information and statistical data the α-contours of the fuzzy posterior probability density are given by (see Frühwirth-Schnatter [10])

$$g_\alpha^L(\theta) = \min_{(\mathbf{x}, \eta \in C(\tilde{\mathbf{x}})_\alpha \times C(\tilde{\eta})_\alpha)} \frac{f(\mathbf{x}|\theta)\, \pi(\theta)}{n(\mathbf{x}, \eta)}, \tag{15}$$

$$g_\alpha^U(\theta) = \max_{(\mathbf{x}, \eta \in C(\tilde{\mathbf{x}})_\alpha \times C(\tilde{\eta})_\alpha)} \frac{f(\mathbf{x}|\theta)\, \pi(\theta)}{n(\mathbf{x}, \eta)}, \tag{16}$$

where $n(\mathbf{x}, \eta)$ is a normalizing constant. Calculations of (15) and (16) are significantly simplified in the case of conjugate priors that depend monotonically both on $\mathbf{x}$ and η. Having these α-contours we can compute the integrals (13) and (14), and hence calculate the membership functions of $\tilde{\mathbf{R}}_0$ and $\tilde{\mathbf{R}}_1$, respectively.

Further generalization may be achieved by assuming a vague character of losses $L(\theta, a_h)$. Let us assume that the fuzziness of $\tilde{L}(\theta, a_h)$ is expressed entirely through fuzzy values of its parameters with the exception of θ. In such a case we may calculate the α-contours of $\tilde{L}(\theta, a_h)$. Let us denote these contours by $L_\alpha^L(\theta, a_h)$ and $L_\alpha^U(\theta, a_h)$, respectively. Then we can calculate the α-cuts of the fuzzy risks $\tilde{\mathbf{R}}_{L,h}$, $h = 0, 1$ using the following formulae:

$$\tilde{\mathbf{R}}_{L,h}^{\alpha,L} = \int_{\Theta_h} L_\alpha^L(\theta, a_h)\, g_\alpha^L(\theta|\mathbf{x})\, d\theta \tag{17}$$

and

$$\tilde{\mathbf{R}}_{L,h}^{\alpha,U} = \int_{\Theta_h} L_\alpha^U(\theta, a_h)\, g_\alpha^U(\theta|\mathbf{x})\, d\theta. \tag{18}$$

This enables us to calculate the membership function of $\tilde{\mathbf{R}}_0$ and $\tilde{\mathbf{R}}_1$, respectively.

2.3 Bayes risk for crisp data and fuzzy hypotheses

In this subsection we present a method for the computation of fuzzy risks related to the test of the fuzzy hypothesis $\tilde{H}_0$ against a fuzzy alternative $\tilde{H}_1$

when all the remaining information (i.e. statistical data, prior information, and loss functions) are crisp.

Let $\widetilde{H}_h : \theta \in \widetilde{\Theta}_h$, $h = 1, 2$ be a fuzzy statistical hypothesis, where $\widetilde{\Theta}_h$ is a fuzzy set described by its membership function $\mu_{\Theta_h}(\theta)$. To simplify the problem let us assume that the fuzzy set $\widetilde{\Theta}_h$ may be presented in a form of a fuzzy interval $\left(\widetilde{\Theta}_{L,h}, \widetilde{\Theta}_{U,h}\right)$, where fuzzy sets $\widetilde{\Theta}_{L,h}$ and $\widetilde{\Theta}_{U,h}$ have α-cuts $\left[\Theta_{L,h}^{\alpha,L}, \Theta_{L,h}^{\alpha,U}\right]$ and $\left[\Theta_{U,h}^{\alpha,L}, \Theta_{U,h}^{\alpha,U}\right]$ such that $\Theta_{L,h}^{\alpha,L} \le \Theta_{U,h}^{\alpha,L}$ and $\Theta_{L,h}^{\alpha,U} \le \Theta_{U,h}^{\alpha,U}$. Denote the membership functions of $\widetilde{\Theta}_{L,h}$ and $\widetilde{\Theta}_{U,h}$ as $\mu_{L,h}(\theta)$ and $\mu_{U,h}(\theta)$, respectively. Using the notation of [8] we may write the membership function of the fuzzy risk $\widetilde{R}_h$ as

$$\mu_{\widetilde{R}_h}(z) = \sup_{u,w:z=\int_u^w L(\theta,a_h)g(\theta|\mathbf{x})d\theta} \min\left(\mu_{L,h}(u), \mu_{L,h}(w)\right) \tag{19}$$

The membership function $\mu_{\widetilde{R}_h}(z)$ may be also determined in a different way.

Let us notice that for a given α-cuts $\left[\Theta_{L,h}^{\alpha,L}, \Theta_{L,h}^{\alpha,U}\right]$ and $\left[\Theta_{U,h}^{\alpha,L}, \Theta_{U,h}^{\alpha,U}\right]$ we have

$$\mu_{\widetilde{R}_h}^{-1}(\alpha) = \int_{\Theta_{L,h}^{\alpha,L}}^{\Theta_{U,h}^{\alpha,U}} L(\theta, a_0) g(\theta|\mathbf{x}) d\theta \tag{20}$$

Hence, the membership function of $\widetilde{R}_h$ is equal to 1 for $R_h = \mu_{\widetilde{R}_h}^{-1}(1)$, and then decreases to 0 for $R_h = \mu_{\widetilde{R}_h}^{-1}(0)$.

2.4 Bayes risk for fuzzy data and fuzzy hypotheses

Finally, let us consider the most general case when we deal with fuzzy statistical data, fuzzy prior information, fuzzy loss function, and fuzzy statistical hypotheses. In this case the fuzzy risk $\widetilde{R}_h$ is given as an integral over a fuzzy set $\widetilde{\Theta}_h$ from a fuzzy function $\gamma(\theta) = \widetilde{L}_\alpha(\theta, a_h) \widetilde{g}_\alpha(\theta|\widetilde{\mathbf{x}})$, i.e.

$$\widetilde{R}_h = \int_{\widetilde{\Theta}_h} \widetilde{L}_\alpha(\theta, a_h) \widetilde{g}_\alpha(\theta|\widetilde{\mathbf{x}}) d\theta. \tag{21}$$

According to Dubois and Prade [8] it is very difficult to find the membership function of a fuzzy number which is calculated as an integral over a fuzzy set from a general fuzzy function. Therefore, we propose to evaluate the fuzzy risk $\widetilde{R}_h$ using the following formulae:

$$\widetilde{\mathbf{R}}_h^{\alpha, L} = \int_{\Theta_{L,h}^{\alpha, L}}^{\Theta_{U,h}^{\alpha, U}} L_\alpha^L(\theta, a_h) g_\alpha^L(\theta | \mathbf{x}) d\theta \qquad (22)$$

and

$$\widetilde{\mathbf{R}}_h^{\alpha, U} = \int_{\Theta_{L,h}^{\alpha, L}}^{\Theta_{U,h}^{\alpha, U}} L_\alpha^U(\theta, a_h) g_\alpha^U(\theta | \mathbf{x}) d\theta \qquad (23)$$

This result seems to be a reasonable upper approximation to the fuzzy risk function $\widetilde{R}_h$ in this very general case.

3 A possibilistic approach in decision making

In the classical (crisp) Bayes verification of statistical hypotheses the problem of decision making is of a secondary importance. If we are able to calculate risks R_0 and R_1 according to (9) and (8), respectively, we choose the action which is related to the smaller risk. Thus, we accept H_0 against H_1 if $R_0 < R_1$, and vice versa. When the respective risks are fuzzy there exists a real problem with taking an appropriate decision, as there is no natural ordering of fuzzy numbers that describe fuzzy risks. Thus, there is a need to use methods which allow us to order the computed fuzzy risks.

When fuzzy statistical data is described using the notion of the fuzzy information system (see [4]) the considered probability measures are weighted with the membership functions. Such an approach provides the decision maker with crisp evaluation of expected risks. A similar method of weighting has been proposed by Taheri and Behboodian [23] in the problem of testing fuzzy hypotheses for crisp data. The problem of defuzzification of fuzzy risks has been considered in the paper by Gil and Lopez-Diaz [12]. They used a very interesting method of ranking fuzzy numbers introduced by Campos and Gonzalez [1]. This particular method seems to be especially useful in decision making as it allows the decision maker to take into account his/hers personal attitude (optimistic, pessimistic or neutral).

In our paper we deal with a relatively simple problem of comparing two fuzzy numbers. Instead of applying any of the defuzzification methods we propose a possibilistic approach introduced by Dubois and Prade [9]. To compare both fuzzy risks $\widetilde{R}_0$ and $\widetilde{R}_1$ we propose to use the concept of the Necessity of Strict Dominance Index (NSD) and Possibility of Dominance Index (PD).

The *Possibility of Dominance* Index (PD) is defined for two fuzzy sets A and B as

$$PD = Poss\,(A \succeq B) = \sup_{x,y;x \geq y}\, \min\{\mu_A\,(x)\,,\mu_B\,(y)\}, \qquad (24)$$

where $\mu_A\,(x)$ and $\mu_B\,(y)$ are the membership functions of A and B, respectively. PD is the measure for a possibility that the set A is not dominated by the set B.

The *Necessity of Strict Dominance*Index (NSD) is used for the comparison of two fuzzy sets A and B and is defined as

$$NSD = Ness\,(A \succ B) = 1 - \sup_{x,y;x \leq y}\, \min\{\mu_A\,(x)\,,\mu_B\,(y)\} =$$
$$= 1 - Poss\,(B \geq A)\,. \qquad (25)$$

NSD represents a *necessity* that the set A dominates the set B.

Calculation of both indices for the relation $\tilde{R}_1 \succ \tilde{R}_0$ allows the decision maker to evaluate his/hers decision. If $NSD > 0$ there exists a strong indication that the acceptance of H_0 (or $\tilde{H}_0$) is preferred over the acceptance of H_1 (or $\tilde{H}_1$), and vice versa. The relation $NSD\left(\tilde{R}_1 \succ \tilde{R}_0\right) > 0$ is equivalent to the relation $\left(\tilde{R}_1^1\right)_L > \left(\tilde{R}_0^1\right)_U$. It means that for the most plausible values of both risks, i.e. for the values with the associated membership function equal to one, the risk of accepting H_1 (or $\tilde{H}_1$) is greater than that of accepting H_0 (or $\tilde{H}_0$). If $NSD\left(\tilde{R}_1 \succ \tilde{R}_0\right) = 1$, then there is no doubt that H_0 (or $\tilde{H}_0$) should be accepted instead of H_1 (or $\tilde{H}_1$). The relation $PD\left(\tilde{R}_1 \succ \tilde{R}_0\right) > 0$ is equivalent to the relation $\left(\tilde{R}_1^0\right)_U > \left(\tilde{R}_0^0\right)_L$. It means that for the least plausible values of both risks, i.e. for the values with the associated membership function equal to zero, the risk of accepting H_1 (or $\tilde{H}_1$) may be greater than that of accepting H_0 (or $\tilde{H}_0$).

4   Discussion and conclusions

The proposed method for the Bayes statistical hypotheses testing in a fuzzy environment differs from other approaches presented by different authors mainly because of the used criterion for making decisions. We believe that there exist at least two advantages of the proposed approach. First, we can assess the impact of vagueness (in statistical data, evaluation of losses or utilities, and stating hypotheses) on our decisions. In contrast to the crisp case the decisions are made basing on the analysis of vague information. Therefore, as the result of the analysis quite different solutions may be proposed with associated levels of possibility. We can find an interpretation of the computed possibility and necessity indices in the framework of the recently rapidly developing theory of preference relations. Let $\mu\,(x,y)$ be a measure of the preference of x over y. The preference relation is complete (see [5] for further references) when

$$\mu\left(x, y\right) + \mu\left(y, x\right) \geq 1 \; \forall x, y. \tag{26}$$

For a complete set of preference relations the measure of the indifference between the alternatives x and y is defined (see [5]) as

$$\mu_I\left(x, y\right) = \mu\left(x, y\right) + \mu\left(y, x\right) - 1, \tag{27}$$

the measure that the alternative x is better than y is defined as

$$\mu_B\left(x, y\right) = \mu\left(x, y\right) - \mu_I\left(x, y\right), \tag{28}$$

and the measure that the alternative x is worse than y is given by

$$\mu_W\left(x, y\right) = \mu\left(y, x\right) - \mu_I\left(x, y\right). \tag{29}$$

Suppose that the PD index is used as the measure of preference of x over y. Then, it is easy to show that

$$\mu_I\left(x, y\right) = PD - NSD, \tag{30}$$

$$\mu_B\left(x, y\right) = NSD, \tag{31}$$

$$\mu_W\left(x, y\right) = 1 - PD. \tag{32}$$

Thus, the knowledge of the indices PD and NSD is sufficient for the complete description of preference relations for two alternatives: to choose the hypothesis H_0 or to choose the alternative hypothesis H_1. We believe that the measure of indifference given by (30) is especially useful for a decision maker. Its value shows how both hypotheses are indistinguishable in the light of available information.

There is also another advantage, in our opinion, of the proposed approach. It is related to the problem which is very rarely considered in the literature on decision making. When we use subjective assessments there is always a possibility that these assessments are interrelated or interconnected. In all mathematical derivations we have tacitly assumed that all considered fuzzy sets are not interconnected. However, it could not be always the case. The interconnections may change the values of respective membership functions. We believe, however, that these changes are relevant only for low α-cuts. The NSD index proposed in this paper for the comparison of the considered hypotheses seems to be robust against possible interconnections of fuzzy sets describing subjectively assessed data.

The results presented in this paper together with the procedures suggested by many other authors provide decision makers with a wide range of supporting tools. There exists, however, the need for further investigations. The most difficult problem - in our opinion - is related to interconnections of fuzzy sets. When we use subjective opinions, vague perceptions and linguistic assessments there is always a possibility that the fuzzy sets used for their description are interconnected. The formal descriptions of these interconnections seems to be an extremely difficult task. Thus, we need methods that are robust to these unknown interdependencies. Another possible area of investigations is related to the fundamental problem of the applicability of expected values in decision making. It seems to be quite possible that in presence of significant vagueness we should use methods from the possibility theory.

In most papers on the decision making in the fuzzy environment it is assumed that the recommended decisions (actions) are crisp. In real problems decision makers ask experts for their advises. When suggested decisions are not obvious the advisors give their opinions in a more or less vague way. This vagueness of opinions reflects the vagueness of the decision problem itself. A user-friendly computer decision support system should communicate with a decision maker using a language which is understandable to him. In the paper of Grzegorzewski and Hryniewicz [14] simple way is proposed of presenting the conclusions of fuzzy statistical tests in a natural language. We believe that the problems of this type should be also investigated in the future.

References

1. Campos L.M., Gonzalez A. (1989) A subjective approach for ranking fuzzy numbers. Fuzzy Sets and Systems, **29**, 145-153
2. Casals M.R. (1993) Bayesian testing of fuzzy parametric hypotheses from fuzzy information. RAIRO, Operations Research, **27**, 189-199
3. Casals M.R., Gil M.A., Gil P. (1986) On the use of Zadeh's probabilistic definition for testing statistical hypotheses from fuzzy information. Fuzzy Sets and Systems, **20**, 175-190
4. Casals M.R., Gil M.A., Gil P. (1986) The Fuzzy Decision Problem: an approach to the Problem of Testing Statistical Hypotheses with Fuzzy Information. European Journ. of Oper. Res., **27**, 3
5. Cutell V., Montero J. (1999) An Extension of the Axioms of Utility Theory Based on Fuzzy Rationality Measures. In: Preference and Decisions under Incomplete Knowledge, J.Fodor, B.De Baets, P.Perny (Eds.), Physica-Verlag, Heidelberg, 33–50.
6. De Groot M.H (1970) Optimal Statistical Decisions. McGraw Hill, New York
7. Delgado M., Verdegay J.L., Vila M.A. (1985) Testing fuzzy hypotheses. A Bayesian approach. In: M.M.Gupta, A.Kandel, W.Bandler, J.B.Kiszka (Eds.), Approximate Reasoning in Expert Systems. Elsevier, Amsterdam, 307-316
8. Dubois D., Prade H. (1980) Fuzzy Sets and Systems. Theory and Applications. Academic Press, New York
9. Dubois D., Prade H. (1983) Ranking fuzzy numbers in the setting of possibility theory. Information Sciences, **30**, 184-244

10. Frühwirth-Schatter S. (1993) Fuzzy Bayesian inference. Fuzzy Sets and Systems, **60**, 41-58
11. Gil M.A. (1988) Probabilistic-Possibilistic approach to some Statistical Problems with Fuzzy Experimental Observations. In: J.Kacprzyk, M.Fedrizzi (Eds.). Combining Fuzzy Imprecision with Probabilistic Uncertainty in Decision Making. Springer-Verlag, Berlin, 286-306
12. Gil M.A., Lopez-Diaz M. (1996) Fundamentals and Bayesian Analyses of Decision Problems with Fuzzy-Valued Utilities. International Journal of Approximate Reasoning, **15**, 203-224
13. Grzegorzewski P. (2000) Testing statistical hypotheses with vague data. Fuzzy Sets and Systems, **112**, 501-510
14. Grzegorzewski P., Hryniewicz O. (1999) Lifetime tests for vague data. In: L.A.Zadeh, J.Kacprzyk (Eds.), Computing with Words in Information/Intelligent Systems, 2, Applications. Physica Verlag,Heidelberg, 176-193
15. Grzegorzewski P., Hryniewicz O. (2001) Soft Methods in Hypotheses Testing. In: D. Ruan, J. Kacprzyk and M. Fedrizzi (Eds.), Soft computing for risk evaluation and management. Physica Verlag, Heidelberg and New York, 55-72
16. Hryniewicz O. (1988) Estimation of life-time with fuzzy prior information: application in reliability. In: J.Kacprzyk, M.Fedrizzi (Eds.), Combining Fuzzy Imprecision with Probabilistic Uncertainty in Decision Making. Springer-Verlag, Berlin, 307-321
17. Kruse R. (1982) The strong law of large numbers for fuzzy random variables. Information Sciences. **28**,233–241
18. Kruse R., Meyer K.D. (1987) Statistics with Vague Data. Riedel, Dodrecht
19. Kwakernaak H. (1978) Fuzzy random variables, part I: definitions and theorems, Information Sciences, **15**, 1–15; Part II: algorithms and examples for the discrete case.Information Sciences, **17**, 253–278
20. Puri M.L., Ralescu D.A. (1986) Fuzzy random variables. J. Math. Anal. Appl., **114**, 409-422
21. Raiffa H., Schleifer R. (1961) Applied Statistical Decision Theory, The M.I.T. Press, Cambridge
22. Tanaka H., Okuda T., Asai K. (1979) Fuzzy Information and Decisions in Statistical Model. Advances in Fuzzy Sets Theory and Applications, North-Holland, 303-320
23. Taheri S.M., Behboodian J. (2001) A Bayesian approach to fuzzy hypotheses testing. Fuzzy Sets and Systems, **123**, 39-48
24. Viertl R. (1987) Is it necessary to develop a fuzzy Bayesian inference. In: R.Viertl (Ed.), Probability and Bayesian Statistics, Plenum Publishing Company, New York, 471-475
25. Viertl R., Hule H. (1991) On Bayes' theorem for fuzzy data. Statistical papers, **32**, 115-122
26. Zadeh L.A. (1978) Fuzzy sets as a basis for a Theory of Possibility. Fuzzy Sets and Systems, **1**, 3-38
27. Zhong C., Zhou G. (1987) The equivalence of two definitions of fuzzy random variables. In: Proceedings of the 2nd IFSA Congress, Tokyo, 59-62.

Fuzzy Comparison of Frequency Distributions

Mark Last [1] - Abraham Kandel[2]

[1]Ben-Gurion University of the Negev, Department of Information Systems Engineering, Beer-Sheva 84105, Israel
[2]University of South Florida, Department of Computer Science and Engineering, 4202 E. Fowler Avenue, ENB 118, Tampa, FL 33620, USA
E-mail: mlast@bgumail.bgu.ac.il , kandel@csee.usf.edu

Abstract. Comparing empirical distributions is one of the fundamental tasks in data analysis. We start with a survey of existing statistical approaches to this problem. The current numeric methods are shown to suffer from several limitations, including restrictive assumptions about the underlying distributions and non-use of available domain knowledge. These limitations can be partially overcome via the time-consuming visual examination of frequency histograms by a human expert. In this paper, we present a fuzzy-based method for automating the process of comparing frequency histograms. Our approach builds upon a novel concept of automated perceptions, introduced in our previous work. We use the evolving approach of type-2 fuzzy logic for representing the domain knowledge of human experts. The proposed method provides an automated interpretation of the differences between histogram plots, based on a cognitive model of human perception. The perception-based approach to comparison of frequency histograms is demonstrated on several samples of real-world data.

Keywords: fuzzy logic, soft computing, comparing distributions, data visualization.

1 Introduction

According to the famous saying "a picture is worth a thousand words", most people find it easier to draw conclusions from graphically presented data than from numeric results of statistical analysis. A frequency histogram (see Mendenhall et al., 1993) is known as a powerful tool of data visualization. The procedure for constructing a frequency histogram is straightforward and it can be implemented by a single pass over data, without using any computerized tools. By observing a histogram, users can see at a glance, what are the most frequent and the most rare intervals of the attribute in question. Moreover, we can compare pairs of histograms, if they are defined on the same range of values. Our visual conclusion may be that the two distributions are identical or one of them is shifted to the right (or to the left) of the other by a small, a moderate or a large magnitude. The problems of comparing two or more distributions include marketing surveys, analysis of engineering experiments, choosing study programs for students, etc.

However, the visualization methods, even supported by the state-of-the-art computer graphics, suffer from a number of serious limitations. First, this is a truly subjective approach: the same data may be represented in different, sometimes misleading, ways, and people may come to different conclusions even from looking at the same presentations of data. The second limitation of the visual data analysis is its poor scalability. Due to the data explosion, we are facing in the last decade, hundreds of frequency distributions (each related to a different attribute) can be extracted from an average modern database. Manual examination of the resulting multi-dimensional multi-color charts is an extremely time-consuming task even for the most experienced statisticians. Consequently, there is a strong need of automated tools for distribution analysis.

Over years, several statistical methods for comparing distributions have been developed. If the data samples are normally distributed, we can compare their mean values by using the t-test (see Minium et al., 1999). The magnitude of the difference between the means (called the effect size) can be measured in the number of pooled standard deviations. The t test assumption of the Normal distribution is justified by the Central Limit Theorem. The main problem with calculating the means of data samples is their sensitivity to outliers: one erroneous value (higher or lower than the others by the order of magnitude) may cause a significant bias in the result. Of course, this problem is easily overcome by the visual analysis: the frequency histogram is an excellent tool for detecting and ignoring outliers.

Several non-parametric methods exist for comparing pairs of distributions (see Hajek et al., 1999). These include the signed-rank (Wilcoxon) test, the median test, and the Kolmogorov-Smirnov test. The non-parametric tests use minimal assumptions about the general shape of the underlying distributions and their purpose is to detect a shift in one distribution with respect to the other, when the distribution densities are identical. To perform the comparison, we need to form a pooled sample of observations from the two distributions and sort the observations by their value, which makes the computational complexity highly dependent on the total number of cases. There is no way to utilize existing knowledge of domain experts in the non-parametric methods.

All parametric and non-parametric tests mentioned above have a binary ("crisp") outcome: the null hypothesis stating that distributions are identical is either rejected, or not. Given a pre-specified significance level (e.g., 1% or 5%), the null hypothesis is rejected only if the test statistic exceeds the corresponding threshold value. The distance between the actual value of the statistic and the threshold value has no importance for the test outcome. Consequently, statistical tests cannot be used directly for measuring the extent of the shift between distributions.

The gap between the limitations of "crisp" statistical methods and the "soft" nature of graphical representation, appealing to the human intuition, may be bridged by using the principles of fuzzy logic (see Klir and Yuan, 1995). Human observations (e.g, "the values of the second distribution tend to be slightly higher than the values of the first one") are usually expressed in qualitative, linguistic terms. Fuzzy logic suggests a mathematical model of linguistic concepts, defined on observed data, by associating a membership function with each linguistic variable. Thus, one can calculate the membership grade of any difference between sample means in the fuzzy set "slightly higher". The form of a membership function (the "linguistic context") is determined by the existing prior knowledge about the phenomenon in question. As demonstrated by Pedrycz (1998), the linguistic context is a powerful tool of data filtering, which can be used for eliminating meaningless (though statistically significant) results. In (Pedrycz, 1996), the fuzzy logic approach is applied to discovering multiple functional dependencies between input and output variables, a task, which is much easier for a human eye than for numeric statistical techniques.

A specific class of linguistic variables, called "linguistic quantifiers", is described by Yager (1996). The linguistic quantifiers express compatibility of proportions, calculated from the raw data, with terms like "most", "few", etc. The concept of linguistic quantifiers seems to be particularly useful for distributions comparison, which is based on comparing pairs of individual proportions. Moreover, the user is usually more interested in discovering linguistic rules (like "engineers are more likely to be credible customers" or "students who are excellent in math, are expected to have higher grades in CS") rather than being presented with numbers and significance levels.

In this paper, we are presenting a novel approach to automating the comparison between frequency distributions by using the Fuzzy Set theory. In Section 2, we describe the cognitive process of comparing frequency histograms. Section 3 presents the fuzzy logic model of applying this process to frequency data. A practical example, based on real-world manufacturing data, is shown in Section 4. Potential directions for integrating the perception-based approach with other statistical tasks are briefly discussed in Section 5.

2 The Cognitive Process of Comparing Distributions

As indicated by (Minium et al., 1999), the key characteristics of a frequency histogram include *central tendency, variability,* and *shape.* For humans, the easiest way of comparing empirical distributions is by observing the distribution histograms. The cognitive process of comparing the central tendency of two different histograms can be summarized as follows:

- Step 1 – If in most intervals there is no significant difference between the proportions conclude that there is no shift in the attribute values. Otherwise, go to step No. 2.
- Step 2 – Find an imaginary threshold point between the intervals, such that below the threshold, most proportions of the first histogram are significantly higher (lower) than the proportions of the second one and vice versa. This means that the values of the first histogram are shifted to the left (right) with respect to the second histogram.

- Step 3 – Make the final conclusion about a positive or a negative shift in the central tendency of the target distribution, based upon the apparent shift in the histogram, the sample size, and the personal expertise (if available).

The cognitive process, discussed above, is not based on any statistical assumptions about the behavior of the underlying distributions. Still, as indicated by us in (Last and Kandel, 1999), the human perception is very efficient when dealing with the uncertainty of visual representations. The human conclusions tend to bear some amount of vagueness and are much easier to be described by words (e.g., "most", "significantly", etc.), rather than by some strict mathematical terms. Thus, the histograms comparison can be seen as a particular case of *Approximate* (or *Fuzzy*) *Reasoning* (see Kandel et al., 1996). Consequently, we are using the Fuzzy Logic approach to model this data analysis process.

3 Comparing Distributions by the Fuzzy Logic Approach

The fuzzy procedure of comparing histograms includes the following steps:
- Determine the primary membership function of the difference between individual proportions.
- Determine the secondary membership function of the above primary membership function based on the number of available observations.
- Evaluate the fuzzy shift between distributions (excluding and including the use of domain knowledge). In this paper, we limit our discussion to the comparison process *without* any use of domain knowledge.

3.1 Evaluating the Fuzzy Difference between Proportions

We assume here that the linguistic variable *proportion change* (denoted by d) can take the following two linguistic values: *bigger* and *smaller*, each being a fuzzy set. Of course, our language contains much more terms for describing a change, like *equal*, *nearly equal*, *much bigger*, etc. However, these two values prove to be sufficient for our purpose of detecting trends in distributions rather than analyzing changes in individual proportions. Since the proportion change is the difference between two proportions (each varying between 0 and 1), it can take any value in the range [-1, 1]. The membership function μ_B associated with the fuzzy set *bigger* should have the following properties:
- Being close to zero, when d is close to -1.
- Being low for $d = 0$.
- Being close to 1, when d is close to 1.

On the other hand, the membership function μ_S (*smaller*) should satisfy the opposite properties (e.g., being close to 1, when d is close to -1).

In our model, the following membership functions will be used for μ_S and μ_B:

$$\mu_S(d) = \frac{1}{1 + e^{\beta d}}, d \in [-1, 1], \beta \geq 0$$

$$\mu_B(d) = \frac{1}{1 + e^{-\beta d}}, d \in [-1, 1], \beta \geq 0$$

Where

d – the difference between measured proportions (relative frequencies) of the same target value in compared distributions.

β - the shape factor, which can change the shape of the membership function from a horizontal line ($\beta = 0$) to the step function ($\beta \rightarrow \infty$). It is associated with the sample size, used for calculating the proportions.

A similar, non-parametric form of membership functions for "bigger" and "smaller" is used by Wang (1997) for defining *gradual rules*. The following lemmas can be easily derived from the above definitions of membership functions:

Lemma 1. When the compared proportions are equal ($d = 0$), the membership functions of "bigger" and "smaller" are equal to each other and can be calculated by:

$$\mu_B = \mu_S = \frac{1}{1 + e^{\beta}}$$

Lemma 2. The membership function of "not smaller" ($1 - \mu_S$) is equal to the membership function of "bigger" (μ_B) and the membership function of "not bigger" ($1 - \mu_B$) is equal to the membership function of "smaller" (μ_S).

3.2 Determining the Form of the Membership Function

The actual membership grade of a given proportion change d depends on the value of the β coefficient. In other words, the membership grade itself is *uncertain* given d. This type of second-order uncertainty in determining membership grades is represented by the *Type-2 Logic System* (Karnik and Mendel, 1998), where primary memberships are the domain-elements of a membership grade and secondary memberships are membership grades of primary memberships.

The effect of the sample size on the membership grade is quite intuitive: the more examples we have the higher is our confidence in the value of a statistical estimate. The extreme case is having no examples at all, implying that both membership functions are constant in d and equal (there is no information about the difference between proportions). These variations in the form of the secondary membership function can be represented by the shape factor β. Thus, we define the factor β to be proportional to the number of examples. Since each compared distribution may be based on a different number of examples, we use the *minimum* between the two sample sizes for estimating β. The linear coefficient relating the number of examples to the shape factor β is denoted by as γ. Consequently, the expression for calculating β is:

$\beta = \gamma n_{min}$

Where n_{min} is the minimum number of examples in one of the two compared distributions.

Furthermore, the expressions for the secondary membership of d as a function of n become:

$$\mu_S(d) = \frac{1}{1 + e^{\gamma n_{min} d}}, d \in [-1, 1], \gamma \geq 0$$

$$\mu_B(d) = \frac{1}{1 + e^{-\gamma n_{min} d}}, d \in [-1, 1], \gamma \geq 0$$

Still, there is a question of how the linear coefficient γ can be determined. As indicated by Karnik and Mendel (1998), we would need an infinite number of approximations to completely represent uncertainty, which, of course, is not practical. Therefore, in this study we have chosen the crisp value of $\gamma = 0.2$ to represent our subjective perception of sample size in comparison of histograms.

3.3 Calculating the Fuzzy Shift between Distributions

After calculating the membership grades of each proportion change in the "smaller" and the "bigger" fuzzy sets, we can evaluate the *shift* between the compared distributions. The following situations are possible:

- There is a *negative (positive) shift*. The values of the second distribution tend to be *lower (higher)* than the values of the first one. This means that there is a threshold point between a pair of histogram intervals. All the proportions below the threshold point are *bigger (smaller)* in the second distribution than in the first one. Above the threshold point we have an opposite situation: the proportions of the second distribution are *smaller (bigger)* than the proportions of the first one.
- There is *no shift*. The distributions have the same central tendency (though they still may differ in their variability and shape). No threshold point can be found for either a negative shift or a positive shift.

According to the above definition of the threshold point, the number of candidate thresholds is $D - 1$, where D is the number of intervals in the histogram of the attribute in question. Each threshold $T \in D$ separates between the intervals $i = 1,..., T$ and $i = T+1,..., D$. We calculate the *net shift* for a candidate threshold T by the following expression:

$$NS(T) = \sum_{i=1}^{T} [\mu_S(d_i) - \mu_B(d_i)] + \sum_{i=T+1}^{D} [\mu_B(d_i) - \mu_S(d_i)]$$

Where d_i is the proportion change for the interval No. i.

Both sum terms of the above expression will be positive if there is a positive shift in the distribution and negative in the opposite case. When there is no shift, both terms will be close to zero. It can be easily shown that the value of the net shift $NS(T)$ over the range of D intervals varies between $-D$ (the lowest possible value) and D (the highest possible value). We use the maximal and the minimal values of NS (T) to normalize the detected shifts, disregarding the number of distribution intervals.

We find the threshold T^* providing the maximal *absolute* value of the net shift by $T^* = arg\ max_T\ |NS\ (T)|$ and then normalize the net shift $NS\ (T^*)$ w. r. t. the number of distribution intervals by $NS_{Norm} = NS\ (T^*)\ /\ D$. The calculated net shifts NS^* may be used to sort the effects in the descending order of distribution shifts (from the highest positive to the lowest negative). With traditional statistical methods, the histogram pairs can only be categorized as "identical" or "different".

4 Case Study: Yield Analysis

We have applied the fuzzy approach to comparing histograms of yield distribution for several families of semiconductor products. The "yield", defined as the proportion of good microchips obtained from a silicon wafer, is the most important performance indicator in semiconductor industry. Today, the analysis of yield data is based mainly on manual inspection of yield histograms by engineers who monitor the manufacturing process.

The normalized net shift NS_{norm} for six histogram-pairs is shown in the left part of Table 1. All histograms are based on nine intervals of equal width. The number of observations in a histogram varies between 200 and 800 batches.

The lowest negative shift (–0.660) has been obtained for the variable *Tol1_2*. The variable *Cur0_1* has the highest positive shift (0.426). The histogram shifts in other variables are closer to zero, but their absolute values are large enough to suggest that the detected shift in central tendency is not random. This conclusion can be further supported (or refuted) by the existing expert knowledge.

Table 1 Summary of Results

			p-values		
Attribute	NS_norm	t-means	rank-sum	median	K-S
Size_C	0.383	0.0004	** 0.0000	** 0.0000	** 0.0000 **
Tol0_1	0.216	0.0882	0.0000	** 0.0000	** 0.0000 **
Tol1_2	-0.660	0.0000	** 0.2802	0.0802	0.0001 **
Size_F	-0.541	0.0001	** 0.0000	** 0.0000	** 0.0000 **
Cur0_1	0.426	0.0012	** 0.0018	** 0.4312	0.0000 **
Cur1_2	-0.404	0.0352	* 0.0008	** 0.0235	* 0.0000 **

We have compared the results of the perception-based approach to several statistical tests, which include t-means (based on Minium et al., 1999) and three non-parametric tests, namely the median test, Wilcoxon rank-sum test, and the Kolmogorov-Smirnov test (all described by Hajek et al., 1999). For each test, we show the *p-value*, which is the probability of obtaining the same or more deviant result of the test statistic under the assumption that the samples were drawn from identical distributions. One and two asterisks denote the 5% and the 1% significance levels respectively.

The statistical tests do not appear to be completely consistent with each other. Thus, the effect of *Tol0_1* is not recognized by *t-means*, while *rank-sum* and *median* have not detected any difference for *Tol1_2*. Generally, we can say that the Kolmogorov-Smirnov (K-S) test is most consistent with the fuzzy approach, since it has detected a significant shift for all attributes. However, as we have indicated above, K-S and other statistical tests are "crisp": they do not provide any "soft" information about the extent of shift between distributions. In the case of yield analysis, this information is critical for planning the efforts to increase the throughput of the manufacturing process.

5 Conclusions

In this paper, we have developed a novel, fuzzy-logic method of comparing frequency distributions. The method provides an automated interpretation of histogram plots, which is based on a general model of human perception and available domain knowledge. The perception-based results are shown to be more consistent and informative than the results of statistical tests. The proposed approach to histogram analysis can be enhanced by associating more linguistic terms with the difference between proportions (e.g., "much bigger", "slightly smaller", etc.). The soft computing methodology can be extended to other graphical forms of data analysis and integrated with several methods of data mining (e.g., feature selection and association rules).

References

J. Hajek, Z. Sidak, P.K. Sen, Theory of Rank Test, Academic Press, 1999.
A. Kandel, R. Pacheco, A. Martins, and S. Khator, The Foundations of Rule-Based Computations in Fuzzy Models. In: Fuzzy Modelling, Paradigms and Practice, W. Pedrycz, Eds., Kluwer, Boston, pp. 231-263, 1996.
N.N. Karnik and J.M. Mendel, An Introduction to Type-2 Fuzzy Logic Systems, Internal Report, 1998.
G. J. Klir and B. Yuan, Fuzzy Sets and Fuzzy Logic: Theory and Applications, Prentice-Hall Inc., Upper Saddle River, CA, 1995.
M. Last and A. Kandel, Automated Perceptions in Data Mining, Proc. of 1999 IEEE International Fuzzy Systems Conference, pages 190-197. IEEE Press, 1999.

W. Mendenhall, J.E. Reinmuth, R.J. Beaver, Statistics for Management and Economics, Duxbury Press, Belmont, CA, 1993.

E.W. Minium, R.B. Clarke, T. Coladarci, Elements of Statistical Reasoning,Wiley, New York, 1999.

W. Pedrycz, Fuzzy Multimodels, IEEE Transactions on Fuzzy Systems, 4, 2, 139-148, 1996.

W. Pedrycz, Fuzzy Set Technology in Knowledge Discovery, Fuzzy Sets and Systems, 98, 3, 279-290, 1998.

L.-X. Wang, A Course in Fuzzy Systems and Control, Prentice-Hall, Upper Saddle River, NJ, 1997.

R.R. Yager, Database Discovery Using Fuzzy Sets, International Journal of Intelligent Systems, 11, 691-712, 1996.

Test of One-Sided Hypotheses on the Expected Value of a Fuzzy Random Variable*

Montenegro, M., Colubi, A., Casals, M.R., and Gil-Álvarez, M.A.

Dpto. de Estadística, I.O. y D.M., Universidad de Oviedo, 33071 Oviedo, Spain

Abstract. In this paper we present a procedure to test one-sided haypotheses about the population expected value of a fuzzy random variable. This procedure is based on a parameterized ranking function making the hypotheses being equivalent to classical ones for the population mean of a real-valued random variable.

1 Introduction

In previous papers (Montenegro *et al.*, 2001, 2002) we have analyzed the problem of testing "two-sided" hypotheses on the population (fuzzy) expected value of a fuzzy random variable. The techniques to test such a type of hypotheses have been based on an operational generalized metric on the space of fuzzy numbers with compact support.

However, these techniques cannot be applied to test one-sided hypotheses on this expected value. In fact, one-sided hypotheses do not make a well-defined sense in case of fuzzy random variables since to rank fuzzy numbers we have to specify an ordering/preordering among them.

For this purpose we can consider a suitable ranking function (like the parameterized one introduced by Campos and González, 1989). López-Díaz and Gil (1998) have proved this function is easy to compute, and when it is applied to the expected value of a fuzzy random variable, we obtain the classical expected value of a real-valued random variable. On the basis of the last result, we can reduce the problem of testing one-sided hypotheses on the population mean of a fuzzy random variable to the problem of testing the mean of a real-valued random variable.

In this paper we first present some possible procedures to test the one-sided hypotheses on the population mean of a fuzzy random variable. We illustrate later one of these procedures with an example. Finally, we will make some remarks to compare the approach in this paper with the one in previous ones (Montenegro *et al.*, 2001, 2002).

---

* The research in this paper has been partially supported by MCYT Grants DGE-PB98-1534 and BFM2001-3494. Their financial support is gratefully acknowledged.

2   Preliminaries

Let $\mathcal{K}_c(\mathbb{R})$ denote the class of nonempty compact intervals. Let $\mathcal{F}_c(\mathbb{R})$ be the space of fuzzy numbers with compact support, that is, $\mathcal{F}_c(\mathbb{R}) = \{\tilde{A} : \mathbb{R} \to [0,1] \mid \tilde{A}_\alpha \in \mathcal{K}_c(\mathbb{R})\}$ for all $\alpha \in [0,1]\}$, where $\tilde{A}_\alpha = \{x \in \mathbb{R} \mid \tilde{A}(x) \geq \alpha\}$ for $\alpha \in (0,1]$ and $\tilde{A}_0 = \mathrm{cl}\{x \in \mathbf{R} \mid \tilde{A}(x) > 0\}$.

Given a probability space $(\Omega, \mathcal{A}, P)$, a mapping $\mathcal{X} : \Omega \to \mathcal{F}_c(\mathbb{R})$ is said to be a *fuzzy random variable* (FRV for short) associated with this space in Puri and Ralescu's sense (1986) iff the α-level mapping defined so that $\mathcal{X}_\alpha(\omega) = \big(\mathcal{X}(\omega)\big)_\alpha$ for all $\omega \in \Omega$, is a random compact convex set whatever $\alpha \in [0,1]$ may be.

A fuzzy random variable $\mathcal{X} : \Omega \to \mathcal{F}_c(\mathbb{R})$ is said to be integrably bounded if, and only if, $\|\mathcal{X}_0\| \in L^1(\Omega, \mathcal{A}, P)$ (with $\|\mathcal{X}_0\|(\cdot) = \sup_{x \in \mathcal{X}_0(\cdot)} |x|$). The *expected value* of an integrably bounded FRV $\mathcal{X}$ is the fuzzy number $\tilde{E}(\mathcal{X})$ such that $\big(\tilde{E}(\mathcal{X})\big)_\alpha =$ Aumann integral (1965) of $\mathcal{X}_\alpha$ (and, because of the convexity of values of $\mathcal{X}$, this is equivalent to say that $\big(\tilde{E}(\mathcal{X})\big)_\alpha = \big[E(\inf \mathcal{X}_\alpha), E(\sup \mathcal{X}_\alpha)\big]$) for all $\alpha \in [0,1]$).

In accordance with Campos and González (1989), given $\tilde{A}, \tilde{B} \in \mathcal{F}_c(\mathbb{R})$ and $\lambda \in [0,1]$, $\tilde{A}$ is said to be *greater than or equal to* $\tilde{B}$ in the λ-*average sense* (and it is denoted by $\tilde{A} \succeq_\lambda \tilde{B}$) if, and only if, $V^\lambda(\tilde{A}) \geq V^\lambda(\tilde{B})$, where $V^\lambda(\tilde{A}) = \int_{[0,1]} \big(\lambda \sup \tilde{A}_\alpha + (1-\lambda) \inf \tilde{A}_\alpha\big) d\alpha$. In case $\tilde{A} \succeq_\lambda \tilde{B}$ but not $\tilde{B} \succeq_\lambda \tilde{A}$, we will write $\tilde{A} \succ_\lambda \tilde{B}$.

The parameterized ranking function V^λ leads to a reasonable ranking and it is quite convenient for computational purposes. In particular, when we deal with the expected value of a FRV, we obtain (see López-Díaz and Gil, 1998) that whenever $\tilde{E}(\mathcal{X})$ exists we have that

$$V^\lambda(\tilde{E}(\mathcal{X})) = E(V^\lambda \circ \mathcal{X}).$$

3   Stating the problem of testing one-sided hypotheses on the expected value of a FRV

Given a random sample of n independent observations, $\mathcal{X}_1, \ldots, \mathcal{X}_n$, from $\mathcal{X}$, the aim of this paper is testing (for a fixed subjective choice of the parameter λ) the null "one-sided" hypothesis $H_0 : \tilde{E}(\mathcal{X}) \succeq_\lambda \tilde{A}$ versus the alternative "one-sided" hypothesis $H_1 : \tilde{A} \succ_\lambda \tilde{E}(\mathcal{X})$ for a given $\tilde{A} \in \mathcal{F}_c(\mathbb{R})$.

On the basis of the definition of the ranking relation and the result by López-Díaz and Gil, the preceding problem is equivalent to that of testing the one-sided hypothesis $H_0 : E(V^\lambda \circ \mathcal{X}) \geq V^\lambda(\tilde{A})$ versus the one-sided hypothesis $H_1 : E(V^\lambda \circ \mathcal{X}) < V^\lambda(\tilde{A})$, which is a classical problem of testing on the mean of the real-valued random variable $V^\lambda \circ \mathcal{X}$ based on the random sample $V^\lambda \circ \mathcal{X}_1, \ldots, V^\lambda \circ \mathcal{X}_n$.

4 Solving the problem of testing one-sided hypotheses on the expected value of a FRV

The problem above can be solved by using classical procedures. In this respect, we can find in the Classical Inference literature different methods.

The methods based on the *normality* of the involved FRVs are now applicable. Thus, if $\mathcal{X}$ is a normal FRV with variance σ^2 in Puri and Ralescu's sense (1985), that is $\mathcal{X}_\alpha(\omega) = X(\omega) \oplus (\widetilde{E}(\mathcal{X}))_\alpha$ for all $\omega \in \Omega$, $\alpha \in [0,1]$, where X is a real-valued random variable having normal distribution $\mathcal{N}(0, \sigma^2)$, then $V^\lambda \circ \mathcal{X}$ is a real-valued random variable having normal distribution $\mathcal{N}(V^\lambda(\widetilde{E}(\mathcal{X})), \sigma^2)$. Consequently, whatever the specified $\lambda \in [0,1]$ and $\widetilde{A} \in \mathcal{F}_c(I\!R)$ maybe, we have that

Theorem 1. *Given a random sample of n independent observations, $\mathcal{X}_1, \ldots, \mathcal{X}_n$, from a fuzzy random variable $\mathcal{X}$ having normal distribution with variance σ^2, to test at the significance level $\alpha \in [0,1]$ the null hypothesis $H_0 : \widetilde{E}(\mathcal{X}) \succeq_\lambda \widetilde{A}$ against the alternative $H_1 : \widetilde{A} \succ_\lambda \widetilde{E}(\mathcal{X})$, the hypothesis H_0 should be rejected whenever*

$$\frac{\overline{V^\lambda \circ \mathcal{X}} - V^\lambda(\widetilde{A})}{\sqrt{\sum_{i=1}^n \left[V^\lambda \circ \mathcal{X}_i - \overline{V^\lambda \circ \mathcal{X}}\right]^2 / n(n-1)}} > t_{n-1,\alpha},$$

where $t_{n-1,\alpha}$ is the $100(1-\alpha)$ fractile of Student's t-distribution with $n-1$ degrees of freedom, and $\overline{V^\lambda \circ \mathcal{X}}$ denotes the sample fuzzy mean of the n associated real-valued random variables (i.e., $\overline{V^\lambda \circ \mathcal{X}} = \sum_{i=1}^n V^\lambda \circ \mathcal{X}_i / n$).

However, the assumption of normality for $\mathcal{X}$ is not too realistic. Only a few practical situations could be properly modeled by this type of FRVs.

In fact, real-life FRVs are commonly *simple fuzzy random variables* (that is, they take on a finite number of different values), whence the associated real-valued random variable $V^\lambda \circ \mathcal{X}$ will be also simple.

To test the considered hypotheses for this type of variables we can make use, for instance, of *asymptotic techniques*. Thus, consider a probability space $(\Omega, \mathcal{A}, P)$. Let $\mathcal{X}$ be a fuzzy random variable associated with it, so that on Ω the fuzzy random variable takes on r different values, $\widetilde{x}_1, \ldots, \widetilde{x}_r$. For each $n \in N$, consider n independent fuzzy random variables having identical distribution that $\mathcal{X}$ on Ω. Let $\mathbf{f}_n = (f_1, \ldots, f_{r-1}) \in [0,1]^{r-1}$, with f_l = relative frequency of $\widetilde{x}_l$ ($l \in \{1, \ldots, r-1\}$) in the performance of the n fuzzy random variables. Let $\overline{V^\lambda \circ \mathcal{X}}$ denote the sample fuzzy mean of the n associated real-valued random variables. Then,

Theorem 2. *Given a random sample of n independent observations, $\mathcal{X}_1, \ldots, \mathcal{X}_n$, from a simple fuzzy random variable $\mathcal{X}$, to test at the significance level $\alpha \in [0,1]$ the null hypothesis $H_0 : \widetilde{E}(\mathcal{X}) \succeq_\lambda \widetilde{A}$ against the alternative*

$H_1 : \widetilde{A} \succ_\lambda \widetilde{E}(\mathcal{X})$, the hypothesis H_0 should be (asymptotically) rejected whenever $2n\left[\overline{V^\lambda \circ \mathcal{X}} - V^\lambda(\widetilde{A})\right] > \gamma_\alpha$, where γ_α is the $100(1 - \alpha)$ fractile of the linear combination of chi-square independent variables $\widehat{\lambda}_1 \chi^2_{1,1} + \ldots + \widehat{\lambda}_k \chi^2_{1,k}$, with $\widehat{\lambda}_1, \ldots, \widehat{\lambda}_k$ ($k \le r - 1$) being the nonnull eigenvalues of the matrix $B^t H\left(\left[\overline{V^\lambda \circ \mathcal{X}} - V^\lambda(\widetilde{A})\right]\right) B$, where $H\left(\left[\overline{V^\lambda \circ \mathcal{X}} - V^\lambda(\widetilde{A})\right]\right)$ is the Hessian matrix

$$
\begin{pmatrix}
\dfrac{\partial^2 \left[\overline{V^\lambda \circ \mathcal{X}} - V^\lambda(\widetilde{A})\right]}{\partial f_{n1} \partial f_{n1}} & \cdots & \dfrac{\partial^2 \left[\overline{V^\lambda \circ \mathcal{X}} - V^\lambda(\widetilde{A})\right]}{\partial f_{n1} \partial f_{n(r-1)}} \\
\vdots & \ddots & \vdots \\
\dfrac{\partial^2 \left[\overline{V^\lambda \circ \mathcal{X}} - V^\lambda(\widetilde{A})\right]}{\partial f_{n(r-1)} \partial f_{n1}} & \cdots & \dfrac{\partial^2 \left[\overline{V^\lambda \circ \mathcal{X}} - V^\lambda(\widetilde{A})\right]}{\partial f_{n(r-1)} \partial f_{n(r-1)}}
\end{pmatrix},
$$

and B is an $(r - 1) \times (r - 1)$ matrix such that $B^t B = \left(I_\mathcal{X}^F(\mathbf{f}_n)\right)^{-1}$, where $\left(I_\mathcal{X}^F(\mathbf{f}_n)\right)^{-1}$ is the inverse of the sample Fisher information matrix $\left[f_{nl}(\delta_{lm} - f_{nm})\right]_{lm}$.

5 Illustrative example

The conclusions in Theorem 2 are now illustrated by means of a real-life example, in which data were supplied by members of the Departamento de Medio Ambiente of the Consejería de Agricultura in the Principado de Asturias in Spain.

Example.
Consider the population of days of a given year, and consider a random sample of 50 days in which *visibility* (variable $\mathcal{X}$) has been observed.

Variable $\mathcal{X}$ takes on the values PERFECT ($\tilde{x}_1$), GOOD ($\tilde{x}_2$), MEDIUM ($\tilde{x}_3$), POOR ($\tilde{x}_4$), and BAD ($\tilde{x}_5$). Experts in the measurement of these values have described them in terms of the fuzzy sets (meaning fuzzy percentages) and based on S- and Π-curves, and triangular and trapezoidal fuzzy numbers, whose support is strictly contained in $[0, 100]$ as follows (see Figure 1):

$$\tilde{x}_1 = \text{Tra}(90, 95, 100, 100),$$
$$\tilde{x}_2 = \text{Tri}(70, 90, 100),$$

$$
\tilde{x}_3 = \begin{cases}
S(40, 50) & \text{in } [40, 50] \\
1 & \text{in } [50, 70] \\
1 - S(70, 80) & \text{in } [70, 80] \\
0 & \text{otherwise,}
\end{cases}
$$

$$\tilde{x}_4 = \begin{cases} S(20,30) & \text{in } [20,30] \\ 1 & \text{in } [30,40] \\ 1 - S(40,50) & \text{in } [40,50] \\ 0 & \text{otherwise,} \end{cases}$$

$$\tilde{x}_5 = S(0,20).$$

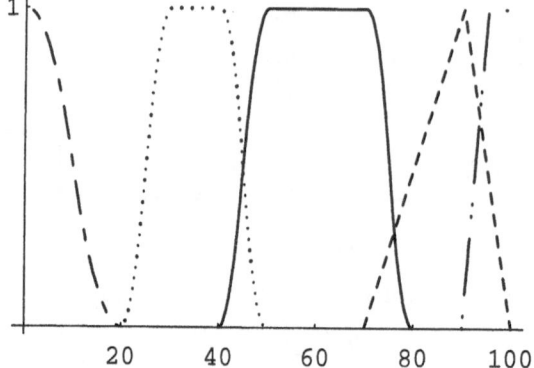 

Fig. 1. Values of the variable *visibility*

For the considered sample the observed fuzzy data have been collected in Table 1.

values of $\mathcal{X}$	$\tilde{x}_1$	$\tilde{x}_2$	$\tilde{x}_3$	$\tilde{x}_4$	$\tilde{x}_5$
absolute frequencies	4	21	12	8	2

Table 1. Data of variable *visibility* from days in the sample

To test the null hypothesis $H_0 : \widetilde{E}(\mathcal{X}) \succeq_{.5} \widetilde{U}$ against the alternative $H_1 : \widetilde{U} \succ_{.5} \widetilde{E}(\mathcal{X})$, where $\widetilde{U})$ denotes the value RATHER GOOD ON THE AVERAGE, which can be assumed to be modeled by the Π-curve $\Pi(60,70,80,90)$, on the basis of Theorem 2 the hypotesis H_0 should be rejected at the significance level $\alpha = .05$ (actually, the p-value of the test is given by .0399).

6 Additional and concluding remarks

An interesting open problem in connection with the study in this paper is that of discussing the effects of choosing the value of parameter λ on the power of tests in Theorems 1 and 2.

In this respect, we have developed a broad introductory analysis. In this way, it should be pointed out that although the test in Montenegro *et al.* (2001, 2002) cannot be applied to deal with one-sided hypotheses, the ideas in this paper can be used also to test two-sided hypotheses. For purposes of comparing these ideas with those in the previous papers, we have performed some simulation studies allowing us to conclude that for a proper choice of λ (the choice depending on the shape of variable values), the power function of the asymptotic test by Montenegro *et al.* (2001, 2002) could be slightly improved by employing the test based on V^λ.

In this way,

Example I. Consider a FRV $\mathcal{X}$ taking on 5 different values in a population. Assume these values are asymmetric triangular ones, whose centers and widths have been obtained by a randomization proccess, and the 5 values are equally likely. Let $\widetilde{U}$ the population expected value of $\mathcal{X}$.

10000 samples of size 300 have been simulated, and 20 different null hypothesis have been considered, namely, [i]: $\widetilde{E}(\mathcal{X}) = \widetilde{U}_i = \widetilde{U} \oplus .02(i-1)$ $(i = 1, \ldots, 20)$.

Test 1 denotes the test in Montenegro *et al.* (2001, 2002) (with the metric using Lebesgue measure on $[0, 1]$ for both, the α-levels and the convex linear combinations).

Test 2 involves the statistic in Theorem 2 applied to test the null hypothesis $V_{0.1}(\widetilde{E}(\mathcal{X})) = V_{0.1}(\widetilde{U}_i)$, Test 3 involves the same statistic applied to test the null hypothesis $V_{0.5}(\widetilde{E}(\mathcal{X})) = V_{0.5}(\widetilde{U}_i)$, and Test 4 involves the same statistic applied to test the null hypothesis $V_{0.9}(\widetilde{E}(\mathcal{X})) = V_{0.9}(\widetilde{U}_i)$.

The following table gathers the percentage of rejections at level $\alpha = .05$ (i.e., when the real percentage of rejections for the null hypothesis being true equals 5%).

Null hyp.	[1]	[2]	[3]	[4]	[5]	[6]	[7]	[8]	[9]	[10]	[11]	[12]	[13]	[14]	[15]	[16]	[17]	[18]	[19]	[20]
Test 1	5.2	6.6	9.2	13.2	18.6	26.2	33.5	44.7	53.3	62.7	71.6	78.7	85.0	89.7	93.1	96.4	97.6	98.6	99.3	99.6
Test 2	5.2	6.3	8.9	12.8	17.7	25.3	32.1	43	51.4	60.8	69.6	77.0	83.6	88.7	92.7	95.7	97.1	98.2	99.2	99.5
Test 3	5.2	6.6	9.2	13.2	18.6	26.2	33.6	44.7	53.3	62.7	71.6	78.7	85.0	89.8	93.1	96.4	97.6	98.5	99.3	99.6
Test 4	5.3	6.8	9.2	13.3	19.1	27.1	34.5	45.7	53.9	63.5	72.5	79.3	85.3	90.3	93.3	96.7	97.7	98.6	99.4	99.6

Example II. Consider a FRV $\mathcal{Y}$ taking on 10 different values in a population. Assume these values are S-, Z- and Π-curves, which have been obtained by a randomization proccess, and the 10 values are equally likely. Let $\widetilde{W}$ the population expected value of $\mathcal{Y}$.

10000 samples of size 500 have been simulated, and 20 different null hypothesis have been considered, namely, [i]: $\widetilde{E}(\mathcal{Y}) = \widetilde{W}_i = \widetilde{W} \oplus .01(i-1)$ $(i = 1, \ldots, 20)$.

Tests 1 to 4 have meanings similar to those in Example I.

The following table gathers the percentage of rejections at level $\alpha = .05$ (i.e., when the real percentage of rejections for the null hypothesis being true equals 5%).

Null hyp.	[1]	[2]	[3]	[4]	[5]	[6]	[7]	[8]	[9]	[10]	[11]	[12]	[13]	[14]	[15]	[16]	[17]	[18]	[19]	[20]
Test 1	5.5	5.8	7.9	12.0	17.7	25.3	35.4	47.2	58.7	70.0	80.1	88.4	93.6	97.0	98.8	99.5	99.9	100	100	100
Test 2	5.1	5.9	6.9	9.4	12.5	16.5	21.4	28.1	33.5	39.8	46.9	55.4	60.9	68.4	74.5	80.5	84.5	89.0	92.1	94.5
Test 3	4.8	6.5	10.5	17.3	26.2	37.2	50.4	62.9	73.3	82.8	89.3	94.2	96.9	99.4	99.8	99.9	100	100	100	100
Test 4	5.0	9.3	24.2	46.9	69.4	86.6	95.4	98.9	99.7	100	100	100	100	100	100	100	100	100	100	100

On the basis of these simulations, we realise there are not general conclusions to get. Thus, in Example I Tests 1 and 3 show a similar behavior, whereas Tests 2 and 4 are worse on the average. On the other hand, in Example II Test 4 is clearly the most powerful one.

Figure 2 includes the graphical representation of the power function of Tests 1 (continuous line) and 3 in Example I, whereas Figure 3 includes the graphical representation of the power function of Tests 1 (continuous line) and 4 in Example II.

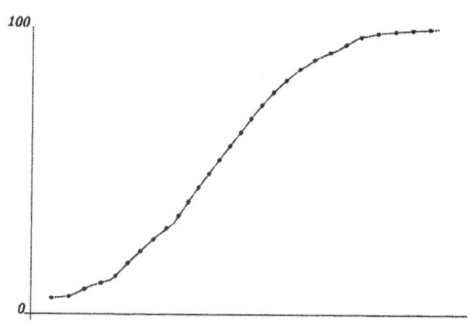

Fig. 2. Figure 2

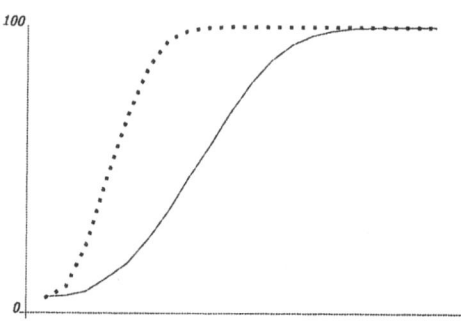

Fig. 3. Figure 3

Anyway, it should be emphasized that Tests 2 to 3 could be valuable to test the null hypothesis $\widetilde{E}(\mathcal{X}) = \widetilde{A}$ in case of rejecting it, but in case of acceptance we know that $\widetilde{E}(\mathcal{X}) \sim_\lambda \widetilde{A}$ is not equivalent to $\widetilde{E}(\mathcal{X}) = \widetilde{A}$, whence the improvement in the power function would not mean a real advantage.

References

1. Aumann, R.J. (1965) Integrals of set-valued functions. $\widehat{J}$. Math. Anal. Appl. **12**, 1–12.
2. Campos, L. de and González, A. (1989) A subjective approach for ranking fuzzy numbers. *Fuzzy Sets and Systems* **29**, 145–153.
3. López-Díaz, M. and Gil, M.A. (1998) The λ-average value and the fuzzy expectation of a fuzzy random variable. *Fuzzy Sets and Systems* **99**, 347–352.
4. Montenegro, M., Colubi, A., Casals, M.R. and Gil, M.A. (2001) Testing the expected value of a fuzzy random variable. A discussion. *Proc. EUSFLAT 2001 Conf.*, 352–355.
5. Montenegro, M., Colubi, A., Casals, M.R. and Gil, M.A. (2002) Asymptotic and Bootstrap techniques for testing the expected value of a fuzzy random variable. (Submitted for publication).
6. Puri, M.L. and Ralescu, D.A. (1985) The concept of normality for fuzzy random variables. *Ann. Probab.* **13**, 1373-1379.
7. Puri, M.L. and Ralescu, D.A. (1986) Fuzzy random variables. *J. Math. Anal. Appl.* **114**, 409–422.

Blackwell Sufficiency and Fuzzy Experiments

Andreas Wünsche

Faculty of Mathematics and Computer Science,
Freiberg University of Mining and Technology,
Agricolastr. 1, 09596 Freiberg, Germany
wuensche@math.tu-freiberg.de

Abstract. Blackwell sufficiency is an accepted instrument for comparison of random experiments.In this paper we discuss, whether Blackwell sufficiency is a suitable instrument to characterize fuzziness and nonspecificity of experiments. The answer will be: Yes in special cases, no in general.

1 Blackwell sufficiency

Assume that we want to perform a statistical experiment with the aim to obtain information on an unknown parameter $\vartheta \in \Theta$. The following definition says that the statistical experiment is represented by the corresponding probability space.

Definition 1 *A statistical experiment $\mathcal{E}$ w.r.t. $\vartheta \in \Theta$ is defined by $\mathcal{E} := [\Omega, \mathcal{A}, \{P^{(\vartheta)}\}_{\vartheta \in \Theta}]$,where for any $\vartheta \in \Theta$ $[\Omega, \mathcal{A}, P^{(\vartheta)}]$ is a probability space.*

Central for Blackwell sufficiency is the the notion of a Markov-kernel.

Definition 2 *Let $[\Omega_1, \mathcal{A}_1]$ and $[\Omega_2, \mathcal{A}_2]$ be two measurable spaces. A mapping*

$$h : \Omega_1 \times \mathcal{A}_2 \to \mathbf{R}$$

is called a Markov-kernel from $[\Omega_1, \mathcal{A}_1]$ to $[\Omega_2, \mathcal{A}_2]$, if:

    1) $\forall \omega \in \Omega_1 :$ $h(w, .)$ *is a probability measure on $[\Omega_2, \mathcal{A}_2]$*

    2) $\forall A \in \mathcal{A}_2 :$ $h(., A)$ *is a measurable function from Ω_1 to $\mathbf{R}$.*

Now we are able to define Blackwell- sufficiency.

Definition 3 *Let $\mathcal{E}_1 = [\Omega_1, \mathcal{A}_1, \{P_1^{(\vartheta)}\}_{\vartheta \in \Theta}]$ and $\mathcal{E}_2 = [\Omega_2, \mathcal{A}_2, \{P_2^{(\vartheta)}\}_{\vartheta \in \Theta}]$ be two experiments, then $\mathcal{E}_1$ is called B-sufficient (Blackwell sufficient) for $\mathcal{E}_2$ (notation: $\mathcal{E}_2 \prec_B \mathcal{E}_1$), if there is $\forall \vartheta \in \Theta$ a Markov-kernel h from $[\Omega_1, \mathcal{A}_1]$ to $[\Omega_2, \mathcal{A}_2]$ with*

$$\forall A \in \mathcal{A}_2: \quad P_2^{(\vartheta)}(A) = \int_{\Omega_1} h(\omega, A) dP_1^{(\vartheta)}(\omega) \quad .$$

An outcome from $\mathcal{E}_2$ can be generated by an outcome from $\mathcal{E}_1$ and a further randomization according to the probability measure $h(\omega, .)$. Since the kernel h does not depend on ϑ the performance of $\mathcal{E}_2$ does not add information about ϑ to what is contained in $\mathcal{E}_1$. In this connection we can say that $\mathcal{E}_2$ is not more informative then $\mathcal{E}_1$. For more details see Blackwell [1] [2] and e.g. Torgersen [11].

Remark 1 (Discrete Case) A discrete finite experiment $\mathcal{E}_1$ is Blackwell sufficient for another discrete finite experiment $\mathcal{E}_2$ if there exists a column stochastic matrix H so that the vector of probabilities $P_2^{(\vartheta)}$ is computable from $P_1^{(\vartheta)}$ like:

$$P_2^{(\vartheta)} = H P_1^{(\vartheta)}. \tag{1}$$

2 Statistical experiments with vague outcomes

Assume that a certain product appears in k qualities, denoted by $1, 2, .., k$ with unknown probabilities $p_1, .., p_k$ of appearance. The quality control procedure, however, diagnoses the true quality only with some vagueness, i.e. for a diagnosis A_i several qualities can be true with some possibility. Thus, we model this procedure by a statistical experiment with n possible outcomes $A_1, ..., A_n$ which are modeled by fuzzy sets on the universe $U = \{1, 2, .., k\}$ with membership values

$$m_{A_i}(j) =: m_{ij} \qquad i = 1, ..., n ; \quad j = 1, ..., k. \tag{2}$$

Furthermore, let us assume that the membership values for fixed j sum up to one, i.e.

$$\sum_{i=1}^{n} m_{ij} = 1 \qquad j = 1, ..., k, \tag{3}$$

which means that the $A_1, ..., A_n$ constitute a fuzzy partition (in Ruspini's sense). Writing (2) in matrix form $M = ((m_{i,j}))$ we have that M is a column stochastic $(n \times k)$-matrix. According to Zadeh [12] for the probability distribution of this fuzzy experiment, we obtain with the unknown parameter $\vartheta = (p_1, ..., p_k)^T$:

$$P^{(\vartheta)}(A_i) = E m_{A_i}(j) = \sum_{j=1}^{k} p_j m_{ij} \qquad ; i = 1, .., n. \tag{4}$$

Note that the $P^{(\vartheta)}(A_i)$ sum up to one (i.e. they build a probability distribution) only if (3) is satisfied. On the other hand (3) is very restrictive and often not natural. Clearly, there are other approaches to probability in fuzzy environment, see e.g Puri & Ralescu [8], Kruse & Meyer [6], but our (restrictive) type of fuzzy experiments is already enough for showing that B-sufficiency is not well suited for comparison. With $P^{(\vartheta)} := (P^{(\vartheta)}(A_1), ..., P^{(\vartheta)}(A_n))^T$ (4) can be written as

$$P^{(\vartheta)} = M\vartheta \qquad \text{for all } \vartheta. \tag{5}$$

Now, consider two fuzzy experiments $\mathcal{E}_1$ and $\mathcal{E}_2$ of the described type both for the same parameter ϑ, but with different M (M_1, M_2) and as a consequence, different $P^{(\vartheta)}$ ($P_1^{(\vartheta)}$, $P_2^{(\vartheta)}$). Let M_1 be ($n_1 \times k$) and M_2 be ($n_2 \times k$). Note that from Remark 1 we have:

$$\mathcal{E}_2 \prec_B \mathcal{E}_1 \iff \forall \vartheta : P_2^{(\vartheta)} = HP_1^{(\vartheta)} \qquad \text{with a column stochastic matrix } H.$$

Using (5) this is equivalent to

$$M_2 = HM_1, \tag{6}$$

i.e. the problem of B-sufficiency for the considered type of fuzzy experiments reduces to an algebraic characterization: $\mathcal{E}_2 \prec_B \mathcal{E}_1$ iff there is a ($n_2 \times n_1$)-column stochastic matrix H which maps the column stochastic matrix M_1 into the column stochastic matrix M_2. Now, the question is whether statements of the following type are valid:

$$M_2 \text{ is "fuzzier" than } M_1 \implies \mathcal{E}_2 \prec_B \mathcal{E}_1. \tag{7}$$

Let us recall the most commonly used notion of fuzziness (see e.g. Klir & Folger [5]).

Definition 4
Let $A^{(1)}$ and $A^{(2)}$ be fuzzy sets on $\{1,..,k\}$. $A^{(1)}$ is called sharper than $A^{(2)}$ (abbreviated: $A^{(2)} <_s A^{(1)}$) iff $\forall j \in \{1,..,k\}$:

$$[m_{A^{(2)}}(j) \le 0,5 \wedge m_{A^{(1)}}(j) \le m_{A^{(2)}}(j)] \bigvee [m_{A^{(2)}}(j) \ge 0,5 \wedge m_{A^{(1)}}(j) \ge m_{A^{(2)}}(j)]$$

A fuzzy set is called maximally fuzzy if $m_A(j) = 0.5$ for all j.

In the special case $n_1 = n_2 = k = 2$ a result of type (7) can be found in Gil [4]. Note that the considered case can be interpreted as comparison of fuzzy Bernoulli experiments. An Bernoulli experiment $\mathcal{E}_1$ is sharper than $\mathcal{E}_2$ if for one of the two possible outcomes $A_1^{(l)}$, $A_2^{(l)}$, say for $A_1^{(l)}$; $l = 1, 2$; it holds

$$A_1^{(2)} <_s A_1^{(1)}.$$

With (3), the same relation holds between $A_2^{(1)}$ and $A_2^{(2)}$. Clearly, an experiment $\mathcal{E}_m$ is maximally fuzzy if for the associated M_m it holds

$$M_m = \begin{pmatrix} 0.5 & 0.5 \\ 0.5 & 0.5 \end{pmatrix}$$

and an experiment $\mathcal{E}_0$ is called crisp (or sharp) if

$$M_0 = \begin{pmatrix} 1 & 0 \\ 0 & 1 \end{pmatrix} \quad or \quad M_0' = \begin{pmatrix} 0 & 1 \\ 1 & 0 \end{pmatrix}.$$

Since $M_m = HM$ with $H = M_m$ for any column stochastic matrix M we have for any fuzzy Bernoulli experiment $\mathcal{E}$: $\mathcal{E}_m \prec_B \mathcal{E}$ (see (6)). Further, any column stochastic matrix M can be generated via (6) by $M = HM_0$ with $H = M$ and by $M = HM_0'$ with $H = MM_0'$ hence $\mathcal{E} \prec_B \mathcal{E}_0$. Thus it holds:

Theorem 1 (Gil)

Condition: $\quad sgn(m_{11}^{(l)} - 0.5) = sgn(m_{22}^{(l)} - 0.5), \quad l = 1,2 \qquad (8)$

1. Let be $\mathcal{E}$ a fuzzy Bernoulli experiment. Then it holds

$$\mathcal{E} \prec_B \mathcal{E}_0, \qquad \mathcal{E}_m \prec_B \mathcal{E}.$$

2. Let be $\mathcal{E}_1$, $\mathcal{E}_2$ two fuzzy Bernoulli experiments. If (8) is satisfied then it holds:
$$A_1^{(2)} <_s A_1^{(1)} \quad \Longrightarrow \quad \mathcal{E}_2 \prec_B \mathcal{E}_1.$$

Thus, Theorem 1 2. presents a statement of the desired form (7). Note that Condition (8) means that only such (2×2)- matrices M_1, M_2 are comparable where both diagonal elements are either larger or smaller than 0.5. Especially (8) is fulfilled if $A_1^{(l)}$ gives main membership to quality 1 and $A_2^{(l)}$ to quality 2; $l = 1,2$; which is really the case in practical situations where the crisp case

$M_0 = \begin{pmatrix} 1 & 0 \\ 0 & 1 \end{pmatrix}$ is fuzzified into $M = \begin{pmatrix} m_1(1) > 0.5 & 1 - m_2(2) \\ 1 - m_1(1) & m_2(2) > 0.5 \end{pmatrix}$.

Note that in Part 2 of Theorem 1 the opposite implication does not hold (see Example 2 in [7]). Note further that without (8) Part 2 of Theorem 1 is not true (see Example 3 in [7]).

Now, let us turn towards non-Bernoulli cases with more than two qualities. Firstly, let us mention that there is no general definition which describes, whether a set of possible outcomes $(A_1^{(1)}, ..., A_{n_1}^{(1)}) =: A_1$ is sharper than $(A_1^{(2)}, ..., A_{n_2}^{(2)}) =: A_2$, abbreviated by $A_2 <_s A_1$. Let us restrict ourselves to $n_1 = n_2 =: n$. For $A_2 <_s A_1$ at least should hold:

$$\forall i \in \{1, 2, ..., n\} : \quad A_i^{(2)} <_s A_i^{(1)} \quad \Longrightarrow A_2 <_s A_1. \qquad (9)$$

Example 1 For example, in the case $k = n = 3$, characterized by

$$M_1 = \begin{pmatrix} 1 & 0.1 & 0 \\ 0 & 0.8 & 0 \\ 0 & 0.1 & 1 \end{pmatrix} \quad and \quad M_2 = \begin{pmatrix} 1 & 0.1 & 0.1 \\ 0 & 0.8 & 0.1 \\ 0 & 0.1 & 0.8 \end{pmatrix}$$

we are intuitively convinced that $A_2 <_s A_1$ since all three rows of M_1 describe fuzzy outcomes which are sharper than the outcomes of M_2, i.e. it holds (7). Unfortunately, we cannot conclude $\mathcal{E}_2 \prec_B \mathcal{E}_1$. A direct solution of (6) yields

$$H = M_2 M_1^{-1} = \begin{pmatrix} 1 & -0.0125 & 0.1 \\ 0 & 0.9875 & 0.1 \\ 0 & 0.0250 & 0.8 \end{pmatrix}$$

which is not a column stochastic matrix. Thus, we cannot prove in general a statement like (14) in the non-Bernoulli case. Thus, Blackwell sufficiency in general cannot be used to characterize different degrees of fuzziness in statistical experiments with vague data.

3 Experiments evaluated by belief and plausibility

Let us consider the case where the uncertainty on the universe U cannot be modeled by a probability distribution. This happens if the state of knowledge on the uncertainty behaviour does not allow to specify a probability distribution. This situation is one of the starting points of Shafer's evidence theory (see Shafer [10]). Let us recall some notions from there. Let $U = \{1, 2, ..., k\}$ be a finite universe. For example, U contains possible answers to the question under consideration. Let P be a probability measure on $\mathcal{P}(U)$, the power set of U. Assume $P(\emptyset) = 0$. The set $A \subset U$ is called a focal set w.r.t. P if $P(A) > 0$. $P(A)$ can be interpreted as the probability that the true answer is in $A \subset U$, but not in any strict subset of A (see e.g. Shafer [10]). The set F_p of all focal sets w.r.t P is called the body of evidence belonging to P. Now, P is used to define for a given $A \subset U$ the degree of belief $Bel_p(A) = \sum_{B \subseteq A} P(B)$ and the degree of plausibility $Pl_p(A) = \sum_{B \cap A \neq \emptyset} P(B)$ that the true answer is contained in A. Clearly, for any P we have $Bel_p(\emptyset) = Pl_p(\emptyset) = 0$ and $Bel_p(U) = Pl_p(U) = 1$. There are two extreme bodies of evidence: the so-called dissonant case where all singletons of U are focal sets and the so-called vacuous case where U is the only focal set. A dissonant body of evidence generates a probability distribution on U via: $P(\{i\}) = p_i$. The interpretation is that we have maximal knowledge on the variability behaviour on U (total evidence) and the uncertainty is totally due to randomness. In the general case we have some knowledge on U expressed by P which is not rich enough to generate a probability distribution on U, but we have

$$\forall A \subset U: \qquad Bel_p(A) \leq Prob(A) \leq Pl_p(A).$$

Note that $Prob$ is a probability measure on U whereas P is a probability measure on $\mathcal{P}(U)$. The length of the interval $[Bel_p(A), Pl_p(A)]$ is related with the so-called specificity of P (see e.g. Körner & Näther [9]).

Definition 5 P_1 *is called to be more specific than* P_2 *(abbr.:* $P_2 \prec P_1$*) iff*

$$\forall A \subset U: \quad [Bel_{p_1}(A), Pl_{p_1}(A)] \subseteq [Bel_{p_2}(A), Pl_{p_2}(A)].$$

Clearly, a dissonant P is the most specific one and a vacuous P is the least specific one. Now, the idea is, to consider a situation described by U and parametric $P^{(\vartheta)}$ as an experiment and to ask for relations between specificity and B-sufficiency. Especially we ask: Let be $P_1^{(\vartheta)}$, $P_2^{(\vartheta)}$ two distributions on $\mathcal{P}(U)$, with parameter $\vartheta \in \Theta$. Is it true that

$$\forall \vartheta \in \Theta : \qquad P_2^{(\vartheta)} \prec P_1^{(\vartheta)} \quad \Longrightarrow \quad \mathcal{E}_2 \prec_B \mathcal{E}_1 \ ? \tag{10}$$

Let us start with the simplest case $U = \{1, 2\}$. Consider a parametric probability distribution $P^{(\vartheta)} = (P^{(\vartheta)}(\{1\}), P^{(\vartheta)}(\{2\}), P^{(\vartheta)}(U))^T$ on $\mathcal{P}(U)$ of the following form:

$$P^{(\vartheta)} = \begin{pmatrix} (1-p_1)\vartheta_1 \\ (1-p_2)\vartheta_2 \\ p_1\vartheta_1 + p_2\vartheta_2 \end{pmatrix} ; \vartheta = \begin{pmatrix} \vartheta_1 \\ \vartheta_2 \end{pmatrix} \in \Theta := \{\vartheta \in [0,1]^2 : \vartheta_1 + \vartheta_2 = 1\}. \qquad p_1, p_2 \in [0,1] \tag{11}$$

Note that we have left out in $P^{(\vartheta)}$ the value $P^{(\vartheta)}(\emptyset) = 0$. Note further that p_1, p_2 model the deviation from the dissonant case $P_d^{(\vartheta)} = (\vartheta_1, \vartheta_2, 0)^T$. The following belief and plausibility is associated with $P^{(\vartheta)}$:

$$Bel_{p^{(\vartheta)}}(\{1\}) = (1-p_1)\vartheta_1 \quad ; \quad Pl_{p^{(\vartheta)}}(\{1\}) = \vartheta_1 + p_2\vartheta_2 \tag{12}$$
$$Bel_{p^{(\vartheta)}}(\{2\}) = (1-p_2)\vartheta_2 \quad ; \quad Pl_{p^{(\vartheta)}}(\{2\}) = \vartheta_2 + p_1\vartheta_1$$

It is easy to check that for the probability distribution $p(\vartheta) = (\vartheta_1, \vartheta_2)^T$ on U which is generated by $P_d^{(\vartheta)}$ we have

$$P^{(\vartheta)} = Ap(\vartheta) \text{ with } A = \begin{pmatrix} 1-p_1 & 0 \\ 0 & 1-p_2 \\ p_1 & p_2 \end{pmatrix},$$

i.e. according to (1), a dissonant experiment $\mathcal{E}_d$ is B-sufficient for any experiment $\mathcal{E}$ evaluated by $P^{(\vartheta)}$ from (11). Hence it holds:

$$\forall \vartheta \in \Theta : \quad P^{(\vartheta)} \prec P_d^{(\vartheta)} \quad \wedge \quad \mathcal{E} \prec_B \mathcal{E}_d.$$

On the other hand, the vacuous P on $\mathcal{P}(U)$ is given by $P = (0, 0, 1)^T$ and it is easy to check that for any $P^{(\vartheta)}$ from (11) we have

$$P = BP^{(\vartheta)} \quad \text{with } B = \begin{pmatrix} 0 & 0 & 0 \\ 0 & 0 & 0 \\ 1 & 1 & 1 \end{pmatrix},$$

i.e. any experiment $\mathcal{E}$ from (11) is B-sufficient for the vacuous experiment $\mathcal{E}_V$. Hence it holds

$$\forall \vartheta \in \Theta : \quad P_V \prec P^{(\vartheta)} \quad \wedge \quad \mathcal{E}_V \prec_B \mathcal{E}.$$

Consider now two experiments $\mathcal{E}_1, \mathcal{E}_2$ characterized by $P_1(\vartheta)$ (with deviations p_{11}, p_{21} from $P_d^{(\vartheta)}$) and by $P_2^{(\vartheta)}$ (with p_{12}, p_{22}). Direct comparison by (12) gives:

$$\forall \vartheta \in \Theta : \qquad P_2^{(\vartheta)} \prec P_1^{(\vartheta)} \quad \Longleftrightarrow \quad p_{11} \leq p_{12} \wedge p_{21} \leq p_{22}. \qquad (13)$$

On the other hand, we have according to (1)

$$\mathcal{E}_2 \prec_B \mathcal{E}_1 \quad \Longleftrightarrow \quad P_2^{(\vartheta)} = HP_1^{(\vartheta)} \text{ with column stochastic matrix } H.$$

Since $P_1^{(\vartheta)} = A_1 P_d^{(\vartheta)}$, $P_2^{(\vartheta)} = A_2 P_d^{(\vartheta)}$ this leads to

$$\mathcal{E}_2 \prec_B \mathcal{E}_1 \quad \Longleftrightarrow \quad A_2 = HA_1 \text{ with column stochastic matrix } H. \qquad (14)$$

Note that B-sufficiency here is characterized by the same algebraic relation as in section 2 (see (6)). Now, we can prove (10), i.e.:

Theorem 2 Let be $U = \{1, 2\}$. $\forall \vartheta :$ $\qquad P_2^{(\vartheta)} \prec P_1^{(\vartheta)} \quad \Longrightarrow \quad \mathcal{E}_2 \prec_B \mathcal{E}_1.$

Thus, in the special case $U = \{1, 2\}$ the relation (10) is satisfied. Note that the opposite conclusion does not hold in general.

Example 2 $P_1^{(\vartheta)} = \begin{pmatrix} \vartheta_1 \\ 0,3\vartheta_2 \\ 0,7\vartheta_2 \end{pmatrix} ; P_2^{(\vartheta)} = \begin{pmatrix} 0,5\vartheta_1 \\ 0,5\vartheta_2 \\ 0,5\vartheta_1 + 0,5\vartheta_2 \end{pmatrix}$

It holds $P_2^{(\vartheta)} = HP_1^{(\vartheta)}$ with column stochastic matrix $H = \begin{pmatrix} \frac{1}{2} & 0 & 0 \\ 0 & 1 & \frac{2}{7} \\ \frac{1}{2} & 0 & \frac{5}{7} \end{pmatrix}$ i.e.

$\mathcal{E}_2 \prec_B \mathcal{E}_1$, but we have not $P_2^{(\vartheta)} \prec P_1^{(\vartheta)}$ since $p_{21} = 0,7 \not\leq p_{22} = 0,5$ (see (13)).

Unfortunately, Theorem 2 does not hold for general $U = \{1, ..., k\}$. This can be seen already in the case $U = \{1, 2, 3\}$ (see [7]).
As a summary, we have the result that comparison of evidences by the criterion of nonspecificity cannot be explained by comparison of the associated probability distribution on $\mathcal{P}(U)$ using the criterion of B-sufficiency. Thus, also for this kind of fuzzy experiments B-sufficiency is not well suited for comparisons.

4   Conclusion

We ask in the first case whether more crisp experiments are Blackwell sufficient for less crisp experiments and in the second case whether more specific evidences are Blackwell sufficient for less specific evidences. We have seen that the answer is "Yes" only in the simplest cases: for fuzzy Bernoulli-experiments and for evidences on universes with two elements. For more general experiments we easily have found counterexamples (see Example 1 and [7]) . Therefore the result is that the notion of fuzziness and the notion of specificity in general do not agree with B-sufficiency.

For both kinds of fuzzy experiments, characterization of B-sufficiency leads to the same algebraic equation: $M_2 = HM_1$ in the case (6) and $A_2 = HA_1$ in the case (14). This equation is equivalent to two further criteria which come from information theory, especially from comparison of channel capacities (see[3]). Note that one of this criteria is constructive for stochastic matrices with two columns but not with more than two columns (see [7]).

For more details and proofs of the theorems see Näther/Wünsche [7].

References

1. D. A. Blackwell (1951). Comparison of experiments. *Proc. 2nd Berkeley Symp. on Math.Statist. Prob.*, pages 93-102. University of Colifornia Press, Berkeley.

2. D. A. Blackwell (1953). Equivalent comparisons of experiments. *Ann. Math. Statist.* **24**: 265-272.

3. Joel E. Cohen and J.H.B. Kemperman and Gh. Zbăganu (1998). *Comparison of stochastic matrices.* Birkhäuser.

4. M. A. Gil (1992). Sufficiency and fuzziness in random experiments. *Ann. Inst. Statist. Math.* **44(3)**:451-462.

5. G.J. Klir and T.A. Folger (1988). *Fuzzy sets, uncertainty, and information.* Englewood Cliffs, NJ: Prentice Hall.

6. R. Kruse and K.D. Meyer (1987). *Statistics with Vague Data.* Dordrecht: Reidel.

7. W. Näther and A. Wünsche, Blackwell sufficiency and fuzzy experiments. *accepted for puplication in Fuzzy Sets and Systems.*

8. M.L. Puri and D.A. Ralescu (1986). Fuzzy random variables. *J. Math. Anal. Appl.* **114**: 409-422.

9. R.Körner and W.Näther (1995). On the specificity of evidences. *Fuzzy sets and systems.* **71**: 183-196.

10. G. Shafer (1976). *A mathematical theory of evidence.* Princeton-London: Princenton University Press.

11. Erik Torgersen (1991). *Comparison of Statistical Experiments.* Cambridge University Press.

12. L.A. Zadeh (1968). Probability measures of fuzzy events. *J. Math. Anal. Appl.* **23**: 421-427.

Optimal Stopping in Fuzzy Stochastic Processes and its Application to Option Pricing in Financial Engineering

Yuji Yoshida

The University of Kitakyushu, 4-2-1 Kitagata, Kokuraminami, Kitakyushu 802-8577, Japan

Abstract. An optimal stopping problem is discussed in a continuous-time process defined by fuzzy random variables. Fuzzy random variables are evaluated by both probabilistic expectation and fuzzy expectation defined by a possibility measure. An optimality equation and an optimal stopping time are given for the process. A numerical example is also given to apply it to a financial model.

1   Introduction and notations

Mathematical modeling of stochastic systems in decision-making has many applications to engineering, economics, etc., and in general, one of the conditions that stochastic modeling works successfully is stability of systems. When the model is applied in an uncertain environment, losses/errors occur between the models and the actual phenomena. For example, losses from time lag is one of the most important elements in optimal stopping problems regarding system. On selling and buying stocks in financial markets when stock prices change radically, losses from time lag by Internet etc. might be more huge. This kind of losses is not only a problem arising from probabilistic sense where something occurs or not, and it is difficult to formulate them by only probabilistic theory(Klir and Yuan [3]). In this paper, probability is applied as the uncertainty such that something occurs or not with probability, and fuzziness is applied as the uncertainty such that we cannot specify the exact values because of a lack of knowledge regarding the present stock market. By introducing fuzziness to stochastic processes in decision-making, we consider a new model with uncertainty of both randomness and fuzziness, which is a reasonable and natural extension of the original stochastic processes.

The optimal stopping problem for random variables has a long history and has been studied by many authors in probability theory. It has many applications in stochastic theory and its related fields, for example, financial engineering, management sciences and so on. This paper discusses the optimal stopping problem with randomness and fuzziness as uncertainty from the viewpoint of fuzzy expectation, taking account of human subjective judgement. In order to describe an optimal stopping model with fuzziness, we need to extend real-valued random variables in the classical probability theory to 'fuzzy random variables'. We formulate an optimal stopping model of stochastic processes with fuzziness, using fuzzy random variables. This paper derives an optimality equation for the 'fuzzy stochastic process' and gives a method to solve the stopping problem without loss of worthy information contained in uncertainty like randomness and fuzziness.

In the next section, we formulate an optimal stopping model for a fuzzy stochastic process and we discuss its optimality condition. In the fuzzy stochastic process, the randomness and fuzziness are evaluated by both probabilistic expectation and 'fuzzy expectation' defined by a possibility measure from the viewpoint of Yoshida [10]. In Section 3, this papers shows that the optimal fuzzy reward is a unique solution of an optimality equation under a differentiability condition, by an approach of dynamic programming. In the last section, we consider a numerical example to apply the results to American put option in a financial model.

Fuzzy random variables were first studied by Puri and Ralescu [5] and have been studied by many authors. It is known that the fuzzy random variable is one of the successful hybrid notions of randomness and fuzziness. In the rest of this section, we give some mathematical notations regarding fuzzy random variables. Let $(\Omega, \mathcal{M}, P)$ be a probability space, where $\mathcal{M}$ is a σ-field of Ω and P is a non-atomic probability measure. $\mathbb{R}$ denotes the set of all real numbers, and let $\mathcal{C}(\mathbb{R})$ be the set of all non-empty bounded closed intervals. A fuzzy number is denoted by its membership function $\tilde{a} : \mathbb{R} \mapsto [0,1]$ which is normal, upper-semicontinuous, fuzzy convex and has a compact support. In this paper, we identify fuzzy numbers with its corresponding membership functions. $\mathcal{R}$ denotes the set of all fuzzy numbers. The α-cut of a fuzzy number $\tilde{a}(\in \mathcal{R})$ is given by $\tilde{a}_\alpha := \{x \in \mathbb{R} \mid \tilde{a}(x) \geq \alpha\}$ $(\alpha \in (0,1])$ and $\tilde{a}_0 :=$ cl$\{x \in \mathbb{R} \mid \tilde{a}(x) > 0\}$, where cl denotes the closure of an interval. We write the closed intervals as $\tilde{a}_\alpha := [\tilde{a}_\alpha^-, \tilde{a}_\alpha^+]$ for $\alpha \in [0,1]$. Hence we introduce a partial order $\succeq$, so called the fuzzy max order, on fuzzy numbers $\mathcal{R}([3])$: Let $\tilde{a}, \tilde{b} \in \mathcal{R}$ be fuzzy numbers. Then $\tilde{a} \succeq \tilde{b}$ means that $\tilde{a}_\alpha^- \geq \tilde{b}_\alpha^-$ and $\tilde{a}_\alpha^+ \geq \tilde{b}_\alpha^+$ for all $\alpha \in [0,1]$. It is known that $(\mathcal{R}, \succeq)$ becomes a lattice ([10]). For fuzzy numbers $\tilde{a}, \tilde{b} \in \mathcal{R}$, we define the maximum $\tilde{a} \vee \tilde{b}$ with respect to the fuzzy max order $\succeq$ by the fuzzy number whose α-cuts are

$$(\tilde{a} \vee \tilde{b})_\alpha = [\max\{\tilde{a}_\alpha^-, \tilde{b}_\alpha^-\}, \max\{\tilde{a}_\alpha^+, \tilde{b}_\alpha^+\}], \quad \alpha \in [0,1]. \tag{1}$$

A fuzzy-number-valued map $\tilde{X} : \Omega \mapsto \mathcal{R}$ is called a fuzzy random variable if the maps $\omega \mapsto \tilde{X}_\alpha^-(\omega)$ and $\omega \mapsto \tilde{X}_\alpha^+(\omega)$ are measurable for all $\alpha \in [0,1]$, where $\tilde{X}_\alpha(\omega) = [\tilde{X}_\alpha^-(\omega), \tilde{X}_\alpha^+(\omega)] := \{x \in \mathbb{R} \mid \tilde{X}(\omega)(x) \geq \alpha\}$. Next we need to introduce expectations and conditional expectations of fuzzy random variables in order to describe an optimal stopping model in the next section. A fuzzy random variable $\tilde{X}$ is called integrably bounded if both $\omega \mapsto \tilde{X}_\alpha^-(\omega)$ and $\omega \mapsto \tilde{X}_\alpha^+(\omega)$ are integrable for all $\alpha \in [0,1]$. Let $\tilde{X}$ be an integrably bounded fuzzy random variable. The expectation $E(\tilde{X})$ of the fuzzy random variable $\tilde{X}$ is defined by a fuzzy number

$$E(\tilde{X})(x) := \sup_{\alpha \in [0,1]} \min\{\alpha, 1_{E(\tilde{X})_\alpha}(x)\}, \quad x \in \mathbb{R}, \tag{2}$$

where $E(\tilde{X})_\alpha := \left[\int_\Omega \tilde{X}_\alpha^-(\omega) \, dP(\omega), \int_\Omega \tilde{X}_\alpha^+(\omega) \, dP(\omega) \right] \ (\alpha \in [0,1])$.

2 An optimal stopping model

In this section, we discuss a continuous-time stopping model of fuzzy stochastic process and derive its optimality conditions. Let $\{\tilde{X}_t\}_{t \geq 0}$ be a family of integrably bounded fuzzy random variables such that $E(\sup_{t \geq 0} \tilde{X}_{t,0}^+) < \infty$, where $\tilde{X}_{t,0}^+(\omega)$ is the right-end of the 0-cut of the fuzzy number $\tilde{X}_t(\omega)$. We assume that the map $t \mapsto \tilde{X}_t(\omega)(\in \mathcal{R})$ is right continuous and has left hand limits on $[0, \infty)$ for almost all $\omega \in \Omega$. $\{\mathcal{M}_t\}_{t \geq 0}$ is a family of nondecreasing sub-σ-fields of $\mathcal{M}$ which is right continuous, i.e. $\mathcal{M}_t = \bigcap_{r:r>t} \mathcal{M}_r$ for all $t \geq 0$, and fuzzy random variables $\tilde{X}_t$ are $\mathcal{M}_t$-adapted, i.e. random variables $\tilde{X}_{r,\alpha}^-$ and $\tilde{X}_{r,\alpha}^+$ $(0 \leq r \leq t; \alpha \in [0,1])$ are $\mathcal{M}_t$-measurable. This paper call $(\tilde{X}_t, \mathcal{M}_t)_{t \geq 0}$ a fuzzy stochastic process. A map $\tau : \Omega \mapsto [0, \infty]$ is called a stopping time if $\{\omega \in \Omega \mid \tau(\omega) \leq t\} \in \mathcal{M}_t$ for all $t \geq 0$. We can extend the stopping range to $[0, \infty]$ by considering $\limsup_{t \to \infty} \tilde{X}_{t,\alpha}(\omega)$ at time $t = \infty$. However we can adopt finite horizons in real applications(see Section 4).

Give a fuzzy goal by a fuzzy set $\varphi : \mathbb{R} \mapsto [0,1]$ which is a continuous and nondecreasing function with $\lim_{x \to -\infty} \varphi(x) = 0$ and $\lim_{x \to \infty} \varphi(x) = 1$. Then we note that the α-cut is $\varphi_\alpha = [\varphi_\alpha^-, \infty)$ for $\alpha \in (0,1)$. For a stopping time τ, we define a fuzzy expectation of the fuzzy numbers $E(\tilde{X}_\tau)$ by

$$\tilde{E}(E(\tilde{X}_\tau)) := \fint_{\mathbb{R}} E(\tilde{X}_\tau)(x) \, d\tilde{P}(x) = \sup_{x \in \mathbb{R}} \min\{E(\tilde{X}_\tau)(x), \varphi(x)\}, \tag{3}$$

where $\tilde{P}$ is the possibility measure generated by the density φ and $\fint d\tilde{P}$ denotes Sugeno integral ([7]). The fuzzy number $E(\tilde{X}_\tau)$ means a fuzzy reward, and the fuzzy expectation (3) implies the degree of decision maker's satisfaction regarding fuzzy rewards $E(\tilde{X}_\tau)$. Then the fuzzy goal $\varphi(x)$ means a kind of utility function for expected payoffs x in (3), and it represents a human subjective judgement from the idea of Bellman and Zadeh [1]. We define an optimal fuzzy reward $\tilde{V}$ as follows: Consider

$$\bigvee_{\tau:\ \text{stopping times}} E(\tilde{X}_\tau), \tag{4}$$

where $\vee$ means the supremum with respect to (1) induced from the fuzzy max order $\succeq$. From (1), we define

$$V_\alpha^\pm := \sup_{\tau:\ \text{stopping times}} E(\tilde{X}_\tau)_\alpha^\pm \tag{5}$$

for $\alpha \in [0,1]$. Then we have $[V_{\alpha'}^-, V_{\alpha'}^+] \supset [V_\alpha^-, V_\alpha^+]$ if $\alpha' < \alpha$. Therefore, we can define a fuzzy number $\tilde{V} \in \mathcal{R}$ by considering a left continuous version of (5)(see [4, Lemma 5]). This paper discusses the following optimal stopping problem with randomness and fuzziness, which is a fuzzy-number-valued extension of continuous-time optimal stopping models in Shiryayev [6].

Problem S. Find a stopping time τ^* such that

$$\tilde{E}(E(\tilde{X}_{\tau^*})) = \tilde{E}(\tilde{V}). \qquad (6)$$

Then, τ^* is called an optimal stopping time, and a real number x^* is called an optimal expected payoff if it attains the supremum of the fuzzy expectation (3), i.e.

$$\tilde{E}(\tilde{V}) = \sup_{x \in \mathbb{R}} \min\{\tilde{V}(x), \varphi(x)\} = \min\{\tilde{V}(x^*), \varphi(x^*)\}. \qquad (7)$$

Next, to analyze the optimal fuzzy reward $\tilde{V}$, we introduce some notation. Let $t \geq 0$ and define

$$Z_{t,\alpha}^\pm := \operatorname*{ess\,sup}_{\tau:\text{ stopping times},\ \tau \geq t} E(\tilde{X}_{\tau,\alpha}^\pm | \mathcal{M}_t) \qquad (8)$$

for $\alpha \in [0,1]$, where $\tilde{X}_{\tau,\alpha}(\omega) = [\tilde{X}_{\tau,\alpha}^-(\omega), \tilde{X}_{\tau,\alpha}^+(\omega)]$ is the α-cut of the fuzzy number $\tilde{X}_\tau(\omega)$. Then we have that $Z_{t,\alpha}^\pm$ are right continuous with respect to t since $\tilde{X}_{t,\alpha}^\pm$ and $\mathcal{M}_t$ are right continuous with respect to t. Further we have $[Z_{t,\alpha'}^-(\omega), Z_{t,\alpha'}^+(\omega)] \supset [Z_{t,\alpha}^-(\omega), Z_{t,\alpha}^+(\omega)]$ if $\alpha' < \alpha$. Therefore, we can define fuzzy random variables $\tilde{Z}_t$ by considering a left continuous version of (8)(see [4, Lemma 5]). The fuzzy random variables $\tilde{Z}_t$ correspond to Snell's envelope in probability theory (Shiryayev [6]). Hence we obtain the following optimality characterization for the fuzzy stochastic process regarding the optimal fuzzy reward $\tilde{V}$ by fuzzy random variables $\tilde{Z}_t$.

Theorem 1. *For $t \geq 0$, the following (i) — (iii) hold:*

(i) *For almost all $\omega \in \Omega$, it holds that $\tilde{Z}_t(\omega) \succeq \tilde{X}_t(\omega)$. Particularly it holds that $\tilde{V} = E(\tilde{Z}_0)$.*

(ii) *For almost all $\omega \in \Omega$, it holds that $\tilde{Z}_t(\omega) \succeq E(\tilde{Z}_s | \mathcal{M}_t)(\omega)$, $s \in [t, \infty)$.*

(iii) *Let $\alpha \in [0,1]$. For almost all $\omega \in \Omega$ satisfying $\tilde{Z}_{t,\alpha}^\pm(\omega) > \tilde{X}_{t,\alpha}^\pm(\omega)$, there exists $\varepsilon > 0$ such that $\tilde{Z}_{t,\alpha}^\pm(\omega) = E(\tilde{Z}_{s,\alpha}^\pm | \mathcal{M}_t)(\omega)$, $s \in [t, t+\varepsilon)$.*

Finally, we discuss an optimality equation for the optimal fuzzy reward process $\{\tilde{Z}_t\}_{t \geq 0}$ by introducing differentials. Let $L^2([0, \infty))$ be the space of continuous functions $u. : [0, \infty) \mapsto \mathbb{R}$ satisfying $\int_0^\infty (u_t)^2\, dt < \infty$ and $\lim_{t \to \infty} u_t = 0$. Let $\mathcal{L}$ be the space of functions $u. \in L^2([0, \infty))$ such that $u.$ is differentiable on $[0, \infty)$ and $du_t/dt \in L^2([0, \infty))$. Then we write $Au_t := -du_t/dt$. For $t \geq 0$, we put a bilinear form on $\mathcal{L} \times \mathcal{L}$ by $\langle u., v. \rangle_t = \int_t^\infty u_s v_s\, ds$ for $u., v. \in \mathcal{L}$.

Assumption A. It holds that $\tilde{Z}^\pm_{\cdot,\alpha}(\omega) \in \mathcal{L}$ and $\tilde{X}^\pm_{\cdot,\alpha}(\omega) \in \mathcal{L}$ for almost all $\omega \in \Omega$ and all $\alpha \in (0,1]$.

Suppose Assumption A holds. Then the following Theorem 2 holds regarding the optimality, using the definition:

$$A\tilde{Z}^\pm_{t,\alpha}(\omega) = \lim_{s\downarrow 0} \frac{\tilde{Z}^\pm_{t,\alpha}(\omega) - \tilde{Z}^\pm_{t+s,\alpha}(\omega)}{s}. \tag{9}$$

Theorem 2 (Optimality equation). *Let $\alpha \in (0,1]$. The optimal reward process $\{\tilde{Z}_t\}_{t\geq 0}$ is a unique solution satisfying the following three inequalities (3.2) – (3.4): For all $t \geq 0$ and almost all $\omega \in \Omega$,*

$$\tilde{Z}^\pm_{t,\alpha}(\omega) \geq \tilde{X}^\pm_{t,\alpha}(\omega); \tag{10}$$
$$A\tilde{Z}^\pm_{t,\alpha}(\omega) \geq 0; \tag{11}$$
$$\left\langle A\tilde{Z}^\pm_{\cdot,\alpha}(\omega), \tilde{Z}^\pm_{\cdot,\alpha}(\omega) - \tilde{X}^\pm_{\cdot,\alpha}(\omega) \right\rangle_t = 0. \tag{12}$$

3 The fuzzy expectation and the optimal stopping

In this section, we discuss the fuzzy expectation of the optimal fuzzy reward $\tilde{V}$, and we give an optimal stopping time for Problem S. Define a grade α^* by

$$\alpha^* := \sup\{\alpha \in [0,1] | \varphi^-_\alpha \leq \tilde{V}^+_\alpha\}, \tag{13}$$

where $\varphi_\alpha = [\varphi^-_\alpha, \infty)$ for $\alpha \in (0,1)$, and the supremum of the empty set is understood to be 0. From the continuity of φ and $\tilde{V}$, we can easily check $\varphi^-_{\alpha^*} \leq \tilde{V}^+_{\alpha^*}$. The following theorem, which implies that α^* is the optimal grade of the fuzzy expectation of the fuzzy rewards, is obtained by a modification of the proofs in [9, Theorems 3.1 and 3.2].

Theorem 3. *It holds that*

$$\alpha^* = \tilde{E}(\tilde{V}) = \sup_{\tau:\ stopping\ times} \tilde{E}(E(\tilde{X}_\tau)). \tag{14}$$

Assumption B. The following (i) and (ii) hold:

(i) The map $t \mapsto \tilde{X}_t(\omega)$ is continuous on $[0,\infty)$ for almost all $\omega \in \Omega$.
(ii) $V^+_{\alpha^*} = \tilde{V}^+_{\alpha^*}$.

Under Assumption B, we obtain the optimality of the following stopping time τ^* by modifying the proof in [10, Theorem 4.1].

$$\tau^*(\omega) := \inf\{t \geq 0 \mid Z^+_{t,\alpha^*}(\omega) = \tilde{X}^+_{t,\alpha^*}(\omega)\}, \quad \omega \in \Omega, \tag{15}$$

where the infimum of the empty set is understood to be $+\infty$.

Theorem 4. *Suppose that Assumption B holds. If τ^* is finite, then τ^* is an optimal stopping time for Problem S. Further, if the function φ is strictly increasing within the grades taken on $(0, 1)$, then τ^* is the shortest in the class of optimal stopping times. The optimal expected payoff is $x^* = \varphi_{\alpha^*}^-$.*

4   Application to option pricing in financial engineering

In this section, we consider American put option in a finance model where there is no arbitrage opportunities ([2]) to illustrate the results of the optimal stopping models in previous sections. Yoshida [11] has discussed European option models from the same viewpoint. Let μ ($\mu \in \mathbb{R}$) be the appreciation rate and σ ($\sigma > 0$) is the volatility. Let K ($K > 0$) be a strike price and let r ($r > 0$) be a discount factor. Let $\{B_t\}_{t \geq 0}$ be a standard Brownian motion on $(\Omega, \mathcal{M}, P)$. $\{\mathcal{M}_t\}_{t \geq 0}$ denotes a family of nondecreasing right-continuous complete sub-σ-fields of $\mathcal{M}$ such that $\mathcal{M}_t$ generated by $B_s (0 \leq s \leq t)$. Let a stock price process $\{S_t\}_{t \geq 0}$ satisfy the log-normal stochastic differential equation: S_0 is a positive constant, and

$$\mathrm{d}S_t = \mu S_t \mathrm{d}t + \sigma S_t \mathrm{d}B_t, \quad t \geq 0. \tag{16}$$

There exists an equivalent probability measure Q such that $e^{rt}S_t$ is a martingale under Q, by setting $\mathrm{d}Q/\mathrm{d}P|_{\mathcal{M}_t} = \exp\left(((r - \mu)/\sigma)B_t - \frac{1}{2}((r - \mu)/\sigma)^2 t\right)$, $t \geq 0([2])$. Under Q, $W_t := B_t - ((r - \mu)/\sigma)t$ is a standard Brownian motion and it holds that $\mathrm{d}S_t = rS_t \mathrm{d}t + \sigma S_t \mathrm{d}W_t$. By Ito's formula, we have

$$S_t = S_0 \exp\left((r - \frac{\sigma^2}{2})t + \sigma W_t\right), \quad t \geq 0. \tag{17}$$

Let c be a constant satisfying $0 < c < 1$ and give a stochastic process $\{a_t\}_{t \geq 0}$ by $a_t(\omega) := cS_t(\omega)$ for $t \geq 0, \omega \in \Omega$. Hence we give a fuzzy stochastic process of the stock price process by the following fuzzy random variables $\{\tilde{S}_t\}_{t \geq 0}$:

$$\tilde{S}_t(\omega)(x) := L((x - S_t(\omega))/a_t(\omega)) \tag{18}$$

for $t \geq 0$, $\omega \in \Omega$ and $x \in \mathbb{R}$, where $L(x) := \max\{1 - |x|, 0\}$ ($x \in \mathbb{R}$) is the triangle-type shape function. Hence, $a_t(\omega) = cS_t(\omega)$ is a half width of triangular fuzzy numbers $\tilde{S}_t(\omega)$ and corresponds to fuzziness in the process. The fuzziness in processes increases as $a_t(\omega)$ becomes bigger. In this financial model, $a_t(\omega)$ should be an increasing function of the stock price $S_t(\omega)$ since it depends on volatility σ and stock price $S_t(\omega)$ of the process from (16). The α-cuts of (18) are

$$\tilde{S}_{t,\alpha}(\omega) = [\tilde{S}_{t,\alpha}^-(\omega), \tilde{S}_{t,\alpha}^+(\omega)] = [S_t(\omega) - (1 - \alpha)a_t(\omega), S_t(\omega) + (1 - \alpha)a_t(\omega)]. \tag{19}$$

By using stopping times τ, we consider a problem to maximize the expected value of the price process $\{\tilde{X}_t\}_{t \geq 0}$ in American put option. Put the optimal fuzzy reward, which is the optimal price of the American put option in this example, by

$$\tilde{V}^y = \bigvee_{\tau:\ \text{stopping times},\ \tau \geq 0} E(\tilde{X}_t \mid S_0 = y) \qquad (20)$$

for an initial stock price y ($y > 0$), where $E(\cdot)$ denotes expectation with respect to the equivalent martingale measure Q, and the fuzzy stochastic process $\{\tilde{X}_t\}_{t \geq 0}$ is defined by

$$\tilde{X}_t(\omega) := e^{-rt}(1_{\{K\}} - \tilde{S}_t(\omega)) \vee 1_{\{0\}} \quad \text{for } t \geq 0,\ \omega \in \Omega, \qquad (21)$$

where $\vee$ is given by (1), and $1_{\{K\}}$ and $1_{\{0\}}$ denote the crisp number K and zero respectively. Their α-cuts are

$$\tilde{X}_{t,\alpha}(\omega) = [\max\{e^{-rt}(K - \tilde{S}^+_{t,\alpha}(\omega)), 0\}, \max\{e^{-rt}(K - \tilde{S}^-_{t,\alpha}(\omega)), 0\}]. \qquad (22)$$

By putting $b(\alpha) := 1 - (1 - \alpha)c$ ($\alpha \in [0,1]$), we put

$$\tilde{V}^{y,+}_{t,\alpha} := \sup_{\tau \geq 0} E(e^{-r(\tau-t)} \max\{K - b(\alpha)S_\tau, 0\} 1_{\{\tau < \infty\}} \mid S_t = y). \qquad (23)$$

Then the right end of the α-cut (20) is $\tilde{V}^{y,+}_\alpha := \tilde{V}^{y,+}_{0,\alpha}$. Define a differential operator

$$\mathcal{L} := \frac{1}{2}\sigma^2 y^2 \frac{\partial^2}{\partial y^2} + ry\frac{\partial}{\partial y} + \frac{\partial}{\partial t}, \quad (y,t) \in \mathbb{R}_+ \times [0,\infty).$$

By taking expectations in (10) – (12) of Theorem 2, we get the following optimality equations (c.f. [2, Theorem 8.5.9]):

$$\tilde{V}^{y,+}_{t,\alpha} \geq \max\{K - b(\alpha)y, 0\};$$
$$\mathcal{L}(e^{-rt}\tilde{V}^{y,+}_{t,\alpha}) \leq 0 \quad \text{in the sense of Schwartz distributions};$$
$$\mathcal{L}(e^{-rt}\tilde{V}^{y,+}_{t,\alpha})(V(y,t) - \max\{K - b(\alpha)y, 0\}) = 0.$$

In a similar proof to Elliot and Kopp [2, Theorem 8.3.1] regarding an American put option model without expiration dates, we obtain a solution $\tilde{V}^{y,+}_\alpha = \tilde{V}^{y,+}_{0,\alpha}$:

$$\tilde{V}^{y,+}_\alpha = \begin{cases} K - b(\alpha)y & \text{if } y \leq s(\alpha) \\ (K - b(\alpha)s(\alpha))(y/s(\alpha))^{-\gamma} & \text{if } y > s(\alpha), \end{cases} \qquad (24)$$

where $s(\alpha) := K\gamma/(b(\alpha)(1+\gamma)) = K\gamma/((1-(1-\alpha)c)(1+\gamma))$ and $\gamma := 2r/\sigma^2$. The optimal stopping time for (5.8), which is called optimal exercise time, is

$$\tau_\alpha(\omega) := \inf\{t \mid S_t(\omega) \leq s(\alpha)\} = \inf\left\{t \mid (r - \frac{\sigma^2}{2})t + \sigma W_t(\omega) = \ln\left(\frac{s(\alpha)}{y}\right)\right\}. \qquad (25)$$

Hence we put a fuzzy goal

$$\varphi(x) = \begin{cases} 1 - e^{-0.2x}, & x \geq 0 \\ 0, & x < 0. \end{cases} \qquad (26)$$

Then we obtain the grade (13) of the fuzzy expectation of the optimal fuzzy reward $\tilde{V}^y$ by $\alpha^* = \tilde{E}(\tilde{V}^y) = \sup\{\alpha \in [0,1]|\ \varphi_\alpha^- \leq \tilde{V}_\alpha^{y,+}\}$, where $\varphi_\alpha^- = -0.2^{-1}\ln(1-\alpha)$ for $\alpha \in (0,1)$. Assumption B is clearly fulfilled. Now put $\sigma = 0.25$, $r = 0.05$, $c = 0.1$, $y = 30$ and $K = 35$. We can easily calculate that the grade of the fuzzy expectation of the optimal fuzzy reward is $\alpha^* \approx 0.805073$, This grade means the degree of writer's's (seller's) satisfaction in pricing. The optimal expected payoff (7) is $x^* = \varphi_{\alpha^*}^- = -0.2^{-1}\log(1-\alpha^*) \approx 8.17566$, which is the optimal expected price of the American put option. Then the optimal stopping time (15), which is the optimal exercise time for (20), is $\tau^*(\omega) = \tau_{\alpha^*}(\omega) = \inf\{t \geq 0 \mid -0.45t + W_t(\omega) = \ln(s(\alpha^*)/10)\}$ with $s(\alpha^*) \approx 21.9667$ (Fig. 1).

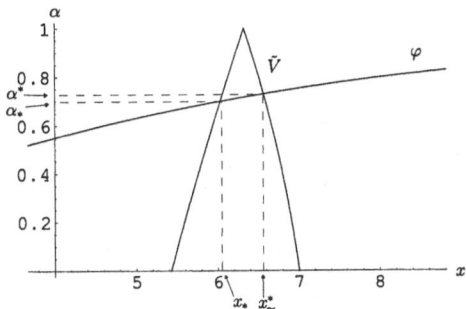

Fig. 1. Optimal fuzzy reward $\tilde{V}^y(x)$ and fuzzy goal $\varphi(x)$.

References

1. Bellman, R. E., Zadeh, L. A. (1970) Decision-making in a fuzzy environment. Management Sci. Ser. B. **17**, 141-164
2. Elliott, R. J., Kopp, P. E. (1999) Mathematics of Financial Markets. Springer, New York
3. Klir, G. J., Yuan, B., (1995) Fuzzy Sets and Fuzzy Logic: Theory and Applications. Prentice-Hall, London
4. Kurano, M., Yasuda, M., Nakagami, J., Yoshida, Y. (1992) A limit theorem in some dynamic fuzzy systems. Fuzzy Sets and Systems **51**, 83-88
5. Puri, M. L., Ralescu, D. A. (1986) Fuzzy random variables. J. Math. Anal. Appl. **114**, 409-422
6. Shiryayev, A. N. (1979) Optimal Stopping Rules. Springer, New York
7. Sugeno, M. (1974) Theory of fuzzy integrals and its applications. Doctoral Thesis, Tokyo Institute of Technology
8. Wang, G., Zhang, Y. (1992) The theory of fuzzy stochastic processes. Fuzzy Sets and Systems **51**, 161-178
9. Yoshida, Y. (1996) An optimal stopping problem in dynamic fuzzy systems with fuzzy rewards. Computers Math. Appl. **32**, 17-28
10. Yoshida, Y. (1997) The optimal stopped fuzzy rewards in some continuous-time dynamic fuzzy systems. Math. and Comp. Modelling **26**, 53-66
11. Yoshida, Y. The valuation of European options in uncertain environment. Europ. J. Oper. Res., to appear
12. Zadeh, L. A. (1965) Fuzzy sets. Inform. and Control **8**, 338-353

Soft Methods in Data Analysis:

Fuzzy, Rough

and Other Approaches

Fuzzy Conjoint Analysis

Malay Bhattacharyya

Decision Sciences Group, Indian Institute of Management
Prabandh Nagar, Off Sitapur Road, Lucknow 226013, INDIA
Email: mb@iiml.ac.in

Abstract. This paper proposes a new methodology for conjoint analysis by fuzzifying the rank or the rating data. The fuzzy conjoint model is solved as a fuzzy regression problem assuming the error term to be a random fuzzy variable. The paper investigates whether or not the proposed fuzzy conjoint model is superior to its non-fuzzy counterpart. A test of superiority is carried out on a real-life example. It appears that the choice of the membership function is extremely critical for the superiority of the fuzzy model. The author conjectures that if the membership functions could be estimated directly from the respondents, the performance of the fuzzy model would definitely improve.

Key words: *Conjoint analysis, fuzzy data, fuzzy regression, marketing research, management.*

1 Introduction

Conjoint analysis also called "trade-off analysis" helps to fathom how people make complex decisions. The basic assumption underlying the technique is that complex decisions such as purchase decisions are based not on a single factor or criterion, but on several factors "considered jointly". This technique enables the researcher to get inside the head of the consumer and observe how decisions are made.

Typically, in a conjoint study, the consumer is presented with a series of choice decisions about products or services. The method uncovers the consumer's "preference structure" from his overall rating or rankings of the products/services/ideas in a realistic manner.

Conjoint analysis computes the relative importance of various product attributes indirectly by asking respondents to make choices similar to those they confront in the marketplace in the context of specified range of variation, and by making trade-offs. Hence, to the extent real differences in importance exist, conjoint analysis more accurately captures those differences.

The basic conjoint methodology uses two broad classes of statistical techniques, viz., (Fractional) Factorial Design and the Multiple Regression with Dummy variables or Monotone Analysis of Variance (MONANOVA). The former is used to design the "Stimuli" which are presented to the respondents for their rating or ranking, while the latter is used for estimating the respondent's utility values for different levels of each factor. For the estimation of the utility values, Multiple Regression is used when the data are collected in the form of rating, while MONANOVA is used when the data are ordinal in the form of ranks, which are converted to continuous variables by some suitable monotonic transformation.

The basic assumption in a conjoint study is that the respondents would be able to rate or rank the "stimuli" with sufficient accuracy. Typically, a labelled rating scale is used with labels such as "Most Preferred", "Moderately Preferred", "Preferred", "Not Very Much Preferred", and "Least Preferred". Similarly, for rank ordering, the respondent is asked to give the highest rank to the "Most Preferred" stimulus, and lowest rank to the "Least Preferred" one. In both the cases it is evident that there is a great deal of ambiguity or vagueness in the way the data are collected. Further, in the depiction of the stimuli the attributes and sometimes their levels are described in linguistic terms that are intrinsically fuzzy. This is the prime motivation for using fuzzy set theory to model the rank or rating data collected in a conjoint study.

The paper is divided into a number of sections. The following section discusses the conjoint models with crisp data. The next section briefly outlines the concepts of fuzzy sets and random fuzzy variables. Section 3 discusses one regression model with fuzzy data. A method for ranking of fuzzy numbers is outlined in section 4. Section 5 describes the steps for conducting the fuzzy conjoint analysis. The new fuzzy conjoint model is compared with its non-fuzzy counterpart with the help of a real-life example in section 6. The last section concludes with a summary and scope for future work.

2 Conjoint model with crisp data

Let

p = number of attributes

q_i = number of levels of the ith attribute

U = overall utility of a stimulus (an alternative)

$$x_{ij} = \begin{cases} 1 & \text{if } j\text{-th level of the } i\text{-th attribute is present in the stimulus,} \\ & i = 1,...,p, \ j = 1,...,q_i \\ 0 & \text{otherwise} \end{cases}$$

W_{ij} = the part-worth contribution or utility associated with the j-th level ($j = 1,...,q_i$) of the i-th attribute ($i = 1,...,p$)

The basic crisp conjoint model may be represented by the following equation:

$$U = \sum_{i=1}^{p} \sum_{j=1}^{g_i} W_{ij} x_{ij} \tag{1}$$

There are several procedures for estimating the basic model. The simplest and the one that is gaining popularity is the OLS regression with dummy variables, wherein the q_i attribute levels are represented by $q_i - 1$ dummy variables. If ratings are obtained, they are assumed to be interval-scaled and form the dependent variable. If the data are in the form of rankings (nonmetric), they are converted to 0 or 1 by making paired comparisons between stimuli. Other procedures that are suitable for nonmetric data include LINMAP, MONANOVA, and the LOGIT model.

3 Fuzzy sets and fuzzy random variables

3.1 Zadeh's fuzzy sets

Definition: A fuzzy subset $A \in \mathcal{F}$ of $\mathbf{R}^d$ is defined by its membership function μ_A: $\mathbf{R}^d \to [0, 1]$. The number $\mu_A(x)$ is called the degree of membership of $x \in \mathbf{R}^d$ belonging to the fuzzy set A. The subset of $\mathbf{R}^d$

$$[A]^\alpha = \{x \in \mathbf{R}^d : \mu_A(x) \geq \alpha\}; \alpha \in (0, 1] \tag{2}$$

is called the α-level set (or α-cut) of A. The support of A is denoted by $[A]^0 = U_{\alpha>0}[A]^\alpha$.

The set of all normal compact convex fuzzy subsets of $\mathbf{R}^d$ is denoted by $\mathcal{F}_c^d$, i.e., any fuzzy set $A \in \mathcal{F}_c^d$ with membership function μ_A: $\mathbf{R}^d \to [0, 1]$ satisfies

i)   A is normal, i.e., $[A]^1 = \{x \in \mathbf{R}^d : \mu_A(x) = 1\}$ is non-empty,
ii) for $\alpha \in (0, 1]$, the α-level sets of A are convex and compact.

Note that α-level sets are nested, i.e., for $0 \leq \alpha \leq \beta \leq 1$, $[A]^\beta \subseteq [A]^\alpha$. Each sequence $\{[A]^\alpha\}_{\alpha>0}$ of nested non-empty compact convex sets defines uniquely a fuzzy set $A \in \mathcal{F}_c^d$ with

$$\mu_A(x) = \sup \{\alpha \in (0, 1] : x \in [A]^\alpha\}, x \in \mathbf{R}^d. \tag{3}$$

3.2 LR-fuzzy numbers

An important subset which can be shown to be a closed subspace of $\mathcal{F}_c$ is the class $\mathcal{F}_{LR}$ of LR-fuzzy numbers

$$<m, l, r>_{LR} = m - lA_L + rA_R, \tag{4}$$

where A_L and A_R are fuzzy numbers with α-cuts $[A_L]^\alpha = [0, L^{(-1)}(\alpha)]$ and $[A_R]^\alpha = [0, R^{(-1)}(\alpha)]$, for $\alpha \in (0, 1]$. Here $L, R: \mathbf{R}^+ \to [0,1]$ are fixed left-continuous and non-increasing functions with $L(0) = R(0) = 1$, and $L^{(-1)}(\alpha) = \sup\{x \in \mathbf{R}: L(x) \geq \alpha\}$, and $R^{(-1)}(\alpha) = \sup\{x \in \mathbf{R}: R(x) \geq \alpha\}$. The functions L and R are called the left and right shape functions, m the modal point and $l, r \geq 0$ are respectively the left

258

and right spreads of the *LR*-fuzzy number. The α-level sets of an *LR*-fuzzy number $A = <m, l, r>_{LR}$ are given by

$$[A]^{\alpha} = [m - lL^{(-1)}(\alpha) + rR^{(-1)}(\alpha)], \alpha \in (0, 1]. \tag{5}$$

If $l > 0$ and $r > 0$, then the membership function of an *LR*-fuzzy number A is

$$\mu_A(x) = \begin{cases} L\left(\dfrac{m-x}{l}\right) & if \quad x < m \\ 1 & if \quad x = m \\ R\left(\dfrac{x-m}{r}\right) & if \quad x > m \end{cases} \tag{6}$$

The advantage of *LR*-fuzzy numbers is that + and . can be defined by simple operations with respect to the parameters m, l, r

$$<m_A, l_A, r_A>_{LR} + <m_B, l_B, r_B>_{LR} = <m_A + m_B, l_A + l_B, r_A + r_B>_{LR} \tag{7}$$

$$\lambda \langle m, l, r \rangle_{LR} = \begin{cases} <\lambda_m, \lambda_l, \lambda_r>_{LR} & if \quad \lambda > 0 \\ <\lambda_m, \lambda_l, \lambda_r>_{LR} & if \quad \lambda < 0 \\ 1_{\{0\}} & if \quad \lambda = 0 \end{cases} \tag{8}$$

3.3 Triangular Fuzzy Number (TFN)

A triangular fuzzy number (TFN) is a special case of LR-fuzzy number with

$$L(x) = R(x) = \max\{0, 1 - x\}. \tag{9}$$

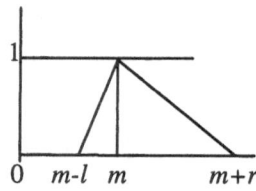

It is easily seen that the α-level sets of triangular fuzzy number $A = <m_A, l_A, r_A>_{LR}$ are given by

$$[A]^{\alpha} = [m_A - l_A(1 - \alpha), m_A + r_A(1 - \alpha)], \alpha \in (0, 1]. \tag{10}$$

The **Steiner point** of a convex set can be viewed as its "middle" point. Using this point some problems can be separated into a problem of location and a problem of shape. This point is in general different from the modal point as well as the point of gravity. The Steiner point of a TFN is given by

$$\sigma_A = m_A + 1/4 \, (r_A - l_A). \tag{11}$$

The best linear estimate approach (discussed later) of fuzzy regression problem is based on the concept of Steiner points of fuzzy numbers.

3.4 δ_2-distance of LR-fuzzy numbers

The δ_2-distance of two LR-fuzzy numbers $A = <m_A, l_A, r_A>_{LR}$ and $B = <m_B, l_B, r_B>_{LR}$ is defined as

$$\delta_2^2(A, B) = (m_A - m_B)^2 + \| A_R \|_2^2 \, (r_A - r_B)^2 + \| A_L \|_2^2 \, (l_A - l_B)^2 \\ + 2(m_A - m_B)\Big(\| A_R \|_1 \, (r_A - r_B) - \| A_L \|_1 \, (l_A - l_B)\Big) \tag{12}$$

where

$$\| A_L \|_2^2 = \frac{1}{2} \int_0^1 \big(L^{-1}(\alpha)\big)^2 \, d\alpha, \quad \| A_L \|_1 = \frac{1}{2} \int_0^1 \big(L^{-1}(\alpha)\big) d\alpha \tag{13}$$

and $\|A_R\|_1$, $\|A_R\|_2^2$ are similarly defined.

For **triangular fuzzy number**, it is easily seen that

$$\| A_R \|_2^2 = \| A_L \|_2^2 = \frac{1}{6} \quad and \quad \| A_R \|_1 = \| A_L \|_1 = \frac{1}{4} \tag{14}$$

3.5 Random LR-fuzzy numbers

Let m, l and r be three real-valued random variables with $P \, [l \geq 0] = P \, [r \geq 0] = 1$. A random LR-fuzzy number is defined by $X = <m, l, r>_{LR}$. By linearity of the expectation, the expectation $E(X)$ of a random LR-fuzzy number X is again an LR-fuzzy number

$$E(X) = E(m) - A_L \, E(l) + A_R \, E(r)$$
$$\text{or} \tag{15}$$
$$E(X) = < E(m), E\,(l), E(r)>_{LR}.$$

And using the δ_2-distance, the variance of random LR-fuzzy number X is given by

$$Var(X) = Var(m) + \| A_L \|_2^2 \, Var(l) + \| A_R \|_2^2 \, Var(r) \\ - 2 \| A_L \|_1 \, Cov(m, l) + 2 \| A_R \|_1 \, Cov(m, r) \tag{16}$$

For a **random triangular fuzzy number** X, the formula for variance reduces to

$$\mathbf{Var}\ (X) = \mathbf{Var}\ (m) + 1/6\ [\mathbf{Var}\ (l) + \mathbf{Var}\ (r)] + \tfrac{1}{2}[\mathbf{Cov}\ (m,\ r) - \mathbf{Cov}\ (m,\ l)]. \quad (17)$$

4 Fuzzy regression

A classical linear crisp regression model is defined as

$$Y = a_1 X_1 + \ldots\ldots + a_m X_m + e, \quad (18)$$

where $Y, X_1, \ldots, X_m$ are all crisp and $e \sim N(0, \sigma^2)$.

When $Y_1, \ldots, Y_n$ are fuzzy numbers but $X_{1i}, \ldots, X_{mi}$ for $i = 1, \ldots, n$ are all crisp, several approaches have been considered. Ping-Teng and Lee considers a model with error term as well as the regression coefficients as fuzzy numbers. In the second approach (extended classical and least square), the error term is random, while the regression coefficients are fuzzy numbers (Körner). The third approach (best linear estimate) considers a model with regression coefficients as crisp numbers but the error term is a random fuzzy variable (Körner). Although the least square approach has some theoretical advantage over the best linear estimate approach, the latter is easier to implement and performs, in terms of coefficient of determination, as well as the former. Further, this approach is statistically comparable to crisp conjoint analysis. Hence it is decided to use the best linear estimate approach for fuzzy regression in this paper. A brief description of this method is given below.

4.1 Best linear estimate approach

In situations where $Y_1, \ldots, Y_n$ are fuzzy numbers but $X_{1i}, \ldots, X_{mi}$ for $i = 1, \ldots, n$ are all crisp, one way to model the regression problem is

$$Y = a_1 X_1 + \ldots + a_m X_m + \xi, \quad (19)$$

where ξ is a random fuzzy variable with $E(\xi) = B \in \mathcal{F}_c$.

If we assume that each Y_i is an LR-fuzzy number, i.e., $Y_i = <y_i, l_i, r_i>_{LR}$ for $i = 1, 2, \ldots, n$ and $E(\xi) = B = <b, l_B, r_B>_{LR}$, it can be shown (Körner) that

$$\hat{a}_i = \frac{\displaystyle\sum_{j=1}^{n} \left(\sigma_{yi} - \sigma_y\right)\left(X_{ij} - \overline{X}_i\right)}{\displaystyle\sum_{j=1}^{n} \left(X_{ij} - \overline{X}_i\right)^2}, \quad i = 1, \ldots, n \quad (20)$$

$$\hat{b} = \sigma_{\overline{y}} - \sum_{j=1}^{n} \hat{a}_i \overline{X}_i - \hat{r}_B \sigma_{A_R} + \hat{l}_b \sigma_{A_l}, \quad \hat{l}_b = \overline{l}, \hat{r}_B = \overline{r} \quad (21)$$

where $\sigma_{\bar{y}}, \sigma_{y_i}, \sigma_{A_R}, \sigma_{A_L}$ denote the Steiner points of the fuzzy numbers Y_i, Y, A_R and A_L respectively. f we further assume that each Y_i, $i = 1,...,n$ and B are triangular fuzzy numbers then

$$\sigma_{A_R} = \sigma_{A_L} = \frac{1}{4}, and \ \sigma_{y_i} = y_i + \frac{1}{4}(r_i - l_i), \sigma_y = \bar{y} + \frac{1}{4}(\bar{r} - \bar{l}) \qquad (22)$$

5 Ranking of Fuzzy Numbers

The final output of the above regression will be fuzzy numbers. We need to convert them to crisp ranks so that they could be compared with the original ranks obtained in a conjoint study. There are several methods for ranking triangular fuzzy numbers. The weighted method (Chui and Park) is not tedious and require simple calculations. Under this method each number is first evaluated by using the expression:

$$E_j = W_{1j}(3m_j - l_j + r_j)/3 + W_{2j} m_j, \text{ for } j = 1, 2, 3, ... \qquad (23)$$

where E_j is the evaluation of the jth number, and W_{1j} and W_{2j} weights. Chui and Park note that "a common practice is to make $W_{1j} = 1$ and $W_{2\ j} = 0.3$ if the parameter m_j is important, otherwise $W_{1j} = W_{2j} = 0.1$". The ranking criterion for this method consists of arranging the E_j in a descending order. The number with the largest E_j is selected as the most preferred number.

6 Conducting fuzzy conjoint analysis

Step1: The ranks obtained through a conjoint experiment are fuzzified. This is done by representing each rank by a TFN. As has been seen above, a TFN can be defined by three numbers, viz., the modal value and the left and right spreads. The modal value and the left and the right spreads are computed using an approach for conversion of ranks into numeric scores.

This approach (Green *et al*, p. 299) assumes that true difference between adjacent objects ranked near the extremes tend to be larger than difference between objects falling near the middle in rank. Relative differences among ranked objects can be viewed as differences between Z values at the boundary points of N-1 equally likely intervals in the midrange of a normal distribution. The interval between each adjacent pair of ranks is defined as an interval corresponding to 100/N of cases in a normal distribution. Finally, 100/2N is arbitrarily set as the percentage of cases in a normal distribution to be cut below the value of the object ranked 1 and above the value of the object ranked N. Operationally, this amounts to converting rank j to $\Phi^{-1}((j - 0.5)/N)$, where $\Phi(.)$ represents the left tail cumulative distribution function of the standard normal distribution.

Rank j is represented by a TFN $A_j = <m_j, l_j, r_j >_{LR}$, where

$$m_j = \Phi^{-1}((j - 0.5)/N), \; m_j - l_j = \Phi^{-1}((j - k - 0.5)/N), \quad (24)$$
$$m_j + r_j = \Phi^{-1}((j + k - 0.5)/N), j = 1, \ldots, N.$$

The value of k can be varied between 0 and N to obtain different TFN representing rank j. These TFNs will have same modal value but varying left and right spreads. Whenever $j - k$ is less than 1, it is replaced by 1, and similarly, when $j + k$ is more than N, it is replaced by N.

Step2: Steiner points are computed for each respondent for the fuzzified ranks for all the concepts using formula (11).

Step3: For each respondent, OLS regression is applied with Steiner points as the dependent variable and the dummy variables of the crisp conjoint model as the predictor variables only for the concepts earmarked for model estimation.

Remark 1: The above regression directly gives the estimates of a_i. The estimate of b can be obtained by adding the last two terms to the estimate of the constant obtained from the regression.

Remark 2: The estimates from the fuzzy regression model can be written as

$$\hat{y}_j = \sum_{i=1}^{n} \hat{a}_i X_{ij} + \left\langle \hat{b}, \bar{l}, \bar{r} \right\rangle_{LR} = \left\langle \sum_{i=1}^{n} \hat{a}_i X_{ij} + \hat{b}, \bar{l}, \bar{r} \right\rangle_{LR} \quad (25)$$

Note that the left and the right spreads are constant (eqn. 25) for all the estimated triangular fuzzy numbers. As these estimated TFNs will be ranked using the weighted method, the left and the right spreads will not affect their ranking. Further note that the last two terms in the estimate of b (eqn. 21) can also be removed without affecting the ranking of the estimates of y_j.

Step4: In view of the above remarks, the crisp predicted values obtained from the regression in step3 are directly ranked to obtain the estimated ranks.

Step5: Kendall's tau, which measures the strength of association between two sets of rank data, is computed to test the goodness of fit of the estimated fuzzy conjoint model and also to compare it with the crisp conjoint model. Obviously, the higher the Kendall's tau coefficient, the better the model fit.

7 A real-life example

The study is concerned with one product category called "health drink" which is described by six attributes. The attributes and their levels are shown in Table 1. A total of 20 product concepts are generated using fractional factorial design, 16 of which are used for estimation of model parameters and

Table 1: Attributes and their levels

Attributes	Levels
Flavor	Chocolate, Elaichi. Plain
Energy	Low, Medium, High
Solubility	Low, High
Color	White, Brown
Nutrition	Low, Medium, High
Price/kg	174, 198, 225, 250

4 for cross validation. A sample of 50 respondents from the institute campus is asked to rank the 20 product concepts on the basis of their purchase intentions. The concepts are presented to them using the "Full profile approach".

The parameters of the crisp conjoint model are estimated for each respondent using the CONJOINT procedure of SPSS. The procedure reconstructs the ranks of all the concepts used for the estimation as well as for cross validation. The reconstructed ranks are compared with the original ranks using Kendall's tau. Respondent wise Kendall's tau values, *Tau_con*, for the crisp conjoint model are given in the second column of Table 2.

Next the fuzzy conjoint analysis is carried out. Seven sets of TFN are used by varying the left and right spreads. This is done by varying the value of *k* in equations (24) between 0 and 6.

The results of comparison of Kendall's tau between crisp and fuzzy conjoint model are shown in Table 2 where : * denotes fuzzy conjoint model performs better than crisp conjoint model (30 cases), ** fuzzy conjoint model is as good as crisp conjoint model (13 cases), *** fuzzy conjoint model is worse than crisp conjoint model (7 cases). Figures in bold indicate the maximum value of Kendall's tau obtained in fuzzy conjoint analysis.

The corresponding Kendall's tau coefficients are given columns 3 to 9 of Table 2 (e.g., *Tau3* refers to the Kendall's tau coefficient corresponding to TFN with $k = 3$).

On the basis of Kendall's tau coefficient, it is observed from Table 2, that in 30 out of 50 cases, fuzzy conjoint model gives better result, 13 cases are as good as the crisp conjoint model, while in the remaining 7 cases it is worse. In the cross validation sample, 66% cases have correct first choice for the fuzzy model as compared to 56% in the crisp model.

8 Conclusion

A new methodology for conjoint analysis by fuzzifying the rank or the rating data has been proposed in this paper. A step-by-step procedure for conducting fuzzy conjoint analysis is given. The fuzzy conjoint model is compared with its non-fuzzy counterpart through a real-life example. As for the estimation of model parameters, the fuzzy conjoint model performs better than the crisp conjoint model in 60% of the cases. It performs equally well as crisp model in 26% of case, while it is worse in 14% of the cases. As expected, it is also observed that the performance of the fuzzy conjoint model is critically dependent on the choice of the membership functions or TFN for the fuzzified rank. Turksen and Willson use a methodology to estimate the individual respondent's membership function directly from the respondents in a similar study. They used the vector conjoint model rather the part-worth or utility model used in this paper. The author conjectures that if the membership functions could be directly elicited from the respondents and their approximations are used in the proposed fuzzy conjoint model, then its performance would be uniformly superior to that of the crisp model. The least square approach should also be explored for conducting fuzzy conjoint analysis to see if it gives better results or not.

Table 2: Comparison of Kendall's tau between crisp and fuzzy conjoint model

Res.No.	Tau_con	Tau0	Tau1	Tau2	Tau3	Tau4	Tau5	Tau6
1	0.867*	**0.883**	**0.883**	**0.883**	**0.883**	**0.883**	0.867	**0.883**
2	0.85*	0.817	0.817	0.850	**0.867**	**0.867**	0.833	0.833
3	0.95***	0.917	0.917	0.917	**0.933**	**0.933**	**0.933**	0.917
4	0.85**	0.800	0.800	0.833	**0.850**	0.833	0.800	0.833
5	0.912*	**0.933**	**0.933**	0.917	0.917	0.883	0.883	**0.933**
6	0.383*	**0.417**	0.400	0.383	0.383	0.367	0.383	0.383
7	0.833**	**0.833**	**0.833**	**0.833**	**0.833**	0.817	0.817	**0.833**
8	0.583**	**0.583**	0.550	0.533	0.533	0.533	0.533	0.517
9	0.983**	**0.983**	**0.983**	**0.983**	**0.983**	**0.983**	**0.983**	**0.983**
10	0.817*	**0.900**	0.883	0.883	0.850	0.883	0.867	0.850
11	0.933**	**0.933**	**0.933**	**0.933**	**0.933**	**0.933**	**0.933**	**0.933**
12	0.7*	**0.733**	0.717	0.717	0.717	0.700	0.683	0.717
13	0.833*	0.833	0.817	0.833	0.833	**0.867**	**0.867**	0.833
14	0.819*	**0.833**	0.800	0.783	0.783	0.783	0.783	0.817
15	0.862*	0.833	0.850	0.850	0.850	**0.867**	0.833	0.833
16	0.807*	**0.817**	**0.817**	0.800	0.783	0.783	**0.817**	0.767
17	0.55*	0.567	0.550	0.567	0.567	**0.583**	0.567	0.567
18	0.95*	0.950	0.950	0.933	0.933	0.933	**0.967**	0.950
19	0.833*	**0.850**	**0.850**	**0.850**	**0.850**	0.833	0.833	**0.850**
20	0.733*	**0.767**	**0.767**	0.750	0.733	0.733	0.750	0.750
21	0.9*	0.883	0.900	0.883	0.883	0.900	**0.917**	0.900
22	0.667*	0.650	**0.683**	**0.683**	0.650	0.617	0.617	0.633
23	0.883**	0.850	0.833	0.833	0.850	0.850	0.867	**0.883**
24	0.941***	**0.917**	**0.917**	**0.917**	0.900	**0.917**	**0.917**	**0.917**
25	0.9*	**0.917**	**0.917**	**0.917**	0.900	0.900	0.900	0.900
26	0.678*	0.667	0.683	0.667	0.683	**0.700**	0.667	0.667
27	0.778*	0.717	0.717	0.717	0.750	**0.783**	0.750	**0.783**
28	0.795*	**0.817**	0.800	**0.817**	0.800	**0.817**	**0.817**	0.783
29	0.85**	0.817	**0.850**	**0.850**	**0.850**	0.833	**0.850**	**0.850**
30	0.867*	0.867	0.850	0.850	0.867	**0.883**	**0.883**	0.867
31	1***	0.950	0.950	**0.967**	0.950	**0.967**	0.950	0.950
32	0.9*	0.833	0.833	0.850	0.867	0.900	**0.917**	0.900
33	0.817***	0.767	0.767	**0.783**	0.767	0.767	**0.783**	**0.783**
34	0.9**	**0.900**	0.867	0.850	0.867	0.850	0.850	0.867
35	0.795***	0.750	**0.767**	**0.767**	**0.767**	**0.767**	0.750	0.733
36	0.517**	**0.517**	0.500	0.483	0.467	0.500	0.500	0.483
37	0.833**	0.817	0.817	0.817	**0.833**	0.800	0.800	0.800
38	0.86***	0.800	0.817	0.817	**0.833**	0.800	0.800	0.783
39	0.917*	0.933	0.933	0.933	0.933	**0.950**	0.917	0.917
40	0.695*	**0.700**	0.650	0.650	0.633	0.650	0.667	0.667
41	0.857*	0.850	0.867	0.850	**0.867**	0.850	0.833	0.833
42	0.933*	0.967	0.967	0.967	**0.983**	**0.983**	0.950	0.950
43	0.783**	0.767	0.767	**0.783**	0.767	**0.783**	**0.783**	**0.783**
44	0.745*	0.750	0.750	**0.767**	**0.767**	**0.767**	**0.767**	**0.767**
45	0.577*	0.567	0.567	0.567	0.583	0.583	**0.600**	**0.600**
46	0.883**	0.867	0.867	0.867	0.867	**0.883**	0.867	0.850
47	0.778*	0.767	**0.783**	**0.783**	**0.783**	**0.783**	0.767	0.733
48	0.9*	0.883	0.883	0.900	**0.933**	0.917	0.917	0.917
49	0.817**	**0.817**	**0.817**	**0.817**	**0.817**	**0.817**	**0.817**	**0.817**
50	0.7***	**0.683**	0.667	**0.683**	**0.683**	0.667	0.633	0.650

References

1. Paul E. Green and Vithala R Rao, Conjoint Measurement for Quantifying Judgmental Data, *Journal of Marketing Research,* **8** (**August 1971**), 355-363.
2. Paul E. Green, Hybrid Models for Conjoint Analysis: An Expository Review, *Journal of Marketing Research,* Vol. **XXI** (**May 1984**), 155-169.
3. Paul E. Green and V. Srinivasan, Conjoint Analysis in Marketing: New developments with Implications for Research and Practice, *Journal of Marketing,* **54**, (**October 1990**), 3-19.
4. Paul E. Green, Donald S. Tull and Gerald Albaum, Research for Marketing Decisions (Fifth Edition), (Prentice Hall of India Private Limited, 1994).
5. I. Burhan Turksen and Ian A. Wilson, A fuzzy set preference model for consumer choice, *Fuzzy Sets and Systems,* **68** (**1994**) 253-266.
6. A. G. Lapiga and V. V. Polyakov, On statistical methods in fuzzy decision-making, *Fuzzy Sets and Systems,* **47** (**1992**) 303-311.
7. D. Dubois and H. Prade, *Fuzzy Sets and Systems: Theory and Applications* (Academic Press. New York, **1980**).
8. A. Kandel, *Fuzzy Techniques in Pattern Recognition* (Wiley, New York, **1990**).
9. G. J. Klir and B. Yuan, *Fuzzy Sets and Fuzzy Logic: Theory and Applications* (Prentice-Hall, New Jersey, **1995**).
10. T. Terano, K. Asai and M. Sugeno, *Fuzzy Systems Theory and its Applications* (Academic Press, London, **1992**) (English edition).
11. L.A. Zadeh, Probability measures of fuzzy events, *J. Math. Anal. Appl.* **23** (**1968**) 421-427.
12. Ping-Teng Chang and E. Stanley Lee, Fuzzy least absolute deviations regression based on the ranking of fuzzy numbers, *IEEE* (**1994**).
13. Ralf Körner, Linear models with random fuzzy variables, *Ph.D. Thesis,* Faculty of Mathematics and Computer Sciences, Freiberg University of Mining and Technology, Freiberg, Germany (1997).
14. Ch. Chui-Yu and Ch. S. Park, Fuzzy cash flow analysis using present worth criterion, *The Engineering Economist,* 39, 113-137 (1994).

Grade Analysis of Repeated Multivariate Measurements

Alicja Ciok[1,2]

[1] Institute of Computer Science PAS,
   ul. Ordona 21, 01-237 Warsaw, Poland
[2] Institute of Home Market and Consumption,
   ul. Al. Jerozolimskie 87, 02-001 Warsaw, Poland

Abstract. The paper shows how the grade methods (correspondence-cluster analysis and overrepresentation maps) can help in analysis of data, which consists of repeated mesurements for a set of objects. Their usefulness will be discussed on a real data example. The analyzed data describe how the Polish retail firms use capital sources and what are their economic conditions.

1 Introduction

This paper tries to call the reader's attention to some features characteristic of grade methods. On one hand, there is a strict formalism on which grade methods are based. Short review of basic ideas and formulas is given in the next section.

On the other hand, this formalism can be applied to data which have a different meaning than the formalism assumes. The data, however, must fulfill the mathematical conditons. One may say that the formalism is applied very "softly". Also the results provided by the grade methods are expressed formally. Nevertheless, they can provide a description of relations (dependence structures) expressed very "roughly" (i.e. without definitions of equations or regression functions). For data which can not be precisely measured (i.e. social and psychological sciences) strict formulation of complex, multivariate relations is often impossible and not always necessary. Very often practitioners are satisfied given a general trend only, which can be very helpful in explanations of complex phenomenon. All these remarks will be illustrated by the results of the application of the grade methods to an analysis of real data concerning Polish retail trade firms and described in Sec. 3.

2 Basic ideas of the grade correspondence-cluster analysis (GCCA)

The basic notions of the grade correspondence analysis and the clustering method is based on it, were presented in several papers, e.g. [1], [2]. Referring interested readers to them we recall now only a few ideas which are necessary for understanding this paper.

- The input data table, after appropriate normalization, must have the *form of a bivariate probability table.* Any two-dimensional table with non-negative values can be easily transformed into this form. The transformed table will be denoted by $P = (p_{i,j})$, $i = 1, ..., m$; $j = 1, ..., k$. Due to the universal form, data structures can be expressed in terms of stochastic dependence between marginal (row and column) variables, say X and Y, irrespective of the data table contents.

- Instead of categorical variables X and Y, the pair of *continuous* variables (X^*, Y^*) defined on $[0, 1] \times [0, 1]$ is considered. The density h of the new distribution is constant and equal to $h_{ij} = p_{ij}/(p_{i\bullet}p_{\bullet j})$ on any rectangle

$$\{(u, v) : S^X_{i-1} < u \leq S^X_i \text{ and } S^Y_{j-1} < v \leq S^Y_j\}$$

where $S^X_i = \sum^i_{l=1} p_{l\bullet}$, $S^Y_j = \sum^j_{l=1} p_{\bullet l}$ and $p_{i\bullet} = \sum^k_{l=1} p_{il}$, $p_{\bullet j} = \sum^m_{l=1} p_{lj}$ for $i = 1, ..., m$; $j = 1, ..., k$. This density is called the randomized grade density of (X, Y). The joint distribution of (X^*, Y^*) is called the copula of (X, Y). Copulas are bivariate distributions on $[0, 1] \times [0, 1]$ with uniform marginals (literature on copulas is enormous - see eg. [4]). Evidently, X^* and Y^* are each uniform on $[0, 1]$.

- Any change in permutations of rows and columns of the data table (categories of X and Y) affects values of S^X_i and S^Y_j and consequently changes the copula.

- The overrepresentation map serves as a very convenient tool for visualisation of data structures. Every cell in the data table is represented by the respective rectangle in $[0, 1] \times [0, 1]$ and is marked by various shades of grey, which correspond to magnitudes of the randomized grade density. The value range of the grade density is divided into several intervals (five are used in the paper). Each colour represents a particular interval; the black corresponds to the highest values, the white - to the lowest. As grade density h measures deviation from *independence* of variables X and Y, dark colours indicate overrepresentation ($h > 1$), light colours show underrepresentation ($h < 1$). Widths of rows (columns) reflect respective marginal sums.

- The randomized grade correlation coefficient $\rho^*(X, Y) = cor(X^*, Y^*)$ measures dependence between variables X and Y; for discrete variables it is equal to Schriever's extension of Spearman's rho (cf. [5]). Coefficient ρ^* may be expressed by various equivalent formulas. The one most convenient in correspondence and cluster analysis is the following:

$$\rho^*(X, Y) = 6 \int_0^1 (u - C^*(Y : X)(u))du = 6 \int_0^1 (u - C^*(X : Y)(u))du,$$

where $C^*(Y : X)(t) = 2 \int_0^1 r^*(Y : X)(u)du$ is called the randomized grade correlation curve and $r^*(Y : X)(t) = E(Y^* \mid X^* = t)$ is the randomized grade regression function.

- The grade correspondence analysis (GCA) maximizes positive dependence between X and Y (measured by ρ^*) in the set of all permutations of rows and columns of the data table (categories of X and Y). Let us note that due to the optimal permutations, there is no need to assume any particular form of this dependency. The important property of the GCA is that *similar rows as well as columns are always placed close to one another*. In this case, the values of the regression functions r^* can serve as a similarity measure, because both regressions: $r^*(Y : X)$ and $r^*(X : Y)$ are nondecreasing for the optimal permutations ([2]). The overrepresentation map of the data table, which has optimally permuted rows and columns with strong dependency, shows tendency to group all dark rectangles along the diagonal of the unit square.
- The grade cluster analysis (GCCA) is based on optimal permutations provided by the GCA. Assuming that numbers of clusters are given, rows and/or columns of the data table (categories of X and/or Y) are optimally aggregated. The respective probabilities are the sums of component probabilities, and they form a new data table. In this case, optimal clustering means that $\rho^*(X, Y)$ is maximal in the set of these aggregations of rows and/or columns, which are *adjacent in optimal permutations*. The rows and columns may be aggregated either separately (i.e. we maximize ρ^* for aggregated X and nonaggregated Y or for nonaggregated X and aggregated Y), or simultaneously. In this paper, only the first method is used. Details concerning the maximization procedure can be found in [1].

3 The use of capital sources by retail trade firms in Poland

The data table analyzed here, came from the research of dr A. Mielczarek from the Institute of Home Market and Consumption in 1999. The aim of this research was to detect:

- which capital sources are preferable by what kind of firms,
- what were the effects of the choices on the firms economic condition,
- whether detected structures (relations) are stable in time.

The questionaire which was used included questions of two kinds. The first group characterized various forms of capital sources obtainable in Poland. The second group provided information about inner capital stuctures of the analyzed firms and their profitability. The list of all questions used in this analysis is included in Tab. 1. For all analysed questions, the firms provided answers for three consecutive years: 1996-1998. Due to technical reasons (no missing data were admissible), the initial data table was restricted to those firms which provided a complete set of answers. In consequense, the sample used is composed of 66 firms (rows in the data table).

Table 1. Analyzed questions

Number	Description	Number	Description
P1	Generation of financial surplus	P7	Bank credit
P3	Profitability level	P11	Credit guaranty fund
P4	Share of firm's own capital	P12	Expansion of partner base
P5	Share of long-term obligations	P15	Leasing
P6	Share of retail obligations	P19	Factoring

4 Typology for pooled, three-year data

The grade correspondence analysis (GCA) applied to the data descibed above, revealed its two specific properties. First, variables which correspond to the same question, however concerning different years, are usually placed one after the other. The only exceptions were questions P12 and P19. This means that the similarity within years is much stronger than the similarity among particular questions (the effect is invariant from the years).

The second observation is that two questions: P19 and P1 provide numerous exceptions from the monotonicity rule (there are many dark rectangles far away from the diagonal). It seems that these questions form outliers in this structure. The deletion of one of them from the data table increases the value of the maximal correlation coefficient ρ^*. After removal of both, the maximal ρ^* increases from the initial value 0.2233 to 0.2502. As the removal of significant variables diminishes dependence in the data table, this fact confirms that both of these questions are not strongly correlated with the revealed structure. Therefore, they are not taken into account in the further analysis.

It is interesting that the optimal permutation of remaining columns (variables) did not change, and was the following: P15(97), P15(98), P15(96), P7(98), P7(96), P7(97), P5(96), P5(97), P5(98), P6(97), P6(98), P6(96), P4(96), P4(97), P4(98), P12(96), P11(96), P11(97), P11(98), P12(97), P12(98), P3(98), P3(97), P3(96). The corresponding overrepresentation map is shown in Fig. 1. It clearly shows, that similar firms are placed close to one another. To an optimally arranged data table, the grade cluster analysis was applied. The horizontal lines in Fig. 1 mark the boudaries of particular clusters. In this case three was chosen as the number of clusters .

The distribution of dark and light rectangles provides an easy interpretation of generated classes. Comparative analysis of the average values of variables in the particular clusters (shown in Tab. 2) confirms this interpretation:

- CLUSTER 1: In this cluster firms are characterized by the greatest number of leasing contracts (P15) and bank credits (P7), and the high share

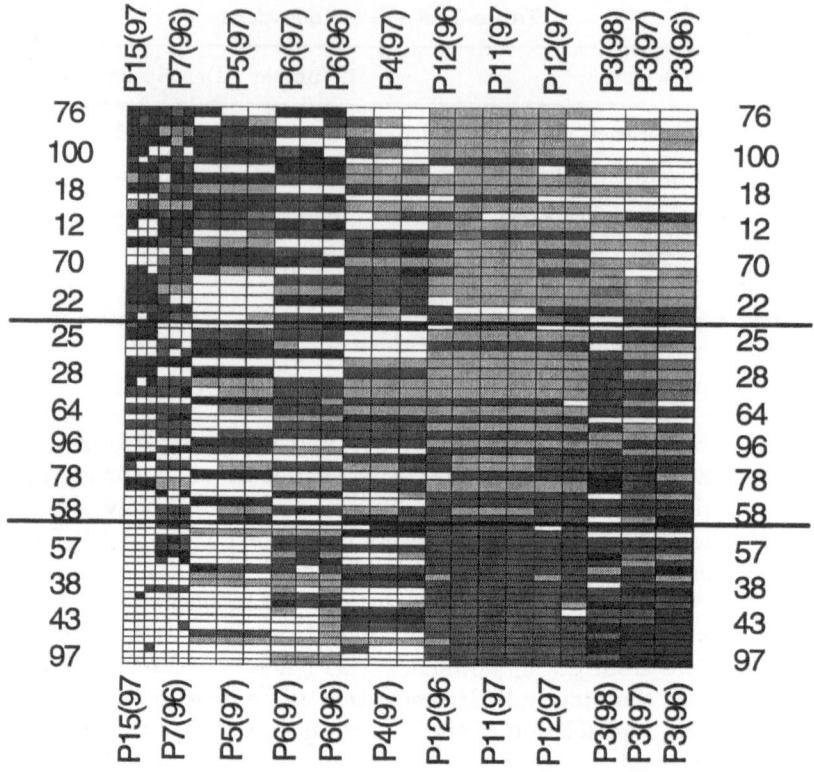

Fig. 1. Overrepresentation map of the analyzed data table optimally arranged

of obligations of both types (P5 and P6). All these characteristics are linked with the lowest average profit (P3).

- CLUSTER 3: Very low number of leasing contracts and bank credits; also very low share of obligation. However, the average profit is the highest.
- CLUSTER 2: This cluster is an intermediary level between cluster 1 and 3. The numbers of leasing contracts and bank credits are lower than those in class 1 and higher than those in class 3. The same observation applies to the share level of retail obligations and profitability. However, the long-term obligations are similar to those in class 1.

Let us note that the cluster interpretation did not include three questions: P4, P11 i P12. The reason is that their averages and distributions are very similar in all classes. Moreover, the averages for questions P11 and P12 are very near to their maximal values. That means that they are almost constant. The maximal values denote in this case, that the respective capital sources were not used. In other words, credit guaranty funds and expansion of partner base are used very rarely by Polish firms. In consequence, all three questions were acknowledged as unsignificant in the generated typology. However, their

Table 2. Average values for the analyzed variables

Questions	Clusters			All	Questions	Clusters			All
	1	2	3	clust.		1	2	3	clust.
P15(97)	1.65	0.78	0.00	0.85	P4(96)	2.26	2.04	1.65	2.00
P15(98)	1.35	0.65	0.05	0.71	P4(97)	2.17	2.04	1.70	1.98
P15(96)	1.39	0.74	0.10	0.77	P4(98)	1.96	2.00	1.70	1.89
P7(98)	1.48	0.70	0.30	0.85	P12(96)	1.87	1.91	1.85	1.88
P7(96)	1.39	0.61	0.40	0.82	P11(96)	1.96	2.00	2.00	1.98
P7(97)	1.30	0.74	0.35	0.82	P11(97)	1.91	2.00	2.00	1.97
P5(96)	2.43	2.09	1.15	1.92	P11(98)	1.91	2.00	2.00	1.97
P5(97)	2.35	2.35	1.10	1.97	P12(97)	1.87	1.96	1.90	1.91
P5(98)	2.39	2.39	1.10	2.00	P12(98)	1.83	1.91	1.90	1.88
P6(97)	2.09	1.61	1.30	1.68	P3(98)	1.65	2.74	2.90	2.41
P6(98)	2.13	1.70	1.30	1.73	P3(97)	1.87	2.65	3.20	2.55
P6(96)	2.00	1.52	1.30	1.62	P3(96)	2.00	2.74	3.40	2.68

behaviour should be interpreted with great care and verified in the future on a greater sample.

Nevertheless, it should be noted that the quality of the presented typology is good, because the aggregation of the data table according to the cluster analysis, diminishes the value of correlation coefficient only slightly - from 0.2502 to 0.2215. That means, that the aggregation caused only a small loss of information.

5 Typologies for annual data

The next step of the analysis was the application of the GCCA method to each of three data subtables. Every table contained data restricted to one year only. The optimal permutations of questions as well as the maximal values of ρ^* are shown in Tab. 3. The values of the correlation coefficients are very similar. The permutations for 1996 and 1997 are identical, the permutation for 1998 differs a little. Due to lack of space the respective tables of the average values for particular variables in each cluster are not presented. However, the interpretation of particular typologies, which follows them, remains identical to the interpretation of that generated for the pooled, three-year data. (As before, three was chosen as the number of clusters.) Therefore, it may be said, that based on the interpretation, the typology provided by the GCCA

Table 3. Results of the GCCA for annual data tables

Year	Optimal permutations								max ρ^*
1996	P15	P7	P5	P6	P4	P11	P12	P3	0.2540
1997	P15	P7	P5	P6	P4	P11	P12	P3	0.2587
1998	P7	P15	P5	P6	P11	P12	P4	P3	0.2620

is stable in time, or in other words, there is the only one typology of firms, invariant from time.

6 Relations between the generated typology and firms profitability

Profitability level appeared to be one of the most important firm characteristics in the typology presented above. It occupies the extremal position in the optimal permutation of variables, what confirms its great discriminating power in firm discrimination. Although this role is constant in the analyzed years, that does not necessary mean, that firm profits are also stable.

Once more, the GCCA method was applied, this time to the data table which included only profit values (P3(96)-P3(98)). An interesting result is that the optimal permutation of variables is identical with the ordering of years. This suggests that some trend across time exists. This trend is described well by the result of cluster analysis. Fig. 2 shows the overrepresentation map of the data table after optimal permutation with marked cluster boundaries. The analysis of this map and the average profit in particular clusters leads to the following typology.

- CLUSTER 1: It is formed by firms with gradual decreasing profits. Moreover, this decrease is rather strong and the starting levels (in 1996) were comparatively high.
- CLUSTER 2: Firms in this class are characterised by differentiated profitability levels, which remain stable.
- CLUSTER 3: It has opposite characteristics to cluster 1. It includes succesful firms with continually increasing profits. Moreover, these firms started from rather low levels. Unfortunately, this cluster contains only seven firms.

Now, let construct a new variable which describes firm memberships to particular clusters for this typology. The similar variable is generated for typology obtained for all questions. The maximal correlation ρ^* between these variables is equal to 0.1867. However, the overrepresentation map of the respective contingency table is regular, without deviations from monotonicity. That means, that these typologies are correlated but dependence is not very strong.

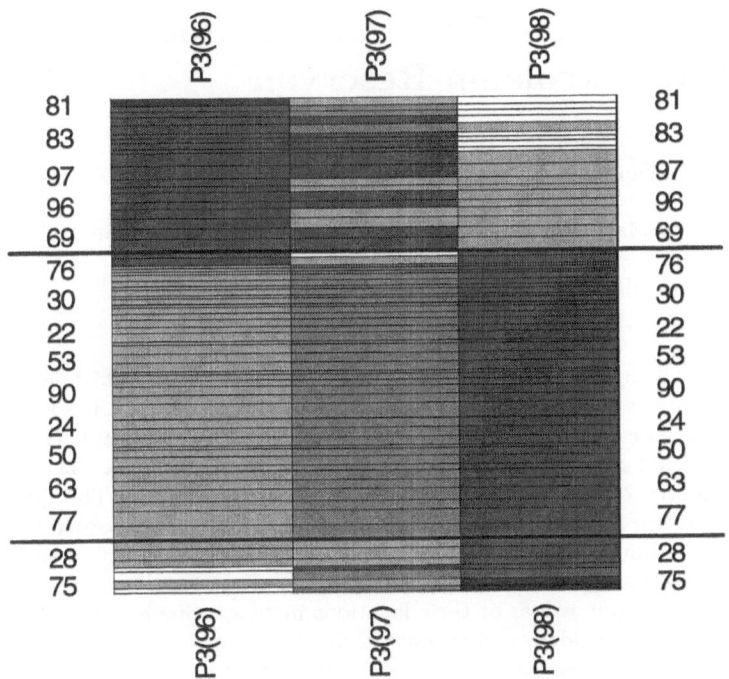

Fig. 2. Overrepresentation map for profitability data table after GCCA

References

1. Ciok, A. (1998) Discretization as a tool in cluster analysis. In: Rizzi A., Vichi M., Bock H.H. (Eds.) Advances in Data Science and Classification. Springer, Berlin Heidelberg, 349–354
2. Ciok, A., Kowalczyk, T., Pleszczynska, E., Szczesny, W. (1995) Algorithms of grade correspondence–cluster analysis. The Collected Papers of Theoretical and Applied Computer Science 7, no 1–4, 5–22
3. Ciok, A., Kowalczyk, T., Pleszczynska, E., (1998) How a New Statistical Infrastructure Induced a New Computing trend in Data Analysis. In: Polkowski L., Skowron A. (Eds.) Rough Sets and Current Trends in Computing, First International Conference, RSCTC'98, Poland, Warsaw, June 1998 Proceedings, Lecture Notes in Artificial Intelligence 1424, Springer, Berlin Heidelberg, 75–82
4. Nelsen, R. B. (1991) Copulas and association. In: Dall'Aglio, G., Kotz, S., Salinetti, G. (Eds.) Advances in Probability Distributions with Given Marginals. Kluwer Academic Publishers, Dordrecht.
5. Schriever, B. F. (1985) Order dependence. Ph.D. Dissertation, Free University of Amsterdam.
6. Szczesny, W. (2001) Grade correspondence analysis applied to contingency tables and questionnaire data. Intelligent Data Analysis, 5, 1–35.

A Contribution to Stochastic Modeling Volcanic Petroleum Reservoir

Sutawanir Darwis

Department of Mathematics , Institut Teknologi Bandung, Jl Ganesa 10, Bandung, Indonesia,
E-mail: e-mail: sdarwis@dns.math.itb.ac.id

Abstract. The hydrocarbon reservoir data do not stem from a statistical designed experiment. Wells are drilled to pump oil and are sampled for data. In some cases the limits of the reservoir are not known beforehand from the geology. They have to be determined from the well data as information becomes available. The uncertainty about geometry of the reservoir introduces a source of error. This justifies the use of stochastic approach, which generate a range of possible reservoir models, respecting the information that is available. Stochastic simulation is an algorithm that allow generating multiple realizations of a spatial process and conditioned to reproduce the sample values at their locations in space. Stochastic simulation can handle sparse, nongridded, and correlated data. This article presents an overview of petroleum reserves estimation methods and proposes a conditional simulation approach to assess the uncertainty of the reserves evaluation on a volcanic reservoir. Conditional simulation seems to be a suitable tool for estimating the volume of hydrocarbon in place and to indicate local anomalies.

1 Introduction

The complexity of reservoir geometry and measurements uncertainty at the wells makes it difficult to perform error calculation of reserves evaluation. Stochastic simulation seems to be suitable tool for hydrocarbon reserves evaluation. Spatial analysis of relevant parameters are provided in a simulation study of reserve estimation. Interpolation techniques can be used for reserves evaluation to calculate the volume of hidrocarbon in place, denoted by Q, is given by the integral (Delfiner, 1979)

$$Q = \int_X \int_Y \int_{\min(WL,Bot(x,y))}^{Top(x,y)} S_0(x,y,z)\phi(x,y,z)dzdydx \qquad (1)$$

where WL = Water Level, Bot = Bottom, ϕ = porosity, S_0 = Oil Saturation, (x,y) = horizontal coordinates, z = depth. Numerical computation would require estimation of the hydrocarbon porosity

$$\phi_h(x,y,z) = S_0(x,y,z)\phi(x,y,z) \qquad (2)$$

at the nodes of a 3-D grid limited by the reservoir boundaries. On account of the vertical heterogeneity of a reservoir, this would be a heavy task. A shortcut consists in reducing the problem to a 2-D by considering cumulated hydrocarbon porosities $\overline{\phi}_h(x, y)$. Suppose

$$[\min[WL, Bot(x, y)] - Top(x, y)] x\overline{\phi}_h(x, y) = H\phi S_0(x, y) \qquad (3)$$

where $H = $ Thickness, $\overline{\phi}_h = $ average hydrocarbon porosity, then

$$Q = \int_X \int_Y H\phi S_0(x, y)dxdy \approx \sum_x \sum_y H\phi S_0(x, y) \qquad (4)$$

It suffices to grid in 2-D the parameters $H\phi S_0(x, y)$ and sum over the grid points either by working on $H\phi S_0(x, y)$ directly or by gridding H, ϕ, S_0 independently and multiplying the grids.

The first approach requires that measurements of H, ϕ, and S_0 are available for all wells. This is usually not the case and the second is preferred. The complexity of reservoir geometry and the fact that several parameters are involved makes it difficult to perform rigorous error estimations.

When the boundaries of the reservoir are very imprecise then one must resort to the conditional simulation approach. Interpolation can provided the input grids for reservoir simulation models. The basic assumption in spatial model was to consider the phenomenon as a realization of a stochastic process. The smoothing effect of kriging will averaged out high and low values. An integration of estimation and simulation is advocated to improve in estimation of recoverable reserves. The idea of simulation is to exhibit other possible realizations of the spatial process. Any interpolation methods gives a smoothed picture of reality, simulations display the same amount of spatial variability that can be expected from the actual phenomenon.

In petroleum geology applications, kriging spatial interpolation has been used to predict porosity, thickness for reserves evaluation. Spatial interpolation can be too smooth to represent variations in porosity. Stochastic simulation is becoming an important tool for uncertainty analysis in reserves evaluation. The basic idea of stochastic simulation is to generate realizations of the process $\{Z(s) : s \in D\}$ that preserves the mean and covariance structure. Let $Z = (Z_1, \ldots, Z_n)'$ denote the values to be simulated and let $EZ = \mu$ and $Var(Z) = \Sigma$. Conditioning the simulation means forcing the simulated surface to pass through the data points. The conditional surface is obtained as

$$Z_{CS}(s) = \widehat{Z}(s) + \left(Z_S(s) - \widehat{Z}_S(s)\right) \qquad (5)$$

where $\widehat{Z}(s)$ is the interpolated surface using data at locations $s_1, \ldots, s_n$, $Z_S(s)$ is the simulated surface and $\widehat{Z}_S(s)$ is a interpolated version of the simulated process using $\widehat{Z}_S(s_1), \ldots, \widehat{Z}_S(s_n)$. Interpolator excact property of kriging implies that $Z_{CS}(s_i) = Z(s_i)$, $i = 1, \ldots, n$.

2 Sequential Gaussian Simulation

Consider the joint distribution of N random variables (RV's) $Z_i, i = 1, \ldots, N$. The N RV's may represent the same attribute at the N nodes of a dense grid discretizing the field D. Consider the conditioning of these N RV's by a set of n data, symbolized by $|(n)$. The corresponding N-variate ccdf is denoted:

$$F_{(N)}(z_1, \ldots, z_N|(n)) = P(Z_i \leq z_i|(n)) \tag{6}$$

Drawing an N-variate sample from the ccdf can be done in N successive steps:

- draw a value $z_1^{(l)}$ from the univariate ccdf of Z_1 given the original data (n). The value $z_1^{(l)}$ is considered as a conditioning datum for all subsequent drawings; $(n+1) = (n) \cup Z_1 = z_1^{(l)}$.
- draw a value $z_2^{(l)}$ from the univariate ccdf of Z_2 given the updated data set $(n+1)$, set $(n+2) = (n+1) \cup Z_2 = z_2^{(l)}$.
- sequentially consider all N RV's Z_i.

The set $\{z_i^{(l)} : i = 1, \ldots, N, \quad l = 1, \ldots, L\}$ represents a simulated joint realization of the N dependent RV's Z_i.

In sequential Gaussian simulation, each variable is simulated sequentially according to its normal ccdf. The conditioning data consist of all original data and all previously simulated values found within a neighborhood of the location being simulated. The conditional simulation of a Gaussian $Z(s)$ proceeds as follows:

1. Determine the univariate cdf $F_Z(z)$ representative of the entire study area.
2. Using the cdf $F_Z(z)$, perform the normal score transform of z-data into a standard normal y-data
3. Check for bivariate normality of the normal score y-data.
4. If a multivariate Gaussian can be adopted for the y-variable, proceed with sequential simulation,
   - Define a random path that visits each node of the grid once. At each node s,retain a specified number of neighboring conditioning data.
   - Use simple kriging (SK) with the normal score semivariogram model to determine the parameters of the ccdf of the $Y(s)$ at location s.
   - Draw a simulated value $y^{(l)}(s)$ from that ccdf
   - Add the simulated value $y^{(l)}(s)$ to the data set
   - Preceed to the next node, and loop until all nodes are simulated.
5. Backtransformed the simulated normal scores $\{y^{(l)}(s) : s \in D\}$ into the original variable $\{z^{(l)}(s) = \Phi^{-1}(s) : s \in D\}$. If multiple realizations are desired $\{z^{(l)}(s) : s \in D, \quad l = 1, \ldots 1, L\}$, the previous algorithm is repeated L times.

3   Reserves Evaluation

There are four techniques used for the calculation of reserves evaluation (Rani, Cheong, 1982). The selection depends on the number of control points, and geometry of the hydrocarbon accumulations. The minimum, most likely, and maximum of each of the basic reservoir parameters such as trap area, trap fill, oil trap fill, porosity, oil saturation are sampled using Monte Carlo simulation to obtain the in-place hydrocarbon. Normal, log normal, triangular, uniform usually used as input to Monte Carlo simulation. The $H\phi S_0$ is used when the reservoir in the development phase. Isoporosity, isopach and isohydrocarbon saturation maps are constructed, and these are then cross-contoured to obtain the $H\phi S_0$ map. Porosity ϕ is the total available pore space between the rock grains which is occupied by oil, gas or water. Oil saturation S_0 is the percentage of the pore space that is filled with oil. The trap area is the total prospect area that might contain oil or gas; the saddle between two structures. Part of the trap is filled with hydrocarbon and expressed as a percentage of trap fill. The total hidrocarbon is the broken down to oil and gas percentages, and expressed as oil trap fill and gas trap fill. The net pay is the vertical thickness of reservoir rock that contains moveable hydrocarbon.

The problem of reserves evaluation is different from classical statistical problem. The data obtained are not the result of a random sample. The measurements are made where it is economically feasible. The resulting measurements are not independent and identically distributed; the spatial correlation can be significant. Local anomalies are indicated on the countur plots by elliptic closed areas. The mapping of these closed areas on each contoured realization of the layer and the intersection of all these areas indicate the optimal location of a structural high. The number of intersecting closures and the variability of their location allows a ranking of favorable areas for drilling. Various attempts have been made in order to map potential oil traps. All methods focus in delineating structural anomalies favorable for oil accumulation. Kriging interpolation methods are able to detect local anomalies. Conditional simulation techniques are suitable to model small scale variations of a spatial process and to indicate local anomalies by conditioning the simulated data to control points. It seems to be possible to localize and to estimate shapes and sizes of structural anomalies. Each realization of a regionalized variable defines areas of positive and negative closure. These areas may be random and disappear in the following simulation run or they may be stable in almost all simulation runs.

The Jatibarang oil field is a fractured volcanic rock with 90 million n^3 estimated recovarable reserves. The productive block covering an area of 5 km × 5 km with 160 wells at 2000 m depth. The study area is divided into 50 × 50 grids of 100 m × 100 m. The data used for the study have been extracted from internal drilling report. Wells were selected with porosity, thickness and oil saturation information (Fig. 1). Porosity is the available

278

pore space which is occupied by water, oil or gas. Oil saturation is the percentage of the pore space that is filled with oil. Porosity and oil saturation are derived from core analysis. The hydrocarbon volume calculation methods are based on nine reservoir parameters : trap area, trap fill, trap fill that is oil, geometric correction factor, thickness, formation volume factor, porosity, oil saturation, barrels per ac.ft. One sample Kolmogorov-Smirnov test of composite normality gives $ks = 0.083$, $p - value = 0.16$ for porosity, $ks = 0.1179$, $p - value = 0.3$ for thickness, and $ks = 0.119$, $p - value = 0.28$ for oil saturation. All test concludes the acceptance of normal distribution for each reservoir parameter.

Figure 4 displays the porosity $\times$ thickness $\times$ oil saturation distribution based on sequential Gaussian simulation on 50×50 grid of 150 m $\times$ 150 m. The plot gives some idea of distribution shape. The simulation result shows a similarity with the porosity-thickness-oil saturation plot (Fig. 2 and Fig. 3). There appears to be a high porosity $\times$ thickness $\times$ oil saturation in the north part of the reservoir and low in the south part. It may be interesting to note that there exist a promising structure in the west-south part of the area with no drillhole in its neighbourhood. The question is to decide whether this structure is real or only a result of simulation. A restriction of the reservoir geometry would yield a better porosity-thickness-oil saturation visualizations and a more realistic reserves evaluation.

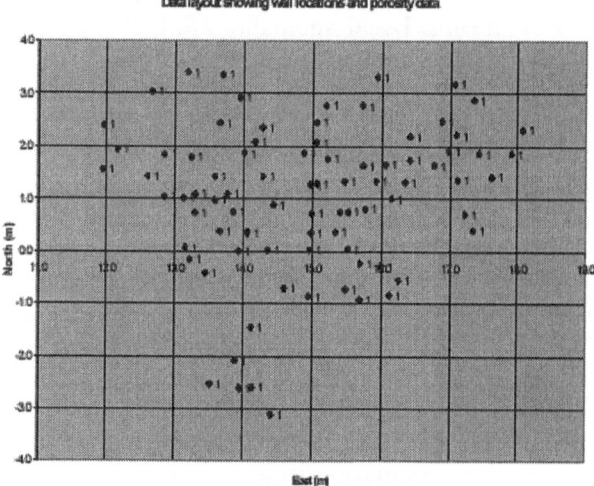

Fig. 1. Well locations and prosity data

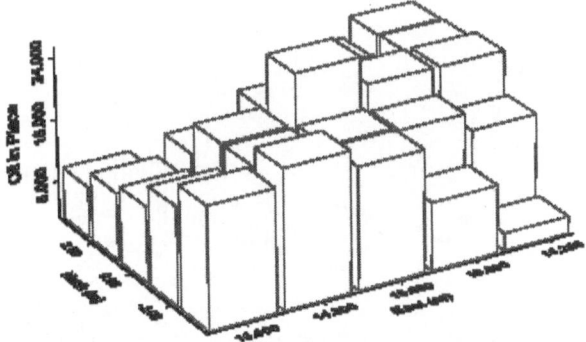

Fig. 2. Porosity-thickness saturation distribution

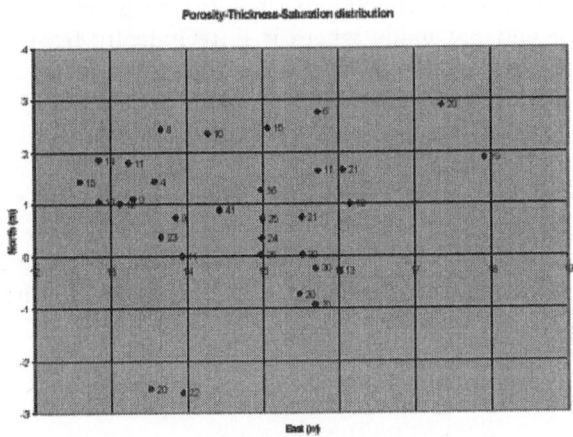

Fig. 3. Porosity-thickness saturation

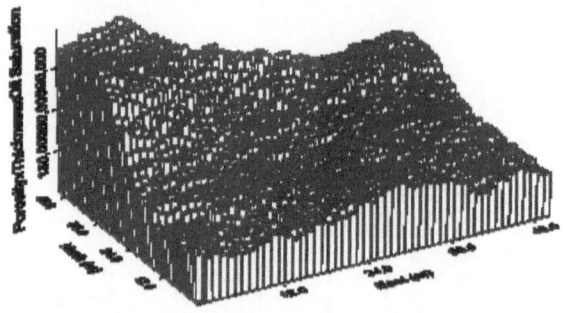

Fig. 4. Sequential Gaussian simulation of porosity-thickness saturation

4 Summary

The study reviews the methodology used in hydrocarbon resource assesment. Wells are drilled to pump oil and are sampled for data. The reservoir data are not the result of a well defined experimental design concepts, measurements are expensive and are made where it is technically feasible. The complexity of geologic structures and the geometry of the trap area are sources of uncertainty in potential reserves estimation. Sequential Gausian simulation seems to be a promising tool for estimating the resource assesment compared with Monte Carlo simulation techniques. A more detailed study on mapping potential oil traps is planned to obtain more quantitative results for volume estimates. The stochastic simulation techniques as used in reservoir characterization is relatively new, many of the statistical properties associated with this method have not been studied; spatial correlation sensitivity, comparison of simulation algorithms

References

1. Delfiner, P. (1979), Basic Introduction to Geostatistics, Ecole d'Ete Fontainbleau CGMM.
2. Deutsch, C. V, Journel, A. G. (1992), GSLIB: Geostatistical Software Library and User's Guide, Oxford University Press.
3. Gotway, C. A. (1994), The Use of Conditional Simulation in Nuclear-Waste-Site Performance Assessment, Journal of the American Statistical Association, 36, 129 - 141.
4. Rani, A. M., Cheong, Y. C. (1982), In-Place Hydrocarbon Volume Calculation Techniques, Proceeding of Monte Carlo Simulation and Related Subjects, Comittee for Co-ordination of Joint Prospecting for Mineral Resources in Asian Offshore Areas (CCOP).

Improving the Quality of Association Rule Mining by Means of Rough Sets

Daniel Delic, Hans-J. Lenz, and Mattis Neiling

Free University of Berlin, Institute of Applied Computer Science,
Garystr. 21, D-14195 Berlin, Germany

Summary. We evaluate the rough set and the association rule method with respect to their performance and the quality of the produced rules. It is shown that despite their different approaches, both methods are based on the same principle and, consequently, must generate identical rules. However, they differ strongly with respect to performance. Subsequently an optimized association rule procedure is presented which unifies the advantages of both methods.

1   Introduction

Data mining methods are part of knowledge discovery from databases. Its objective is to extract nontrivial, beforehand unknown and potentially useful information. There exist several data mining procedures which differ in their methodology as well as in their data types.

In our work we focus on a comparison of the *association rules* with the *rough sets*. The association rule method was developed particularly for the analysis of large databases, whose attributes posses only Boolean values. The rough set method, however, investigates databases. Its attributes can posses several values with the constraint, that a predefined set of attributes must exist on which the generated rules are based on. In order to be able to make a fair comparison, both procedures should operate on the same data type. Our aim is to find out, which method is more efficient, and whether a combination of both methods improves the overall quality of unscrambled rules.

2   Association Rules

The association rule method was developed originally for the analysis of large databases of customer transactions. Each transaction consists of items purchased in a supermarket. In order to apply this method to data multi-value attributes, we defined the term *item* in a different way.

2.1   Transforming Attributes into Items

An item combines the name (label) of an attribute with one of its possible values. Each attribute has a finite set of values which is called domain.

282

Within a database each combination of attribute name and attribute value is distinguishable. Consequently, there exists a set $A = \{a_1, a_2, a_3, ..., a_p\}$ with $p \in N$ items in a database.

Example 1. To represent an attribute with binary domain we only need one item $a_i \in \{0, 1\}$. For example, for attribute 'spaghetti' with domain(spaghetti) $= \{bought, not bought\}$ $(i = 1)$ we set $a_1 = 1$ if spaghetti are bought (see Fig. 1).

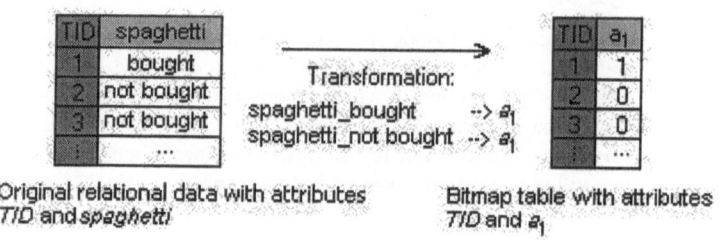

Original relational data with attributes *TID* and *spaghetti*

Bitmap table with attributes *TID* and a_1

Fig. 1. Transformation of relational data into an efficient bitmap representation for attributes with binary domains.

Example 2. For an attribute with no-binary domain, each attribute value corresponds to one item. For example, for attribute 'blood_pressure' with domain(blood_pressure) = {high, low, normal} $(i = \{1, 2, 3\})$ the following items result: a_1 = "blood_pressure_high", a_2 = "blood_pressure_low" and a_3 = "blood_pressure_normal" (see Fig. 2).

TID	blood press.		Transformation:		TID	a_1	a_2	a_3
1	high		*blood pressure_high* --> a_1		1	1	0	0
2	low		*blood pressure_low* --> a_2		2	0	1	0
3	low		*blood pressure_normal* --> a_3		3	0	1	0
⋮	...				⋮	...	...	

Original relational data with attributes *TID* and *blood pressure*

Bitmap table with attributes *TID*, a_1, a_2 and a_3

Fig. 2. Transformation of relational data into an efficient bitmap representation for attributes with no-binary domains.

Finally an association rule can be defined as follows: Given the subsets $X \subset A$ and $Y \subset A$ with $X \cap Y = \emptyset$, an association rule is an implication in the form $X \rightarrow Y$, with a confidence c $(0 \leq c \leq 1)$ and a support s $(0 \leq s \leq 1)$.

The confidence is a measure of how many tuples that contain X also contain Y. The support is the relative frequency of the set $X \cup Y$.

2.2 Data Structure 'Bitmap'

It is known from Database Theory that bitmaps are efficient data structures (Jürgens/Lenz, 2001). Therefore the original relational schema has to be transformed into a bitmap table. The first column holds the primarykey attribute *TID*, each of the remaining columns contain a binary bitmap-attribute i.e. an item from the original data set (see Fig. 1 and 2). This structure is efficient, because the column-address of a bitmap-attribute doesn't change. Therefore it is possible to check very fast each tuple, by direct access for the appropriate column, whether the searched item is or is not available in the corresponding tuple of the original data set. If its present there is a '1', otherwise a '0' in the bitmap.

2.3 Rules Derivation

The procedure to generate association rules is based on the multi-pass-algorithm *Apriori*[1]. The principle idea of this algorithm is to scan several times a database while searching for sets of items that occur in a sufficient large number. After each pass the number of items in those "large sets" is increased by one until all existing sets in the database are found. Subsequently the association rules are derived from these item sets.

3 Rough Sets

The rough set theory was introduced by Zdzislaw Pawlak [2]. It is a method for uncovering dependencies in data, which are recorded by relations. A detailed introduction to rough set theory can be found in *Munakata* (1998).

3.1 Model

The rough set method operates on data matrices, so called "information tables" (see Table 1). It contains data about the *universe U* of interest, *condition attributes* and *decision attributes*. The goal is to derive rules that give information how the decision attributes depend on the condition attributes.

A prerequisite for rule generation is a partitioning of U in a finite number of blocks, so called equivalence classes, of same attribute values by applying an equivalence relation. For example, the equivalence relation R_1 $= \{(u, v) | u(blood\,pressure, temperature) = v(blood\,pressure, temperature)\}$

---

[1] See *Agrawal/Srikant* (1994).

[2] See Pawlak (1982).

Table 1. Information table.

universe person	condition attributes temperature	blood pressure	decision attribute heart problem
Adams	normal	low	no
Brown	normal	low	no
Carter	normal	medium	jes
Ford	high	high	jes
Gill	high	high	no
Bellows	medium	medium	jes

leads to a partition of U into three equivalence classes $U_1 = \{Adams, Brown\}, U_2 = \{Carter\}$ and $U_3 = \{Ford, Gill\}$ (see Table 1). Given these classes, rules like e.g. "If temperature *normal* and blood pressure *low* then heart problem *no.*" can be derived . Generally, no unique rule derivation is possible. For example, *Ford* and *Gill* have identical values of the condition attributes, but differ in their values of the decision attribute. In order to analyze such data, the concept of approximation spaces is used to determine the confidence of the derived rules.

The quality of the extracted rules depends also strongly on the possibility of attribute reduction. Reducing the number of attributes in a dataset by removing the redundant ones is one of the main objectives of rough set theory and at the same time one of the main problems. Finding a minimal reduct is a NP-hard[3] problem. Finding all reducts has exponential complexity[4].

In our approach we try to reduce the computing time by applying the concept of reduct extraction directly to the produced rules, not to attributes. First, in order to generate strong rules, all rules which support and confidence values don't reach the given minimum threshold, are deleted. Second, the reducts are extracted. Suppose, there are two rules with same decision item and same confidence value, and their only difference is the set of condition items. The condition itemset of rule no 1 is a subset of the condition itemset of rule no 2. In this special case rule no 2 is redundant and can be deleted without having loss of information.

3.2 Rough Set Based Rule-Generation with *RS-Rules+*

Besides the "classic" version of rule generation with a fixed decision attribute (cf. Table 1), our algorithm *RS-Rules+* offers the possibility of varying the selected decision attributes of a table. Each attribute of the table will be included either as a decision or condition attribute. *RS-Rules+* is explained on the basis of Table 2: The attributes of the original data set are $\{A\}, \{B\}$ and

---

[3] See Rauszer (1991).
[4] See Skowron/Rauszer (1992).

$\{C\}$. Each attribute has two non *null* values $A = \{a_1, a_2\}, B = \{a_3, a_4\}$ and $C = \{a_5, a_6\}$, so that there are six items (bitmap-attributes) for the resulting bitmap-table: $\{a_1\}, \{a_2\}, \{a_3\}, \{a_4\}, \{a_5\}, \{a_6\}$. In the first step all possible rules are constructed from all bitmap-attributes of the table. For example, for $\{A\} \rightarrow \{B\}$ four item-related rules can be set up: $\{a_1\} \rightarrow \{a_3\}, \{a_1\} \rightarrow \{a_4\}, \{a_2\} \rightarrow \{a_3\}$ and $\{a_2\} \rightarrow \{a_4\}$. All rules not fulfilling the minimum support and minimum confidence are deleted. The condition items from the remaining rules ($\{a_1\}$ and $\{a_4\}$), build pairwise the condition itemsets for the next run[5] (i. e. $\{a_1, a_4\}$). Again all possible rules are produced, their support and confidence is measured and the 'weak' ones are deleted. The set of the remaining rules $((\{a_1, a_4\} \rightarrow \{a_5\}))$ is compared with the set of produced rules in the previous run in order to remove the unnecessary ones.

Suppose, $\{a_1, a_4\} \rightarrow \{a_5\}$ has the same confidence as $\{a_4\} \rightarrow \{a_5\}$. In this case $\{a_1, a_4\} \rightarrow \{a_5\}$ is dominant and will be deleted. Since there are no more rules left to extract from the itemsets of the next generation, the algorithm stops.

In case that $\{a_1, a_4\} \rightarrow \{a_5\}$ is necessary it is stored as a valid rule and its condition itemset is part of the basis for next generation condition itemsets. Since there is no other new rule left, no new condition itemset(s) can be setup and the algorithm stops.

Table 2. Rule-Generation with *RS-Rules+*.

	1st run	2nd run
New condition attribute-sets	$A = \{a_1, a_2\}$ $B = \{a_3, a_4\}$ $C = \{a_5, a_6\}$	$\{a_1, a_4\}$
Possible Rules	$\{a_1\} \rightarrow \{a_3\}$ $\{a_1\} \rightarrow \{a_4\}$ $\{a_1\} \rightarrow \{a_5\}$ $\{a_1\} \rightarrow \{a_6\}$ $\{a_2\} \rightarrow \{a_3\}$ $\{a_2\} \rightarrow \{a_4\}$ $\{a_2\} \rightarrow \{a_5\}$ $\{a_2\} \rightarrow \{a_6\}$ $\{a_3\} \rightarrow \{a_5\}$ $\{a_3\} \rightarrow \{a_6\}$ $\{a_4\} \rightarrow \{a_5\}$ $\{a_4\} \rightarrow \{a_6\}$	$\{a_1, a_4\} \rightarrow \{a_2\}$ $\{a_1, a_4\} \rightarrow \{a_3\}$ $\{a_1, a_4\} \rightarrow \{a_5\}$ $\{a_1, a_4\} \rightarrow \{a_6\}$
Rules with minsupport (valid rules)	$\{a_1\} \rightarrow \{a_3\}$ $\{a_4\} \rightarrow \{a_5\}$	$\{a_1, a_4\} \rightarrow \{a_5\}$
Condition attribute-sets with minsupport	$\{a_1\}, \{a_4\}$	$\{a_1, a_4\}$

---

[5] The formation of condition itemsets is based on the same principle as the k-itemset construction in 2.3.

4 Comparison of the Procedures

The transformation of the underlying data set into a bitmap-representation and the modification of the rough set rules induction scheme allow an objective comparison of both methods. Three benchmark data sets have been selected, which differ in size and number of attributes[6]:

1. *Car Evaluation Database*: 1728 tuples, 25 Boolean attributes in the bitmap-table.
2. *Mushroom Database*: 8416 tuples, 12 original attributes selected, 68 boolean attributes in the bitmap-table.
3. *Adult*: 32561 tuples, 12 original attributes selected, 61 boolean attributes in the resulting bitmap-table.

The computing times of the rough set algorithm for the benchmark data were between 2 minutes and 4 hours, those of the association rule algorithm between 40 seconds and 45 minutes[7] (c.f. Table 3). Both procedures supplied similar results for all examined tables, except for few cases due to reducts of the rough set algorithm[8]. However, not creating redundant rules does not lead to any principal difference of rule production. Hence we conclude that both procedures are equivalent with respect to the produced rules.

Theorem. Let $c, s \in (0, 1]$ values for minimum confidence and minimum support. Then the algorithms *Apriori* and *RS-Rules+* induce identical rules from given data D w.r.t. c and s if the extraction of reducts by *RS-Rules+* does not take place.

Proof. Let $X \rightarrow Y$ be any rule derived with *RS-Rules+* and S the approximation space for Y. S is spanned by all the items, which are involved in X. The restriction of S on the lefthand-side X of a rule selects from D exactly those records which inherit a condition itemset equal to X. For this set D_X of the extracted records $\frac{|D_X|}{|D|} \geq s$ holds. Since only those rules are derived by *RS-Rules+*, which fulfill the minconfidence c and the minsupport s, both criterions are satisfied for the set $D_{X \cup Y} \subset D_X$. Hence the *Apriori*-algorithm will induce this rule also.

Conversely, let $X \rightarrow Y$ be any rule derived with *Apriori*. Then $D_{X \cup Y}$ the set of all records inheriting the large itemset $X \cup Y$ fulfills the inequalities $\frac{|D_{X \cup Y}|}{|D|} \geq s$ and $\frac{|D_{X \cup Y}|}{|D_X|} \geq c$. Since *all* approximation spaces S are derived by *RS-Rules+*, there exist a S', such that S' is an approximation space for Y, and hence the rule $X \rightarrow Y$ is induced, too. $\square$

---

[6] The benchmark data can be found in *UCI Repository of Machine Learning Databases and Domain Theories* (URL: ftp.ics.uci.edu/pub/machine learning-databases/Adult, /car, /mushroom).

[7] All tests were executed on a PC with an AMD K6-2/400 processor.

[8] A more detailed description can be found in Delic (2001).

5 Hybrid Association Rule Procedure *Apriori+*

Since the generated sets of rules of both procedures are almost equivalent, the selection of the procedure depends on the role of reducts. The best solution would be an additional extension of the faster association rule method with functions of the rough set procedure, such as formation of reducts and the ability to refer to a fixed decision attribute if needed. This hybrid method was realized with our algorithm "Apriori+": The derived rules are examined on reducts, unnecessary rules are deleted. If there is a given fixed decision attribute, all itemsets without this attribute can be ignored for rule generation.

A further benchmark test confirmed the advantages of *Apriori+*: With all examined data sets the same rules (including reducts) as with *RS-Rules+* have been extracted, whereby the necessary computing times corresponded to almost 100% to the Apriori association rule procedure (see Table 3).

Table 3. Complexity (CPU time) grouped by *database, parameters and the use of the decision attributes* for the algorithms *Apriori* (Apr), *Apriori+* (Apr+)and *RS-Rules+* (RS+).

Database	Car Evaluation					Mushroom					Adult				
Minsupport	10%					35%					17%				
Minconfidence	75%					90%					94%				
Fixed Decision Attribute	Yes		No			Yes		No			Yes		No		
Method	RS+	Apr+	Apr	RS+	Apr+	RS+	Apr+	Apr	RS+	Apr+	RS+	Apr+	Apr	RS+	Apr+
CPU Time [min]	1,15	1,12	1,10	3,15	1,12	3,32	2,02	2	15	2,02	64	44	44	233	44

From the examination of reducts on the rules and Theorem 1 follows directly:

Corollary. Let $c, s \in (0,1]$ values for minimum confidence and minimum support. Then both algorithms *Apriori+* and *RS-Rules+* induce identical rules from given data D w.r.t. c and s, if the extraction of reducts does take place.

In this case, the same redundant rules are left out by both algorithms.

6 Summary

The association rule procedure, which was originally developed for processing attributes with Boolean domains, can be based upon bitmap tables in order to analyze attributes with multi-value domains. The Rough set procedure originally was for generating rules with a fixed decision attribute. Evidently, the need of an a-priori-definition was removed.

On the basis of different data sets we compared the quality of produced rules and the necessary computing times of the algorithms. It turned out, that the rules of *RS-Rules+* and *Apriori+* do not differ. The computing times were in favor of the association rule procedure.

Even so, any final judgement should be stated carefully, since many factors have a great impact. The needed time of computation e.g. depends also strongly on the implemented algorithms. Improvements with the rough set algorithm could lead to reduction of computing time. Another factor is the methodology of the rule generation. It cannot be excluded that there exists another rough set based procedure which is more efficient for rule generation than the one introduced here. The investigation on these and other related questions will be subject of further work.

References

1. Agrawal, R., Imielinski, T., Swami, A. (1993). Mining Association Rules between Sets of Items in Large Databases. In: *Proceedings of the 1993 ACM SIGMOD International Conference on Management of Data*, Washington D.C., USA. 207–216. ACM Press.
2. Agrawal, R. and Srikant, S. (1994). Fast Algorithms for Mining Association Rules in Large Databases. In: *VLDB'94*, 487–499. Morgan Kaufmann.
3. Delic, D. (2001). Data Mining - Abhängigkeitsanalyse von Attributen mit Assoziationsregeln und Rough Sets. MS thesis. Free University of Berlin, Institute of Applied Computer Science, Berlin, Germany.
4. Jürgens, M. and Lenz, H.-J. (2001). Tree Based Indexes Versus Bitmap Indexes: A Performance Study. *International Journal of Cooperative Information Systems*, **10**, 355–376.
5. Munakata, T. (1998). Rough Sets. In: *Fundamentals of the New Artificial Intelligence*, 140–182. New York: Springer-Verlag.
6. Pawlak, Z. (1982). Rough Sets. *Int. J. Computer and Information Sci*, **11**, 341–356.
7. Skowron, A. and Rauszer, C. (1992). The discernibility matrices and functions in information systems. In R. Słowiński (ed.): *Intelligent Decision Support. Handbook of Applications and Advances of Rough Sets Theory*, 331–362, Dordrecht: Kluwer.
8. Rauszer, C. (1991). Reducts in information systems. *Fundamenta Informaticae*, **15**, 1–12.

Towards Fuzzy c-means Based Microaggregation

Josep Domingo-Ferrer[1] and Vicenç Torra[2]

[1] Dept. Comput. Eng. and Maths - ETSE, Universitat Rovira i Virgili
Av Paisos Catalans 26, 43007 Tarragona (Catalonia, Spain)
e-mail: jdomingo@etse.urv.es
[2] Institut d'Investigació en Intel·ligència Artificial - CSIC
Campus UAB s/n, 08193 Bellaterra (Catalonia, Spain)
e-mail: vtorra@iiia.csic.es

Abstract. National Statistical Offices collect data from respondents and then publishes them. To avoid disclosure, data is protected before the release. One of the existing masking methods is microaggregation. This method is based on obtaining a set of clusters (clustering stage) and then aggregating the values of the elements in the cluster (aggregation stage). In this work we propose the use of fuzzy c-means in the clustering stage.

1 Introduction

National Statistical Offices (NSO) collect data from respondents and then disseminate them. Due to legal restrictions, data has to be protected so that no disclosure of sensitive data is possible. This is, it should not be possible to link sensitive data from a particular respondent to this respondent.

To avoid disclosure, data is masked before their release. Statistical Disclosure Control studies masking methods for applying some distortion to data in such a way that the data is still analytically valid. This is, the published data is valid for researchers and users because they can reach to similar conclusions than the ones inferred from the original data.

At present, there exist a large set of micro-data protecting methods. For example, among the most widely used [4], we find the following ones: sampling, top and bottom coding, recoding, data swapping, microaggregation. See [2] and [8] for an extensive review of micro-data methods and [1] for a detailed analysis comparing information loss and disclosure risk for existing methods.

This work is devoted to one of the methods for numerical micro-data protection: microaggregation.

1.1 Microaggregation

Given a datafile to be protected, microaggregation consists on obtaining microclusters of similar records (at least k records have to be included in each cluster) and then publishing the averages of each cluster instead of publishing the original data. According to this, the method is defined by the following two steps (see [3] for details):

Step 1: Partition of the records into a set of disjoint groups so that the number of records in each group is at least k.

Step 2: For each record, the value of each variable is replaced by the aggregation of the corresponding values of all records in the same group.

Thus, this method consists on applying a clustering method to the original data and then returning for each record the prototype of the cluster instead of returning the record itself. This method can be formalized as follows:

1. Let $X = \{x_1, x_2, \cdots, x_n\}$ be a set of given data for n respondents for which p variables are observed. Thus, x_i is a p-dimensional vector $x_i = (x_{i1}, \cdots, x_{ip})$. Let $d(x, y)$ be a p-dimensional distance. Then, if we denote a partition of X into g groups by $G = \{G_1, \cdots, G_g\}$, with $|G_i|$ being the cardinality of groups G_i and $\hat{x}_i$ being the average data vector of G_i, the optimal partition G for microaggregation is the one that minimizes:

$$\sum_{i=1}^{g} \sum_{x_j \in G_i} d(x_j, \hat{x}_i) \tag{1}$$

   while satisfying $|G_i| \geq k$ for all $G_i \in G$.

2. Given an optimal solution G, x_j is replaced by $\hat{x}_i$ if x_j belongs to group G_i. elements in G_i).

Finding the optimal solution in Equation 1 is not an easy task. The following difficulties are found:

1. The size of all clusters has to be at least k and as similar to k as possible. Clusters of less records are not allowed. This constraint, not usually considered in clustering, required the development of specific clustering methods.

2. The problem of finding the best k-size partition of the domain is a NP problem. To have approximated solutions, heuristic approaches have been developed.

3. Classical microaggregation techniques assign the same values to all records in a cluster. This give clues to attackers, specially in the case of applying multivariate microaggregation to several groups of variables for the same records.

In this work, we introduce the use of fuzzy c-means for microaggregation. We define an heuristic method based on fuzzy c-means to obtain partitions of at least size k. Then, for each record, the mean value of all variables is not necessarily taken from the same cluster. Instead, using fuzzy membership values, they can be taken from different clusters. An additional advantage of fuzzy c-means is the existence of efficient computational algorithms (as the one described [6]).

The structure of this work is as follows. In Section 2, we review fuzzy c-means. Then, in Section 3 we detail our approach. The paper finishes with the conclusions.

2 Fuzzy c-means

While classical clustering methods partition the records of a given domain into a disjoint set of clusters, fuzzy clustering methods build a set of clusters in which elements can belong at the same time to several of them. When an element belongs to more than one cluster their membership is partial. This is modeled considering membership degrees in the $[0, 1]$ interval in such a way that 0 means no membership and 1 means full membership. In this case, it is commonly assumed that for a given element in the set, the summation of the membership of this element to all clusters is 1. Formally, this is defined as follows:

Let $X = \{x_1, \cdots, x_n\}$ be the set of elements (records or respondents in our application), then a set of membership functions $A = \{A_1, \cdots, A_c\}$ is a fuzzy partition of X into c clusters if and only if:

$$\sum_{i=1}^{c} A_i(x_k) = 1 \qquad \text{for all } x_k \in X$$

Here, $A_i(x_k)$ is interpreted as the membership of the k-th element to the i-th set.

Note that restricting $A_i(x_k)$ in $\{0, 1\}$ this definition corresponds to a crisp partition.

Among existing fuzzy clustering methods, a well known one is Fuzzy c-means. This method is a generalization of the k-means clustering method. This latter method consists on finding the set of c clusters such that the following objective function is minimized:

$$J(A, V) = \sum_{k=1}^{n} \sum_{i=1}^{c} A_i(x_k) \cdot ||x_k - v_i||^2$$

subject to the following constraints: $A_i(x_k) \in \{0, 1\}$ and $\sum_{i=1}^{c} A_i(x_k) = 1$ for all $x_k \in X$. Here, $|| \cdot ||$ corresponds to the Euclidean distance, and $V = \{v_i\}_{i=1,\cdots,c}$ are the centers of the clusters.

Defining,

$$M = \{(A_i(x_k)) | A_i(x_k) \in \{0, 1\}, \sum_{i=1}^{c} A_i(x_k) = 1 \text{ for all } k\}$$

the problem to be solved is to find A and V such that $min_{A \in M} J(A, V)$.

For computing the fuzzy c-means, the constraints M are replaced by:

$$M_f = \{(A_i(x_k))|A_i(x_k) \in [0, 1], \sum_{i=1}^{c} A_i(x_k) = 1 \text{ for all } k\}$$

However, solving the same problem with these new *fuzzy* constraints lead to the same crisp solution because the objective function is linear with respect to $A_i(x_k)$, and the optimization problem is solved using linear programing. Therefore, the optimal solution for $A_i(x_k)$ is found at an extremal point in the $[0, 1]$ interval. Thus, it is a crisp value.

To obtain a fuzzy solution, the following objective function was proposed (see e.g. [7] and its references for details):

$$J(A, V) = \sum_{k=1}^{n} \sum_{i=1}^{c} (A_i(x_k))^m \cdot ||x_k - v_i||^2$$

Here, m is a real number $m \geq 1$ that influences the membership values. With $m = 1$, the solution is crisp and, then, the larger is m, the more fuzzy the clusters we obtain.

To find A and V that minimize this objective function constrained in M_f, the following algorithm is used (see [7] or [5]):

Step 1: Generate an initial A and V
Step 2: Solve $min_{A \in M_f} J(A, V)$ computing (A_{ik} stands for $A_i(x_k)$):

$$A_{ik} = \Big(\sum_{j=1}^{c} \Big(\frac{||x_k - v_i||^2}{||x_k - v_j||^2} \Big)^{\frac{1}{m-1}} \Big)^{-1}$$

Step 3: Solve $min_V J(A, V)$ computing:

$$v_i = \frac{\sum_{k=1}^{n} n(A_{i,k})^m x_k}{\sum_{k=1}^{n} (A_{i,k})^m}$$

Step 4: If the solution does not converge, go to step 2; otherwise, stop

3 At least k Fuzzy c-means

As the fuzzy c-means do not assure that clusters have at least k elements, we have defined an algorithm for clustering based on the fuzzy c-means. Being k the minimum number of records allowed in a cluster, the algorithm is defined in the Algorithm 1. Note that in this algorithm, $fcm(X, c, m)$ corresponds to standard fuzzy c-means (as described in Section 2).

Once the clusters have been obtained, the values for the variables are (randomly) replaced by the values of the center of the class. When an element belongs to several clusters, the corresponding center is selected by a random distribution proportional to the membership degree.

Algorithm 1 micro-fuzzy-c-means

Algorithm *At least k* fuzzy c-means (X, k, m) **is**
begin
  $c := n/(3 \cdot k)$;
  $P = fcm(X, c, m)$;
  $cardMin = \min_{p \in P} |p|$
  if $cardMin < k$ **then**
    decrease c, increase m and start again
  end if
  for $p \in P$ **do**
    nothing
    begin
      if $|p| > 2 \cdot k$ **then**
        define m' in terms of m and $|p|$;
        micro-fuzzy-c-means (p, k, m');
      end if
    end
  end for
end

---

4 Conclusions

In this work we have shown the applicability of fuzzy c-means to microaggregation. Future work is on the comparison of our approach to existing techniques following the approach in [1].

Acknowledgements

Josep Domingo-Ferrer and Vicenç Torra are partially supported by the EU project CASC: Contract: IST-2000-25069 and CICYT project STREAMOBILE (TIC2001-0633-C03-01/02)

References

1. Domingo-Ferrer, J., Torra, V., (2001), A Quantitative Comparison of Disclosure Control Methods for Microdata, 111-133, in Confidentiality, Disclosure, and Data Access: Theory and Practical Applications for Statistical Agencies, P. Doyle, J. I. Lane, J. J. M. Theeuwes, L. M. Zayatz (Eds.), Elsevier.
2. Domingo-Ferrer, J., Torra, V., (2001), Disclosure Control Methods and Information Loss for Microdata, 91-110, in Confidentiality, Disclosure, and Data Access: Theory and Practical Applications for Statistical Agencies, P. Doyle, J. I. Lane, J. J. M. Theeuwes, L. M. Zayatz (Eds.), Elsevier.
3. Domingo-Ferrer, J., Torra, V., (2002), Aggregation techniques for statistical confidentiality, in "Aggregation operators: New trends and applications", (Ed.), R. Mesiar, T. Calvo, G. Mayor, Physica-Verlag, Springer.

4. Felso, F., Theeuwes, J., Wagner, G. G., (2001), Disclosure Limitation Methods in Use: Results of a Survey, 17-42, in Confidentiality, Disclosure, and Data Access: Theory and Practical Applications for Statistical Agencies, P. Doyle, J. I. Lane, J. J. M. Theeuwes, L. M. Zayatz (Eds.), Elsevier.
5. Klir, G., Yuan, B., (1995), Fuzzy Sets and Fuzzy Logic: Theory and Applications, Prentice-Hall, U.K.
6. Kolen, J. F., Hutcheson, T., (2002), Reducing the time complecity of the Fuzzy C-Means Algorithm, IEEE Trans. on Fuzzy Systems, April, 263-267.
7. Miyamoto, S., Umayahara, K., (2000), Methods in Hard and Fuzzy Clustering, pp 85–129 in Z.-Q. Liu, S. Miyamoto (Eds.), Soft Computing and Human-Centered Machines, Springer-Tokyo.
8. Willenborg, L., De Waal, T., (1996), Statistical Disclosure Control in Practice, Springer LNS 111.

Statistical Profiles of Words for Ontology Enrichment

Andreas Faatz, Cornelia Seeberg, Ralf Steinmetz

Multimedia Communications Lab
Darmstadt University of Technology
Merckstrasse 25, 64283 Darmstadt, Germany
{afaatz, seeberg, rst}@kom.tu-darmstadt.de

Abstract. The following paper focuses on an enrichment method for ontologies. We define similarities of possible new concepts and base the similarity and dissimilarity of concepts on the usage statistics in large corpora. The method is soft in the sense, that we define a semantically motivated heuristics for the influence of different linguistic properties influencing the similarity definition.

1 Introduction

The following paper focuses on the semi-automatic enrichment of ontologies based on statistical information.

An ontology is a structured network of concepts from an knowledge domain and interconnects the concepts by semantic relations and inheritance [1]. Ontologies give a formal representation and conceptualisation of a knowledge domain. For a given ontology we find propositions automatically, which could extend the ontology by new concepts. This means we

- use a text corpus
- detect a set of candidate concepts from the corpus
- finally select a subset of those candidate concepts by ranking their similarity to concepts already
- existing in the given ontology.

The final selection ends up in possible new concepts for the ontology to be proposed to a (human) ontology engineer.

The concepts in the ontology have one or more descriptors, which are words or phrases from natural language. On the other hand, the extractable information from large text corpora are words or phrases. For our technique this means that we develop a method of finding suitable definitions for the semantic similarity and dissimilarity of words. Throughout the paper we treat the ontological concepts and their descriptors as the same objects.

The paper is organised in the following way:

- we give the definitions in need
- we formally explain the enrichment algorithm
- we focus on properties of the algorithm which extend and systematically treat the linguistic properties we take under consideration
- we point out related work from the area of word clustering
- finally we summarise our results and open research issues

2.1 Definitions

An *ontology* is a 4-tuple $\Omega := (C, \text{is\_a}, R, \sigma)$, where C is a set we call *concepts*, *is_a* is a partial order relation on C, R is a set of relation names and $\sigma: R \rightarrow \wp(C \times C)$ is a function [1].

Throughout this paper we assume that a concept has a character string as a descriptor. This character string may be a word or a phrase.

A *distance measure* on Ω is a function $d: (C \times C) \rightarrow [0,1]$. Examples of distance measures are:

1) $d(x,y) = e^s$, where e is Euler's constant and s denotes the number of steps along the shortest relational path between the concepts x and y
2) $d(x,y) = 1$, if there exists a relational path between the concepts x and y and $d(x,y) = 0$, if there does not exist a relational path between the concepts x and y.
3) [3] defined criteria for distance measures in thesauri, which can be applied to the restriction of an ontology $\Omega := (C, \text{is\_a}, R, \sigma)$ to the pair $(C, \text{is\_a})$.

Moreover one can show, that there is an infinite number of distance measures fulfilling more restrictive characteristics than 1) and 2).

A *text corpus* ζ is a collection of text documents written in exactly one natural language. We assume ζ to be electronically available. From a text corpus we define a set of words or phrases to be the candidate concepts. A *proposition* for the ontological enrichment is a word or a phrase from ζ, which is used similarly to the concepts from the given ontology. Candidates are to be predefined, for example as all nouns occuring in ζ. Note that ζ might be extended during or after the application of the enrichment algorithm.

A *rule set* ρ is a finite set of linguistic properties, each of which can be tested in terms of its fulfilment frequency in the text corpus.

The entries m_{ij} of a *representation matrix* $M(C, \rho, \zeta)$ list, how often the j-th property from ρ was fulfilled in ζ for the descriptor the i-th concept from C.

2.2 The basic optimisation for ontology enrichment

The enrichment algorithm processes information available from ζ and Ω. It computes an optimal solution for the problem of fitting the distance information among the concepts expressed by Ω and the dissimilarity information between words or phrases to be extracted from the word usage statistics considering ζ.

Let us assume a given $M(C, \rho, \zeta)$. We search for a set $k = \{k_1, \cdots, k_n\}$ of non-negative reals with $|k| = |\rho|$, which will be called *configuration* of the rule set ρ. Each k_i corresponds to a rule ρ_i.

The configuration k decides about the quantities of dissimilarity we derive from $M(C, \rho, \zeta)$.

The *Kullback-Leibler divergence* generally measures the dissimilarity between two probability mass functions [2] and was applied successfully to statistical language modelling and predicition problems in [4]. The Kullback-Leibler divergence $D(x,y)$ for two words x, y is defined as

$$D(x, y) = \sum_w P(w|x) \log \frac{P(w|x)}{P(w|y)} \tag{1}$$

In the basic version of the Kullback-Leibler divergence, which is expressed by formula (1), w is a linguistic property and $P(w|x)$ ist the probability of this property being fulfilled for the word x. In the sum indicated by formula (1), w ranges over all liguistic properties one includes in a corpus analysis. In our case the frequencies of observing the linguistic properties are denoted by $M(C, \rho, \zeta)$. For our purposes we change (1) in such a way, that k weighs the influence of each property w:

$$D_k(x, y) = \sum_w k(w) P(w|x) \log \frac{P(w|x)}{P(w|y)} \tag{2}$$

with $k(w) \in k$ in our case

Considering our representation matrix notation $M(C, \rho, \zeta)$ we obtain

$$P(w|x_i) = \sum_{l=1}^{|\rho|} \left[\frac{m_{il}}{\sum_{n=1}^{|\rho|} m_{in}} \right] \tag{3}$$

Let us clarify the notation of formula (3):

x_i denotes the i-th concept from C. Correspondingly in (3) the m_{il} are the matrix entries in $M(C, \rho, \zeta)$ in the row expressing the linguistic properties of x_i. With this notation $k(x_i) = k_i$ holds. In that sense, we will be able to determine an optimal $k = \{k_1, \dot{}, k_n\}$.

Taking the distances from the ontology Ω as an input, which should be approximated by the $D_k(x, y)$ as well as possible, the question of finding an optimal configuration k reduces to the question:

which configuration k minimises the average squared error expressed by the differences

$$(d(x, y) - D_k(x, y))^2 \ ?$$

Finally we present a formulation of this question in terms of a quadratic optimisation formula. Searching for an optimal k means searching for a minimum of the following fitness expression

$$\min_k \sum_{i=1}^{|C|} \sum_{i=1}^{|C|} (d(x_i, x_j) - D_k(x_i, x_j))^2 \tag{4}$$

where $k = \{k_1, \ , \ , k_n\}$ and $k_l \geq 0$ for all $k_l \in k$. Note that we minimise over the set of all configurations, that means over all possible k. We now explain, which words phrases are propositions for the ontological enrichment. Once we optimised formula (4) we obtain the configuration in need to compute all the distance measures between all the concepts from Ω and the candidates. We apply an enrichment step starting with the optimal similarity measures $D_k(x, y)$.

We only take into concern the $D_k(x, y)$ with $x \in C$ and a candidate y. If such a distance between a formerly known concept (i.e. its descriptor) and a candidate (i.e. a word from the corpus) formerly unknown to Ω is lower than a predefined threshold, y proposition to enrich Ω. A suitable threshold can for example be defined from the average of the distances $d(x, y)$ where $x \sim y$ holds for some $\sim \ \in R$.

Additionally the $D_k(x, y)$ with $x \in C$ and a candidate y carry even more information, namely an optimal placement of the candidate concepts. The candidate concepts and the concepts from Ω can be presented together, if a candidate turns out to be a proposition. This simplifies the knowledge engineer's understanding of how the candidate concepts evolved and in which semantic area of Ω they might belong.

3 Discussion of the algorithm - extending ρ

The application of the algorithm needs a rule set ρ as one of the parameters given. The following section focuses on a technique which systematically selects and constructs ρ We will use the fact, that - for several distance measures d - our experiments showed, that the optimisation (4) ended up in supressing most of the rules from ρ setting the corresponding k_i to zero.

Applying the optimisation to derive a configuration k we chose a ρ in the following way: for each concept from C (Ω given) we collect the collocators occuring in the same sentence at a distance of at most five words in ζ, create a list of all these collocators and finally check for each $c \in C$, how often it was collocated in the same sentence, but at maximum distance of five words within ζ.

We choose this particular ρ, as for the German language, with which we carry out our experiments. The property is a standard configuration of the German online corpus analysis tools [5] and [6].

We obtained two general observations, which characterise all of our ongoing enrichment experiments and which imply interesting further developments of the algorithm, because they point to a compression of the property set ρ while applying the algorithm. The vast majority of the $k_i \in k$ are zero (in our first experiments about 90% of the $k_i \in k$) and there are many minor influences of nonzero k_i, if we also consider the fact, that that similarity must exceed a threshold T for a candidate concept to become a proposition.

The fact that we observe many zeros in the solution k is related to the sparse structure of the optimisation problem (4). But even if we cannot predict the exact structure of $M(C, \rho, \zeta)$ beyond sparsity, the sparsity of the data belonging to candidate concepts additionally leads to properties with minor influence. Although we admit, that this observation needs a further strict systematic fundament, we use it as a working hypothesis.

Our first experiments also point to a fact, which we already expected intuitionally: if a concept becomes a proposition and its similarity was only determined by exactly one feature ρ_i from ρ (leaving $k_j = 0$ for $i \neq j$) we detected a higher risk of bad propositions in the sense of a semantic mismatch or an overgeneral proposition.

Another complication may arise, if we extend the initial corpus while applying the algorithm. Such a strategy is useful, if we start with a small specialised corpus and a few concepts in Ω. In that case the initial corpus may contain not enough information, consequently information must be added by including new texts in the initial corpus. Only in that sense it is true, that specialised corpora perform well in domain dependent conceptual clustering or ontology enrichment problems like [7] stated. But extending a corpus goes along with introducing a fulfilment of properties, which we did not observe in the initial corpus [4], which gives a bias to our computation of important properties via the optimal configuration k.

For a systematic treatment of the difficulties we discussed in this section - arbitrary choice of ρ, bias with propositions guided by exactly one linguistic property, bias after extending corpus - we give a prospect on a stepwise application of the algorithm: after applying the algorithm once we make up a new property set $\tilde{\rho}$ keeping the properties, which turned out to be influential, and adding new properties.

An example for new properties is a larger context, in which a collocation may occur. As long our data remain sparse, several extensions of the linguistic property set can result in a rich representation making the selection of our the modified $\tilde{\rho}$ less arbitrary. In the special case of word-cooccurence in contexts the extension of ρ by a stepwise application of the enrichment algorithm becomes systematic, if we start with narrow contexts (as the distance of five words we used in the first experiments) and broaden the contexts monotonically in every step.

4 Related work

Similarity between words is a topic from the theory of word clustering algorithms and requires statistical information about the contexts, in which the words are used. Many approaches test and count collocation features of the words in large text corpora, such that a word is represented by a large vector, which has entries communicating, how often a collocation feature was fulfilled in the corpus. The vectors are sparse [8].

The notion of similarity definitions by vector representations normally does not weight every single dimension of the vectors. In the paper we stated that this is possible by a soft method using the information already defined in the given ontology. The method is soft in the sense, that semantic information from an ontology is quantified and thereafter guides the selection of relevant linguistic properties.

The influence of the ontological structure on the word (-vector) similarities results in an optimisation problem, which gives an answer to the question which dimension in the word (-vector) representation is influential for the similarity computation.

Automatic thesaurus and ontology construction dates back from the 1970s [9]. Our approach is a further development of methods, which try to construct the whole ontology. The soft method of introducing heuristics for the ontological information given can only be applied, if we enrich an existing ontology instead of fully constructing the ontology. Besides the question about the heuristics guiding the optimisation procedures described above, a number of other interesting questions arises along with the approach. Among them are the fol-

lowing: how do we construct a suitable corpus to learn from, which linguistic preprocessing is necessary or helpful, do we need absolute, relative or probabilistic vector entries, how does the approach scale for larger ontologies.
The question of evaluating the results is interesting for related areas such as Kohonen maps for documents and word clustering algorithms [10].

5 Summary and future work

We presented an algorithm, which returns propositions for the enrichment of an ontology. The algorithm selects from a set of linguistic properties regarding the information encoded in the ontology, for wich we wish an enrichment: at this point, our soft method is based on a modified Kullback-Leibler divergence for each single given enrichment problem.
We also gave a prospect on a more systematic setup of the algorithm, which has to undergo genuine evaluation to overcome its merely constructive status. The area of evluation methods for the algorithm together with further experiments will be the focus of our future work on the subject.

References

[1] Stumme, G., Mädche A.:
FCA-Merge: A Bottom-Up Approach for Merging Ontologies JCAI '01 - Proceedings of the 17th International Joint Conference on Artificial Intelligence, Seattle, USA, August, 1-6, 2001, San Francisco/CA: Morgan Kaufmann, 2001
[2] Kullback, S.: *Information Theory and Statistics.* John Wiley and Sons, New York, 1959.
[3] Resnik,P.: *Semantic Similarity in a Taxonomy: An Information-Based Measure and its Application to Problems of Ambiguity in Natural Language*, Journal of Artificial Intelligence Research, vol. **11**, 1999
[4] Dagan I., Perreira F., Lee L.: *Similarity-based Estimation of Word Co-occurence Probabilities*, Proceedings of the 32nd Annual Meeting of the Association for Computational Linguistics, ACL'94, New Mexico State University, June 1994
[5] the Cosmas corpus querying service, http://corpora.ids-mannheim.de/~cosmas/
[6] the Wortschatz corpus querying service, http://wortschatz.uni-leipzig.de/
[7] Bisson, G. and Nedellec, C. and L. Cañamero: *Designing clustering methods for ontology building - The Mo'K workbench*, Proceedings of the Ontology Learning ECAI-2000 Workshop, August 2000
[8] Sahlgren, M.: *Vector-Based Semantic Analysis: Representing Word Meanings Based on Random Labels,* Proceedings of the ESSLLI 2001 Workshop on Semantic Knowledge Acquisition and Categorisation, Helsinki, Finland, 2001
[9] Spark-Jones K.: *Readings in Information Retrieval,* Morgan Kaufmann, 1997
[10] Lagus, K.: *Studying similarities in term usage with self-organizing maps.* Proceedings of NordTerm 2001, Tuusula, Finland. pp. 34-45, 2001

Quest on New Applications for Dempster-Shafer Theory: Risk Analysis in Project Profitability Calculus

Mieczyslaw A. Klopotek[12] and Slawomir T. Wierzchon[13]

[1] Institute of Computer Science, Polish Academy of Sciences, Warsaw, Poland
[2] Institute of Computer Science, University of Podlasie, Siedlce, Poland
[3] Department of Computer Science, Bialystok University of Technology, Bialystok, Poland

Abstract. This paper is concerned with seeking new applications for the Dempster-Shafer Theory that are by their nature better suited to the axiomatic framework of this theory. In particular, wafer processing on a integrated circuits production line, chemical product quality evaluation etc. are considered. Some extensions to basic DST formalism are envisaged.

1 Introduction

The methods for decision analysis based on Dempster-Shafer theory have been already studied for along time, see for example works of Jaffray [1], Strat [8] and Yager [13]. A number of studies has been carried out within the framework of VBS (Valuation-based systems, comprising among other the probability theory and the Dempster-Shafer theory) can also represent and solve Bayesian decision problems [7,6]. Shenoy [5] has shown that the solution method of VBS for decision problems is more efficient than that of decision trees and of influence diagrams. Xu [12] proposed a method for decision analysis using belief functions in the VBS. This is done by generalizing the framework of VBS for decision-making under Dempster-Shafer theory (DST). Another variant of DST, the Transferable Belief Model has also been studied in this context [11].

All these studies concentrate essentially on an attempt to extend the traditional Bayesian decision-making problem under uncertainty by modeling incomplete knowledge. It is usually assumed that an outcome of a test is definite and does not influence the tested process, and that for belief functions the removal operator is applicable, that is there exist conditional belief functions.

We think that these assumptions prevent from exploiting the true potential behind belief functions. As demonstrated in [2], the traditional approach to belief functions as interval probabilities simply fails if we want to work with general belief functions (and not ones with very special properties). It has been shown there that the intrinsic nature of combining belief functions is "destructive" that is if DST should model a process on real objects these objects must be occasionally destroyed in the process. The other feature is "blind" application of steps of a process.

Hence when looking for application areas of DST in decision-making we should concentrate on situations where we have to do with destructive processing.

In this paper we will recall the current approach to decision making in DST and its shortcomings and then we will point to actual challenges and new areas of potential applications of DST.

Many books and papers can provide with a detailed introduction to DST. We use generally accepted denotations, explained e.g. in [4,10].

2 Decision Making with DST in the VBS Formalism

Shenoy [7,6,5] introduced a formalism for decision making within his framework of VBS (valuation based systems) which includes also decision making in DST.

VBS representation for a decision problem is denoted by a 6-tuple $\Delta = (U_D, U_R, \{\Theta_X\}_{X \in U}, \{\pi\}, \{Bel_1, Bel_2, \ldots, Bel_n\ldots\}, \rightarrow)$ representing decision variables, random variables, frames of the variables, utility valuations, belief valuations, and precedence constraints respectively. The VBS representation of a canonical decision problem $\Delta_c = \{\{D\}, \{R\}, \{\Theta_D, \Theta_R\}, \{\pi\}, \{Bel[R|D]\}, \{D \rightarrow R\}\}$ is illustrated in Fig. 1. A graphical description is called a valuation network.

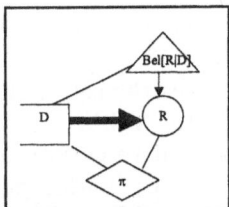

Fig. 1. A graphical representation of VBS for DST decision making

The set of variables U consists of decision variables U_D and random variables U_R. The possible values of a decision variable represent the acts available at that point. The possible values of a random variable represent the states of nature. Graphically decision variables are represented by the rectangles and random variables are represented by the circles. In Δ_c, R is a random variable and D is a decision variable.

Let $A \subseteq U$. A utility (or payoff) valuation π for A is a function from Θ_A to the set of real numbers. The values of utility valuations are utilities. The utility function is deemed to be a non-normalized belief function.

To solve the decision problem in DST Shenoy introduced his fusion algorithm [7]. He also gave some restriction for VBS representation in order to avoid divisions during computation. Since removal of two belief functions (reverse operator to $\oplus$) does not always result in a. belief function, generally with VBS the decision calculus in the case where the removal can be avoided, is discussed. Therefore, frequently one states the following assumption for the VBS representation. (Strat, 1990):

Fundamental Assumption In the VBS representation, at least one of the following two conditions should be satisfied. (1) There is only one utility valuation. (2) We only have a conditional for each random variable such that the variables on which the belief function is conditioned always precede the random variable.

Solving Bayesian decision problem s in VBS is based on the criterion of maximizing expected payoff for combination of all decisions and all random situations with utility function. In the context of DST, we can think of combining (conditional) belief functions representing relationship between variables and representing the utility function using the Dempster-rule operator. Then we can obtain the interval expectation of minimal and maximal utility over the random variables R. But such an interval would be very hard to be used as a ground for decision making. Hence Strat proposes, to obtain a unique decision, to introduce a lambda factor and to consider optimal decision d* with respect to this factor such a decision that $(\pi \oplus Bel)^{\downarrow D}(d*) = (\pi \oplus Bel)^{\downarrow D})^{\downarrow \{\}}(\{\}))$ with {} meaning an empty set.

As for Bayesian decision problems, in order to use VBS solution method, the VBS representation of a decision problem need to be well-defined for case of belief functions. The conditions are as follows:

- $U_D \subseteq \cup H_D$ where H_D denotes the set of subsets of U_D for which payoff valuations exist in the VBS.
- $U_R \subseteq \cup H_R$ where H_R denotes the set of subsets of U_R for which belief function valuations exist in the VBS.
- For the operator $\rightarrow$ the transitive closure over U is a partial order (irreflexive and transitive) over U; in this partial order for any decision variable D and any random variable R either R precedes D or D precedes R; if there is a conditional for R given A-{R} and the decision variable D is in A, then D precedes R in the partial ordering; in there is a belief valuation for A and a decision variable D is in A, then D precedes some random variable R in A. ; .
- Suppose D is the subset of decision variables included in the domain of the joint belief function $Bel_1 \oplus ... \oplus Bel_n$. Then $(Bel_1 \oplus ... \oplus Bel_n)^{\downarrow D}$ is a vacuous belief function.

The fusion algorithm consists essentially in deleting all variables from U. If variable X precedes variable Y then Y must be deleted before X. The deletion means marginalization. The important thing here is that it can be executed by local computation. See the papers of Shenoy, and also [9].

3   Objections to Traditional DST Applications in Decision Making

Notice that the Fundamental Assumption along with usage of proper conditional belief functions (that is ones with non-negative mass values) assures that DST decision making is applied when only one of the neighboring variables in the dependence network is a true DST (multi-valued) variable, and the other are traditional (single-valued) ones. The proper conditional belief functions can always be transformed to conditional probability functions (with transformation of variables such that atomic events being disjoint sets of atomic events of the original variables) so that the "DST decision making" turns to be in fact Bayesian decision making and means predominantly adding some "sugar" of formalism than solving more general problems. The only advantage is that in case of many zero probability conditional events the information can be stated more compactly.

We think, however, that decision-making applications should be sought for which traditional Bayesian representation falls short of capturing the intrinsic nature of the problem. In this case it may only be productive to try to check the applicability of DST.

By an appropriate empirical model for DST we understand the following one (see Fig. 2) [4].

- We observe ("measure") a real world $state_1$, encode it as a belief function Bel_1, know that a real
- world process may be represented by a DST reasoning process Bel_p, run the reasoning while the real world
- process runs in the real world transforming it into the $state_2$. We observe (in exactly the same way as before) the real world and encode the $state_2$ as belief function Bel_2. Bel_2 shall coincide or at least be one of alternative predictions of the reasoning process.

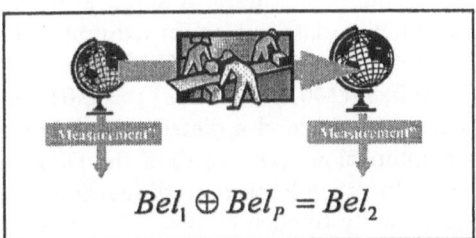

$$Bel_1 \oplus Bel_p = Bel_2$$

Fig. 2. Proper DST model of the real world

As the only true models of DST turn to be "destructive models" [2], we shall look for applications where losses of material, energy etc. takes place in processes we should make decisions about.

4   A New Challenge: Integrated Circuits Case Study

One of the areas of applications, where the traditional approach to DST decision making is not applicable and where the destructive processes, we were talking about in the introduction, take place, is a charge-based production of products with differentiated grades of quality.

For example the production of integrated circuits as run once in the Semiconductor Center CEMI in Warsaw. The integrated circuits had a multistage quality lattice with products higher in the lattice sold usually more expensively. For example a product matching requirements for UCA6401N could also be sold as UCY7401N (lower requirements) or UC7401N (still lower), or 401N (the lowest), or the given chip could be completely useless, see Fig. 3 (the openness of the TBM model of Smets would be useful here). The quality lattice was in many cases even more complicated, forming a lattice (instead of linear ordering).

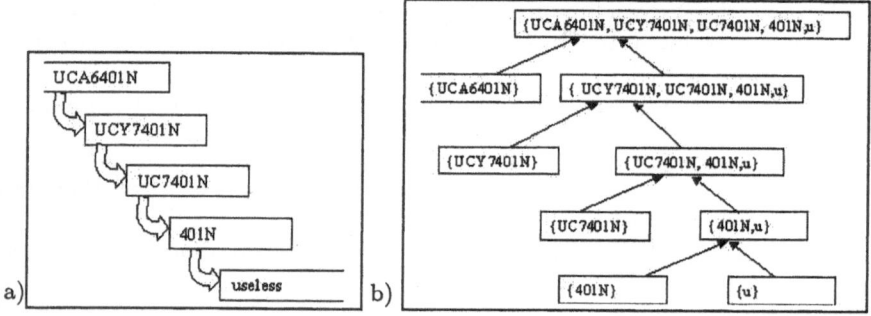

Fig. 3. Products based on falling quality (a) and the representation in DST (b) (arrows mean subset relationship)

The first fragment of the production process consisted in application of various processes to a charge of wafers. A single stage yielded results similar to a Dempster-Shafer belief distribution for example a belief distribution basic belief assignment had the form $m(\{UCA6401N, UCY7401N, UC7401N, 401N, useless\}) = 0.95$, $m(\{401N, useless\}) = 0.01$, $m(\{useless\}) = 0.015$ etc. Two subsequent stages yielded a distribution of quality result similar to application of the combination operator $\oplus$ of the DST: $Bel_{12} = Bel_1 \oplus Bel_2$. The reason was that due to microscopic dimensions one had no chance to apply processes individually to each chip on the wafer and hence the quality disturbances of the next stage applied independently of the quality disturbances of the preceding stage.

In the process there were several checkpoints ("test" facilities) that had to decide on the quality of the chips. However, not the individual chips but only a few representatives (out of several hundreds) were checked. Hence the results of the tests never yielded a definite decision that the quality is good or bad but rather suggested a belief function distribution. A decision had to be made based on each test whether to proceed with the production process or to abandon the charge altogether. The decision was to reject if the quality function distribution was worse than an assumed one. The comparison has to be based on *Bel* function to be rational (a worse distribution has consistently higher *Bel* function values). Notice that a decision influences the belief distribution in a way that is difficult to predict from the traditional belief function operators. We do not have currently a handy mathematical expression for the belief function distortion. The calculations have to be carried out in a simulated manner (sampling a set of samples from the distribution, then evaluation, and then restoring distributions from data).

The most difficult problem is the combination of the utility function with the belief distribution for the charge. The utility function reflects the distribution of orders of chips of each type (the focal points set on singletons: e.g. UCY7401N, UCA6401N etc.). The operator $\oplus$ is hardly applicable, because the utilization is not independent of the resulting distribution. For example, if there are too many products of UCA6401N type, they will not be thrown away, but they will be sold as UCY7401N. So the combination of the charge belief distribution cannot be considered as "combination of independent evidence" and therefore instead of Dempster rule a discrete optimization procedure has to be applied.

Usually, several alternative technologies are available consisting of same or different steps in different order. Each process step and each testing step imposes some costs. With these restrictions the goal is to find the most profitable scenario of managing the production line. In order to rich satisfactory management results, one has to fix the precise organization of the charge management. It has to be decided, given the customer order portfolio, which technology is to be applied, and within each technology, whether or not particular testing steps are to be included or not.

Under these circumstances the actual decision making procedure requires in fact the computation intense analysis of all the paths possible in the decision making graph to find the set of the most profitable way of processing the charge.

5 Other Decision Making Situations Where Dempster-Shafer Theory Is Applicable

Another situation when the decision making under intrinsic belief functions takes place, is production planning in a chemical plant. Here we find also frequently a kind of quality lattice that may lead to belief function decision making problem: A hair shampoo is quite well suited as bath foam agent (though a bit too expensive in production). More important is here, however, the issue of quality control. The quality control is in fact a destructive process: a product sample taken for quality control will be in fact not usable thereafter.

308

The quality control step can be described for example by a belief function Bel1 with $m_1(\{good, bad\}) = 0.95$ and $m_1(\{bad\}) = 0.5$. Hence if the potential for quality distribution in the sample was described before quality control by Bel, thereafter it is described by Bel$\oplus$Bel1. Decisions made at tests may lead either to continuation of the process, to total abandoning of the process or to reclassification of the product (downwards the quality lattice). The practical issue here is then the amount of tests that are carried out during the production process and the balancing of costs and savings of carrying out or dismissing tests at various stages of the process. Again, a strong dependence between the customer order portfolio and the utility function is present.

Dempster-Shafer evidence combination model describes well the situation of bringing to the market a set of multifunctional non-durable goods (see Fig. 4). We may have knowledge of general requirements in a market of goods fitting some functional requirements A,B,C,D in terms of Bel1 function: Bel1(A,B,C)=0.8 saying that 80% of customers need goods with properties A,B,C (but not D) etc. We produce therefore goods with property profile described by Bel2 and deliver them to a set of shops, but without knowing the particular requirements distribution in that shop: just delivering Bel_2 to each shop. Then our expected sale amounts would be described by $Bel_1 \oplus Bel_2$. (combination without normalization). The decision making under such circumstances would concern choice of the distribution Bel_2 from among the set of technologies available to us.

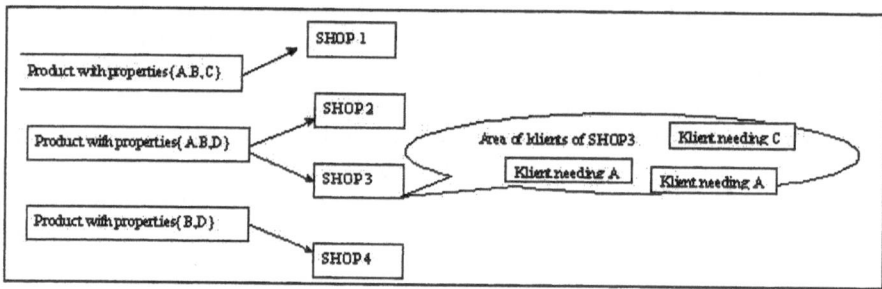

Fig. 4. Random distribution of products matching agains client needs. Product with properties {A,B,D} will turn to be useless for a client seeking property C.

6   Concluding Remarks

We feel that the research run so far on applications of the Dempster-Shafer theory in decision-making processes has avoided the intrinsic problems. DST has long ago been shown not to be a model of interval probabilies except for very special cases when only one of the neighboring variables in the dependence network is a true DST (multi-valued) variable, and the other are traditional (single-valued) ones. Problems with applicability of true belief functions should be sought.

We have pointed to a few sample applications here. Then one shall understand that the combination of "evidence" cannot be run only using the Dempster-rule because not always the evidence is independent. The alternative rules derived so far in the literature do not fit also some typical situations in DST related decision making. One important example is the apriori known customer order portfolio. Another is the issue of modifying belief distribution by application of some tests (when test results are described by a belief function also).

We continue an intensive research to resolve the problem of special purpose evidence combination operators. New evidence combination operators need to be derived to replace the expensive simulations carried out currently.

References

1. Jaffray J. Y., Linear utility theory for belief functions. *Operations Research Letters* 8 (1989) 107-112.
2. Klopotek M.A: *Methods of Identification and Interpretations of Belief Distributions in the Dempster-Shafer Theory* (in Polish). Monograph. Publisher: Institute of Computer Science, Polish Academy of Sciences, Warsaw,1998.
3. Klopotek M.A.: Dempster-Shafer theory.*Encyclopedia of Mathematics*. Supplement III. M.Hazenwinkel (ed). Kluwer Academic Publishers, 2002,
4. Klopotek M.A., Wierzchon S.T.: Empirical Models for the Dempster-Shafer Theory. In Srivastava, R.P.; Mock, T.J., (Eds.): *Belief Functions in Business Decisions*. Springer-Verlag, volume 88 of the series "Studies in Fuzziness and Soft Computing, 2002
5. Shenoy P. P., A Comparison of Graphical Techniques for Decision Analysis. *European Journal of Operational Research* (1994)
6. Shenoy P. P., Valuation-based Systems for Bayesian Decision Analysis. *Operations Research* 40 (1992) 463-484.
7. Shenoy, P.P.: A Fusion Algorithm for Solving Bayesian Decision Problems. In *Uncertainty in Artificial Intelligence*, Proc. of the 7th Conf. 1991, pp 361-369.
8. Strat T.: Decision analysis using belief functions. *Int. J. of Approximate Reasoning* 4 (1990) 391-417. 24
9. Wierzcho S.T.: *Methods of representing and processing of uncertain information within the framework of Dempster-Shafer Theory*. Monograph. Publisher: Institute of Computer Science, Polish Academy of Sciences, Warsaw, 1996.
10. Wierzchon S.T., Klopotek M.A.: *Evidential reasoning. An interpretative investigation*. University of Podlasie Publishing House, Siedlce, 2002 304 pages,
11. Xu H., Hsia Y-T., Smets Ph.: Transferable Belief Model for Decision Making in Valuation-Based Systems. *IEEE Trans. Systems, Man and Cybernetics* (1995)
12. Xu H.: A Decision Calculus for Belief Functions in Valuation-Based Systems. In Dubois D., Wellman M. P., D'Ambrosio B. and Smets Ph. eds. Proc. 8th Uncertainty in Artificial Intelligence (San Mateo, Ca.:Morgan Kaufmann, 1992) 352-359.
13. Yager R. R.: Decision Making under Dempster-Shafer Uncertainties, Iona College Machine Intelligence Institute Tech. Report MII-915, (1989).

Decision Making Based on Informational Variables

Juliusz L. Kulikowski

Institute of Biocybernetics and Biomedical Engineering
4, Ks. Trojdena St., 02-109 Warsaw
E-mail: jlkulik@ibib.waw.pl

Abstract. It is presented a method of decision making based on a concept of informational variables. Information variable is defined as a generalisation of random variable used in the case when probability distribution functions are not available. Instead of them a preference relation being a particular case of semi-ordering is introduced into the Borel family of subsets on a real axis. Instead of using numerical probabilities preference relationships among events are established according to the general principles of a topological logic, and the rules specified in the paper. The preference relationships can be established by operations performed on the graphs of preference or on their matrix representations. It is shown how a decision problem analogous to this of statistical hypothesis verification, usually solved by Bayesian methods, in the case of a lack of conditional probability distributions can be formulated in the terms of informational variables and solved using the method described in the paper.

Key words: decision making, topological logic, preference relation, informational variables

1 Introduction

The concept of *informational variable* has been introduced in [1] as a generalisation of the well known concept of a random variable. However, it was inspired by a growing interest to the methods of decision making under uncertainty, and, in particular, to the possibilities offered in this domain by the concept of a *topological logic*, originally published in 1937 by C. G. Hempel [2] and continued by Ch. A. Vessel [3] and by the author [4,5,6]. In this paper we would like to show possible applications of information variables in decision making based on limited preliminary information. Such situations arise often in various application areas where the number of past observations is not sufficient for building a strong statistical basis for decisions' optimisation.

2 Basic assumptions

Let us take into account an ordered triple:

$$V = [\Omega, C_\Omega, \prec]$$ (1)

where Ω denotes a non-empty (at lest two-element) set of *elementary events*; C_Ω is a sigma-algebra of subsets of Ω (called *events*), and $\prec$ is a *preference relation* in C_Ω that will be described below. V will be called an *informational space*. Its similarity to the well known concept of a *probabilistic space* introduced by A. Kolmogorov (see [7], § 1.4)) is evident; the main difference consists in the term $\prec$, in a probabilistic space taking the form of a *probabilistic measure*.

Preference relation is a specific type of semi-ordering introduced into C_Ω. To make it clear, let us take into account a non-empty (at least two-element) set A. Any binary reciprocal, symmetrical and transitive relation described in A is called an *equivalence relation* ([8], chapt. 1). Let $\approx$ be an equivalence relation and let us take into account a binary reciprocal, asymmetrical and transitive relation $\prec$ (read: "is less preferred than") described in A, such that any two elements $e', e'' \in A$ satisfy both, $e' \prec e''$ and $e'' \prec e'$, if and only if $e' \approx e''$. So defined relation $\prec$ is called a *weak semi-ordering* relation. A weak semi-ordering relation $\prec$ becomes a *strong semi-ordering* relation if any two elements e', e'' satisfy both, $e' \prec e''$ and $e' \prec e''$, if and only if $e' \equiv e''$. Otherwise speaking, a strong semi-ordering relation is antisymmetrical: for a pair of different elements no more but one of the relationships $e' \prec e''$ or $e' \prec e''$ may be valid. Each strong semi-ordering relation is a weak one, as well. Any pair of elements $e', e'' \in A$ satisfying neither $e' \prec e''$ nor $e'' \prec e'$ will be called *mutually incomparable* and denoted as $e' ? e''$. The relation ? is non-reciprocal, symmetrical and non-transitive. Sometimes, instead of the notation $e' \prec e''$ its inverse form $e'' \succ e'$ (read: "e'' is more preferred than e'") will be used.

A typical example of a strong semi-ordering relation is the one induced in the family C_Ω of subsets by the inclusion $\subseteq$ of subsets. It is easy to show that it satisfies the above-mentioned conditions of reciprocity, anti-symmetry and transitivity. We shall call it a *natural semi-ordering of subsets*.

For a given set A and a binary relation r described on it any pair of elements e', $e'' \in A$ satisfying r is called a *syndrome of r*. A notion $e' r e''$ indicating both, a syndrome and the corresponding relation, will be called a *relationship*.

Let us suppose that there are two relations, r and q, described in A. The relation q is called a *sub-relation of r* if each syndrome of q is, at the same time, a syndrome of r. In such case r will be called an *over-relation of q*.

Each weak semi-ordering relation $\prec$ described on the family C_Ω of subsets and satisfying the conditions:

a/ for any two subsets $\alpha, \beta \in C_\Omega$ the inclusion $\alpha \subseteq \beta$ implies $\alpha \prec \beta$;

b/ for any $\gamma \in C_\Omega$ such that $\gamma \cap \alpha = \gamma \cap \beta = \emptyset$, if $\alpha \prec \beta$ then

312

$(\alpha \cup \gamma) \prec (\beta \cup \gamma)$;

c/ under the conditions mentioned in b/, if $\alpha ? \beta$ then $(\alpha \cup \gamma) ? (\beta \cup \gamma)$;

d/ for any subsets $\alpha, \beta, \gamma, \delta \in C_\Omega$ if $\alpha \prec \beta$ and $\gamma \prec \delta$ then

$(\alpha \cup \gamma) \prec (\beta \cup \delta)$,

will be called a *preference relation* in C_Ω.

The preference relation $\prec$ is thus an over-relation of the natural semi-ordering in C_Ω; in this context natural semi-ordering may be called a *trivial preference relation in C_Ω*. Let us also remark that in preference relation a couple of relationships $\alpha \prec \beta$ and $\beta \prec \alpha$, in general, does not imply an identity $\alpha \equiv \beta$, but rather an equivalence $\alpha \approx \beta$.

Let us take into account a finite family of subsets $\alpha, \beta, \gamma,...,\iota, \kappa \in C_\Omega$ and a corresponding linearly ordered series of relationships:

$$\Lambda = [\alpha \ r' \beta, \ \beta \ r'' \gamma,..., \iota \ r^* \kappa].$$

Such a series will be called a *logical chain* and it will lead to a question: what type of relationship $\alpha \ r \ \kappa$, called a *conclusion of the chain,* can be deduced from it? The answer is given by the following *inference rules of preference*:

1/ if in the given logical chain Λ there is $r', r'',..., r^* = r$ where $r = \approx, \prec$ or $\succ$ then, correspondingly, the conclusions are: $\alpha \approx \kappa, \alpha \prec \kappa$ or $\alpha \succ \kappa$;

2/ if for some $\varepsilon, \eta \in C_\Omega$ there is $(\varepsilon \approx \eta) \in \Lambda$ then it has no influence on the conclusion of Λ;

3/ if for some $\varepsilon, \eta \in C_\Omega$ there is $(\varepsilon ? \eta) \in \Lambda$ then $\alpha ? \kappa$;

4/ if for some $\varepsilon, \eta, \mu, v \in C_\Omega$ there is $(\varepsilon \prec \eta) \in \Lambda$ and $(\mu \succ v) \in \Lambda$ then $\alpha ? \kappa$.

Logical chains whose conclusions are $\alpha \prec \kappa$ (or $\alpha \succ \kappa$) are called, correspondingly, *directed up-going* (correspondingly, *down-going*) *logical chains.* Logical chain whose conclusion is $\alpha \approx \kappa$ is called *a chain of logical equivalence.* A logical chain Λ for which $\alpha \equiv \kappa$ is called a *logical loop.* The following additional inference rules will be introduced:

5/ any up-going or down-going logical loop consisting of some events can be replaced by a loop of logical equivalence consisting of the same events;

6/ if the logical chain Λ contains a proper sub-chain Λ', $\Lambda' \subset \Lambda$, then all relationships belonging to Λ' can be neglected in the inference of a conclusion of Λ.

The following example will illustrate the above-given concepts.

Example 1: Let us take into account a finite set $\Omega = \{a,b,c\}$. The corresponding family of subsets C_Ω consists of 8 elements (including Ω and an empty set $\emptyset$). Its natural semi-ordering can be illustrated by a directed graph plotted in Fig. 1.

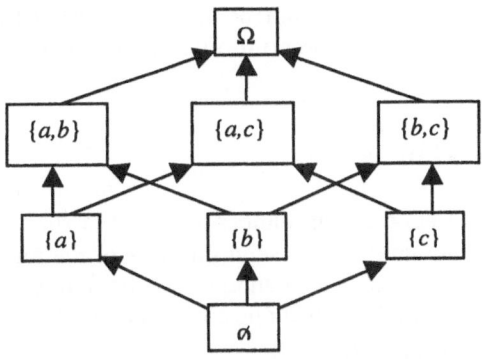

Fig. 1. Natural semi-ordering of subsets of a three-element set.

Let us consider the subsets $\{a,b\}$, $\{a,c\}$ and $\{b,c\}$. In the sense of natural semi-ordering they are mutually incomparable. However, let us assume that an additional logical preference of $\{a,b\}$ with respect to $\{a,c\}$ takes place. This means that in the graph an arc directed from $\{a,c\}$ to $\{a,b\}$ should be introduced. Then, according to the condition b/ of the definition of preference relation an additional arc directed from $\{c\}$ to $\{b\}$ should be also introduced. The graph of preference will thus contain three types of arcs: 1/ the ones indicating natural preferences, 2/ introduced according to additional assumptions, and 3/ deduced from the former ones according to the general properties of preference relations. Finally, the preference relation considered here will be represented by a modified graph shown in Fig. 2. Introduced arc is plotted by a bold line, while the deduced one by a dotted line.

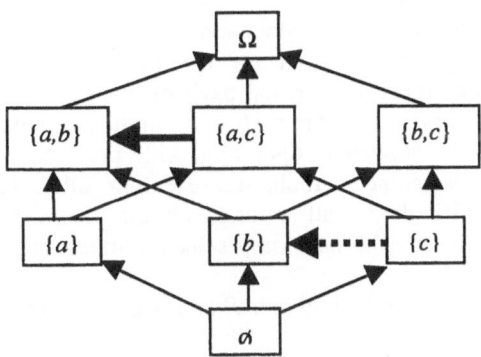

Fig. 2. A modified directed graph representing a preference relation.

Let us remark that a graph representing a preference relation is, in general, oriented and contourless. It makes it possible to compare logically such events only that can be connected by an ordered path directed from a less to a more preferred event. Such ordered pairs correspond to logical chains whose conclusions are not of the type α ? κ. In fact, only those ordered paths in the graph that have not been induced by natural semi-ordering of events indicate non-trivial preferences.

3 Informational variables

Let us take into consideration a real-number axis R. We shall define the set of elementary events Ω as a family of all semi-intervals of the form $(-\infty, x]$, for $x \in R$. The corresponding sigma-algebra of subsets will be defined as a Borel family B_R of such semi-intervals and their algebraic combinations. Then the quadruple

$$X = [R, \Omega, B_R, \prec] \tag{2}$$

where $\prec$ is a preference relation, will be called an *informational variable* described on R.

It is evident that the concept of information variable is an extension of this one of a random variable, the main difference consisting in replacement of a probability measure by a preference relation. Probability measure is a function assigning probabilities to the measurable sets of B_R; in practice it is given in the form of an *integral probability distribution* (ipd) or in an equivalent form of a *probability density function* (pdf). As it has been shown above, it induces in B_R an equivalence based on the equality of probabilities and a weak ordering relation based on the inequalities of probabilities. On the other hand, the preference relation is a rule that to each pair of events $\alpha, \beta \in B_R$ assigns one of four possible relationships: 1° $\alpha \prec \beta$, 2° $\beta \prec \alpha$, 3° $\alpha \approx \beta$ (when both 1° and 2° are valid) or 4° $\alpha ? \beta$ (when neither 1° nor 2° is valid). In general, no numerical probabilities for this purpose are necessary. Operations on probabilities are based mainly on the set algebra and on integral calculus, while, as it will be shown below, operations on preferences can be based mainly on set algebra, linear algebra and on the theory of graphs. Probabilities can be calculated (according to some preliminary assumptions) or can be evaluated on the basis of statistical observations. Preferences can be assigned to selected pairs of events on the basis of statistical observations, of experience and/or of less or more conscious and sophisticated intellectual processes. However, operating with preference relations it is not necessary to use preference graphs representing all events of B_R and all relationships (in particular – all natural) among them. Instead, their partial subgraphs containing only the necessary selected nodes and selected arcs can be used.

Let us take into account the real axis R and two finite intervals D_1 and D_2 in it, as shown below.

$$D_1 \qquad D_2 \qquad\qquad R$$

Fig. 3. Two (partially overlapping) finite intervals on a real axis.

Considering both intervals as events of B_R one can construct for them a graph of natural preference shown in Fig. 4. Let us remark that if non-overlapping intervals are considered, then the graph will be modified so that the relationships $(D_1 \cap D_2) \prec D_1$, $(D_1 \cap D_2) \prec D_2$, $(D_1 \setminus D_2) \approx D_1$ and $(D_2 \setminus D_1) \approx D_2$ are taken into account.

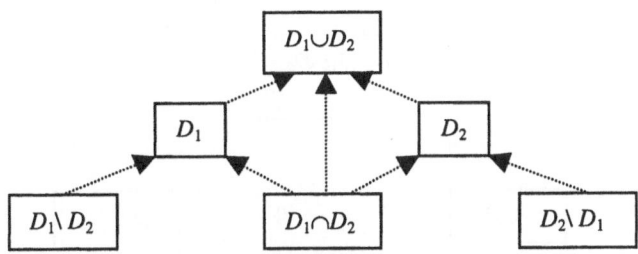

Fig. 4. A graph of natural preferences for two overlapping intervals.

Let us assume that additional preference $D_1 \prec D_2$ is established. Then the graph of preferences takes the form shown in Fig. 5, where non-natural preferences are shown by continuous arrows.

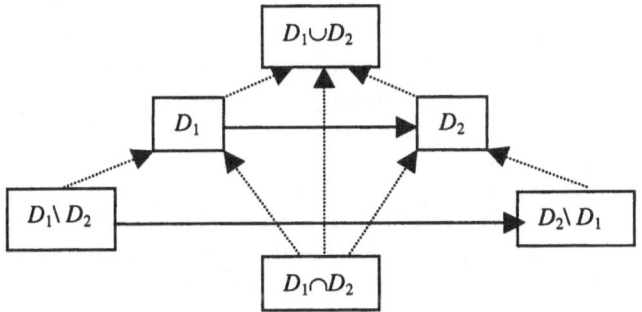

Fig. 5. A graph of preference for two overlapping intervals with assumed preference $D_1 \prec D_2$.

However, if D_1 and D_2 are mutually disjoint, the graph in Fig. 5 should be modified: the pairs of nodes corresponding to: D_1 and $D_1 \setminus D_2$, D_2 and $D_2 \setminus D_1$, should be replaced by single nodes, as representing mutually equivalent events. In similar way, the nodes corresponding to $D_1 \cap D_2$ and to ø should be combined together as representing empty sets. The corresponding graph of preferences is shown in Fig. 6 a. In Fig. 8 b a graph corresponding to a situation when $D_1 \approx D_2$, and in Fig 6 c a one for $D_1 \subset D_2$ are plotted.

316

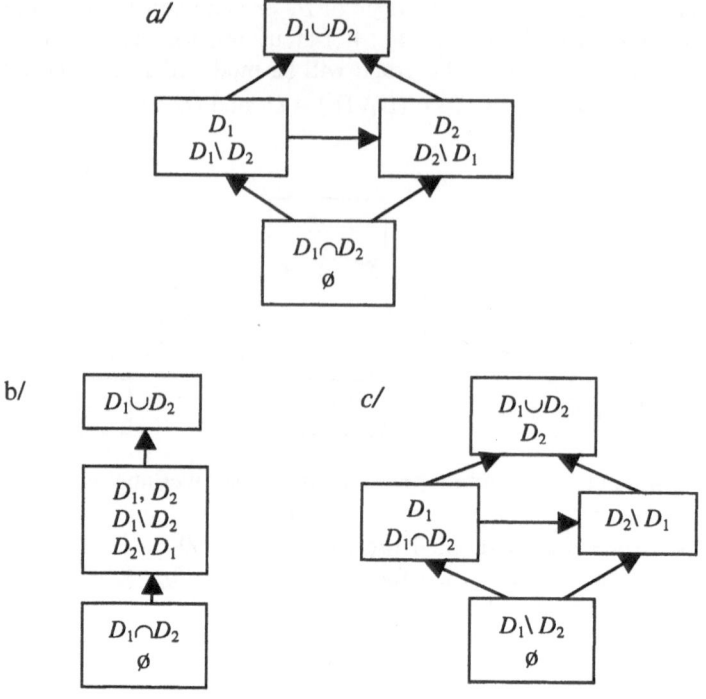

Fig. 6. Graphs of preference for two intervals: a/ mutually disjoint, $D_1 \prec D_2$, b/ as before, $D_1 \approx D_2$, c/ for $D_1 \subset D_2$.

The above-given considerations have shown how to establish preferences among two events represented by intervals on the real axis R. An extension of this approach on any finite number of events is no more but a technical problem.

The case of a set of pair-wise disjoint intervals covering the real axis R is of special interest. Let us take into account a countable set of points $\xi_0, \xi_1, ..., \xi_i, ... \in R$ such that $-\infty < \xi_0 < \xi_1 < ... < \xi_i < ...$ Then there will be considered a series of intervals $D_0 = (-\infty, \xi_0]$, $D_i = (\xi_{i-1}, \xi_i]$ for $i = 1, 2, 3,...$ covering the real axis. Taking into account that the intervals are mutually disjoint one can observe that each pair of consecutive intervals will be represented by a graph of the form shown in Fig.6 a or b. However, it will reasonable to represent graphs of preference in a more convenient for computer calculations form, for example, in the form of graph-representing matrices. For this purpose let us define a set of symbols $F = \{\prec, \succ, \approx, ?\}$. Then a *matrix of preference* will be defined as a square matrix $Q = [q_{ij}]$ such that $q_{ij} \in F$ and $D_i q_{ij} D_j$ represents the preference established among D_i and D_j. The matrix Q has the following properties:

a/ its diagonal elements are $q_{ii} \equiv \approx$;

b/ if $q_{ij} = \prec$ then $q_{ji} = \succ$ (and *vice versa*);

c/ if $q_{ij} = \approx$ or ? then $q_{ji} \equiv q_{ij}$.

Example 2: Let us take into account four consecutive intervals on the real axis: D_1, D_2, D_3, D_4, for which a graph of preferences takes the form shown in Fig. 7.

Fig. 7. A graph of preferences for four intervals.

The same relationships represented by a matrix of preference takes the following form:

	D_1	D_2	D_3	D_4
D_1	$\approx$	$\prec$	$\prec$	?
D_2	$\succ$	$\approx$	$\approx$	$\succ$
D_3	$\succ$	$\approx$	$\approx$	$\succ$
D_4	?	$\prec$	$\prec$	$\approx$

The matrix of preference in its primary form represents the introduced preferences only. However, it can be easily completed by the natural and induced ones. Being given the statements: "$\xi \in D_i$" for $i = 1,2,...,k$, in the case of limited information one cannot say, in general, which of them are "true" or "false" or even what are the probabilities of the corresponding random events. Instead of this, the concept of informational variable gives us the possibility to compare the statements without assigning to them logical, probabilistic or any other numerical weights. So, the extended matrices of preferences in informational variables play a role similar to this of ipdfs or pdfs in random variables.

4 Decision making based on informational variables

First, let us remind the well known statistical decision problem: there are given two conditional pdfs of two alternative random variables: $v(x|1)$ for X^1 and $v(x|2)$ for X^2. It was observed a value ξ. The problem consists in answering with a minimum probability of error, whether ξ represents the random variable (statistical population) X^1, or X^2? Assuming that both statistical populations are of equal prior probabilities the answer is given by calculation and comparison of the values $v(\xi|1)$ and $v(\xi|2)$. Then the greater of them indicates the right answer. Let us remark that for making the decision, in fact, it is not necessary to know exact values of the probabilities; it is enough to know, which of them is greater. This suggests that for making the decision redundant information has been used. It will be shown below that the above-formulated problem can be solved using the concept of informational variable needing reduced primary information.

It will be considered the case of a finite set of informational variables:

$$X^\mu = [R, \Omega, B_R, \prec^\mu], \quad \mu = 1,2,...,m, \tag{3}$$

where R, Ω, and B_R have the same sense as in formula (2) while $\prec^\mu$ denotes a preference relation (specific for X^μ). It will be here assumed that all relations $\prec^\mu$ are described on the same system of intervals of R. Therefore, all they will be represented by matrices of preference Q_μ, $\mu = 1,...,m$, of the same size.

Let us consider μ as an additional component of informational variable. Therefore, the pairs $[\mu, D_i]$ will be considered as new informational variables. Then it arises the following problem:

Let us assume that it was observed a ξ, $\xi \in D_i$, D_i being an interval in R. Establish a decision rule for answering the question: what is the μ^* such that with the highest logical preference ξ can be stated being a realisation of $[\mu^*, D_i]$?

For this purpose a matrix of preference P for the pairs $[\mu, D_i]$ should be considered. P can be constructed as a block-matrix $[P_{\mu\nu}]$, $\mu,\nu = 1,2,...,m$, where $P_{\mu\mu}$ are matrices of preference of the variables X^μ and $P_{\mu\nu}$ for $\mu \neq \nu$ are filled by symbols corresponding to additional relationships introduced among selected events according to the above-given rules of the matrices of preference construction. Lacking symbols should be represented by ? as the symbol of incomparability.

P being given, a sub-matrix of preference P_i for the events $[\mu, D_i]$, $\mu = 1,2,...,m$, should be drawn out from it and taken into account. It gives us the possibility to find out the solution. For this purpose it is enough to select all such rows of P_i that contain no symbol $\prec$.; the corresponding indices μ form a subset of those, maximising the logical preference. This will be illustrated by the following example.

Example 3: There will be taken into account three informational variables, X^1, X^2 and X^3 on a real axis R, described by a matrix of preference Q. For the sake of simplicity it will be assumed that in all cases the same set of intervals: $D_1 = (0, \xi_1]$, ..., $D_4 = (\xi_3, \xi_4]$, $D_5 = (\xi_4, \infty)$, will be used for an approximation of the variables. Let us assume that it was observed $\xi \in D_2$. The question is what is the informational variable represented by ξ ?

In this case the matrix of preference has the following block structure:

$$Q = \begin{bmatrix} Q^{11} & Q^{12} & Q^{13} \\ Q^{21} & Q^{22} & Q^{23} \\ Q^{31} & Q^{32} & Q^{33} \end{bmatrix}$$

Each sub-matrix $Q^{\mu\nu}$ is of 5×5 size and its element $r^{\mu\nu}{}_{ij}$ represents the assumed relationship between the events D_i of X^{μ} and D_j of X^{ν}. However, for the solution of the above-formulated problem it is enough to take into account the following elements: $r^{12}{}_{22}$, $r^{13}{}_{22}$ and $r^{23}{}_{22}$. Let them be:

$$r^{12}{}_{22} = \approx, \quad r^{13}{}_{22} = \prec, \quad r^{23}{}_{22} = ?$$

By symmetry they can be completed by

$$r^{21}{}_{22} = \approx, \quad r^{31}{}_{22} = \succ, \quad r^{32}{}_{22} = ?$$

Then the following graph of preference can be plotted:

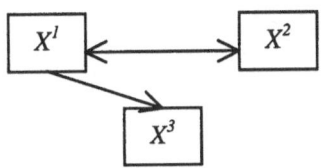

Fig. 8. Graph of preference for a triple of informational variables.

From the graph it becomes clear that the statement "ξ represents X^3" has the highest logical preference.

It may happen that some of so selected rows consist only of symbols $\approx$ and ?. This means that the set of introduced relationships is not sufficient for making an univocal decision. It is evident that any system of logical inference tells us how to manage the available information, not how to create it in the case of its acute shortage.

Bibliography

1. Kulikowski J.L. *"Computers in experimental research"* (in Polish). PWN, Warsaw, 1993.
2. Hempel C.G. *"A purely topological form of non-Aristotelian logic"*. Journ. Symb. Logic, vol. 2, nr 3, 1937.
3. Vessel Ch. A. *"On the topological logic"*. In: "Non-classical Logic" (in Russian), Nauka, Moscow, 1970.
4. Kulikowski J.L. *"Application of the non-classical logical models in operations research"*. In: "Modern Management Problems" (in Polish), PWN, Warsaw, 1974.
5. Kulikowski J.L. *"Decision making in a modified version of topological logic"*. In: "Proceedings of the Seminar on "Nonconventional Problems of Optimization", part 1. Prace IBS PAN, nr 134, Warsaw, 1986.

6. Kulikowski J.L. *"Topological logic and rough sets. A comparison of methods"*. In: "Rough Sets and Their Applications" (in Polish, eds J. Chojcan, J, Łęski). Wyd. Politechniki Śląskiej, Gliwice, 2001.
7. Wilks S.S. *"Mathematical statistics"*. Russian translation: Nauka, Moskva, 1967.
8. Rasiowa H., Sikorski R. *"The mathematics of metamathematics"*. PWN, Warsaw, 1968.

An Algorithm for Identifying Fuzzy Measures with Ordinal Information

Pedro Miranda[1] and Michel Grabisch[2]

[1] Departamento de Estadística, I.O. y D.M., Calvo Sotelo s/n, 33007 Oviedo, Spain, pmm@pinon.ccu.uniovi.es

[2] Université Pierre et Marie Curie- LIP6, 8, rue du Capitaine Scott, 75015 Paris, Michel.Grabisch@lip6.fr

1 Introduction and basic concepts

Consider a decision problem in which the preferences of the decision maker can be modelled through the Choquet integral [2] with respect to a fuzzy measure [10]. This means that he follows some behavioural rules while making decision (see [1], [9]). Next step consists in obtaining such a measure. The problem of identifying fuzzy measures from learning data has always been a difficult problem occurring in the practical use of fuzzy measures [5], [6].

Our starting point is a set of experimental data from which we want to identify the fuzzy measure. Our choice criterion will be the mean squared error, as this is the criterion usually considered in practical problems, due to its good properties.

When dealing with cardinal information, it is well known that the minimization of a squared error criterion leads to a quadratic program. Grabisch and Nicolas [6] have developed an algorithm for the identification of fuzzy measures from sample information. In this algorithm, they solve the quadratic problem whose variables are the coefficients of the fuzzy measure over each subset of the universal set N. A similar algorithm using Möbius inverse and Shapley interaction was developed in [7]. Of course, it can be argued that this situation is not realistic, as in most practical situations the decision maker is not able to provide numerical values but he will be only able to range the objects following his preferences, i.e. in many practical problems one has often to deal with non numerical, qualitative information, coming from any source providing information in natural language. If this addresses in the large the problem of modelling knowledge, we address here in particular the problem of dealing with *ordinal* information, that is, information given on some ordinal scale. The consequence is that any manipulation of the data is forbidden, unless these manipulations involve only order, and then, the algorithm developed for the cardinal case cannot be directly used.

If many powerful tools exist when information is quantitative (or cardinal), the practitioner is devoid of adequate tools to deal with problems involving ordinal information. Usually, an arbitrary mapping on a numerical scale is performed to come back to the cardinal world.

However, in the last years some methods treating with ordinal information have been developed; among them, we will use the MACBETH method [3].

In this paper, we try to build a meaningful numerical scale, keeping the ordinal information. To make this, we do not deal with only one numerical scale but a family of possible cardinal representations. Then, we look for a fuzzy measure that best represents the data, i.e. that best conserves the ordinal information. We will propose as solution a quadratic problem under constrains. Finally, an algorithm leading to a linear problem is proposed.

In the sequel, we will consider a finite set of criteria $N = \{1, ..., n\}$. Subsets of N are denoted with capital letters A, B, and so on. The set of all subsets of N is denoted $\mathcal{P}(N)$.

A fuzzy measure is denoted by μ. The set of fuzzy measures is denoted by $\mathcal{FM}$. Vectors are denoted with bold letters $\mathbf{u}, \mathbf{b}$, while matrices are denoted with bold capital letters $\mathbf{H}, \mathbf{A}$ and so on.

We suppose then that we are given l objects or alternatives, from which we want to derive the fuzzy measure. Let us denote the set of alternatives by $X = \{x^1, ..., x^l\}$. We also suppose that the decision maker is able to compare (in an ordinal way) these objects for each criterion and also that he is able to express a preference between any two alternatives, so as to obtain a ranking of the alternatives.

2 Identification of fuzzy measures with cardinal data

Let us now recall the method developed in [6] and [7]. We suppose that we are given l numerical values of Choquet integral $e_1, ..., e_l$ and also the numerical values $f^1(x_1), ..., f^1(x_n), ..., f^l(x_1), ..., f^l(x_n)$ of the scores of each alternative over each criterion. Our goal was to determine a fuzzy measure μ minimizing the quadratic error

$$\sum_{k=1}^{l}(\mathcal{C}_\mu(f^k(x_1), ..., f^k(x_n)) - e_k)^2.$$

Of course, we have restrictions over μ, namely the restrictions of monotonicity.

It has been proved in [6] that if we consider the quadratic error criterion, the problem reduces to solve the quadratic problem:

$$\begin{aligned} \text{minimize} \quad & \tfrac{1}{2}\mathbf{u}^T \mathbf{D}_\mu \mathbf{u} + \Gamma_\mathbf{u}^T \mathbf{u} \\ \text{under the constraints} \quad & \mathbf{A}_\mu \mathbf{u} - \mathbf{b}_\mu \geq \mathbf{0} \end{aligned} \tag{1}$$

where $\mathbf{u}$ is the vector containing the values of $\mu(A)$ for all $A \subset N$. In [7], other expressions of this problem in terms of Möbius transform and Shapley and Banzhaf interactions have been developed.

3 A method for identifying fuzzy measures

Let us now develop a method for the ordinal case. The idea is to use the learning process with cardinal data studied in Section 2. To obtain these cardinal data we are going to use the MACBETH algorithm, as it is a theoretically well-founded method to build interval scales from ordinal information. Once cardinal data are obtained, we obtain the fuzzy measure μ by solving a quadratic problem as it was done in Section 2.

MACBETH is based on two fundamental assumptions: the fact that the decision maker is able to express in an ordinal way the intensity of preference between two objects, and the existence on each criterion i of two particular elements, denoted $\mathbf{0}_i$ and $\mathbf{1}_i$, which have the meaning of "neutral" and "satisfactory", respectively. These two elements convey an absolute meaning in the sense that the degree of satisfaction is the same for $\mathbf{0}_i$, $\forall i$, and for $\mathbf{1}_i$, $\forall i$.

These assumptions enable the construction of interval scales $\nu_1, ..., \nu_n$ over each criterion, which are made commensurable by putting $\nu_i(\mathbf{0}_i) = 0$, $\forall i$, and $\nu_i(\mathbf{1}_i) = 1$, $\forall i$. Also, an overall interval scale ν can be built similarly for the overall score (see [3] for details). Doing this, we are back to the cardinal case and μ can be obtained by solving the following problem:

$$\min_{\mu \in \mathcal{FM}} \sum_{i=1}^{l} (\mathcal{C}_\mu(\nu_1(x_1^i), ..., \nu_n(x_n^i)) - \nu(x^i))^2.$$

However, before solving this problem, we have to make ν commensurable with all scales ν_i.

In the sequel, we will denote $\mathcal{C}_\mu(\nu_1(x_1^i), ..., \nu_n(x_n^i))$ by $\mathcal{C}_\mu(x^i)$ in order to simplify notation. We will also denote $\nu_j(x_j^i)$ by x_j^i.

In next paragraphs, we will develop some approaches in which we consider different degrees of information about the overall scale.

3.1 Method 1

Let us start with the simplest solution. For the overall scale, the "overall neutral" and the "overall satisfactory" can be defined as follows: the overall neutral is a synthetic object whose criteria scores are $\mathbf{0}_1, ..., \mathbf{0}_n$, denoted $(\mathbf{0}_N, \mathbf{1}_\emptyset)$. On the other hand, we define the overall satisfactory as another synthetic object whose criteria scores are $\mathbf{1}_1, ..., \mathbf{1}_n$, denoted $(\mathbf{1}_N, \mathbf{0}_\emptyset)$. We then define $\nu((\mathbf{0}_N, \mathbf{1}_\emptyset)) = 0$, $\nu((\mathbf{1}_N, \mathbf{0}_\emptyset)) = 1$, which is consistent with $\nu_i(\mathbf{0}_i) = 0$, $\nu_i(\mathbf{1}_i) = 1$, $i = 1, ..., n$, thanks to the relation $\mathcal{C}_\mu(0, ..., 0) = 0$ and $\mathcal{C}_\mu(1, ..., 1) = 1$.

Like this, we have some values of Choquet integral $\nu(x^1), ..., \nu(x^l)$ and the values $\nu_1(x^1), ..., \nu_1(x^m), ..., \nu_n(x^1), ..., \nu_n(x^l)$; thus we are in the conditions of the cardinal case. Applying then the results of Section 2, we obtain the fuzzy measure or the set of fuzzy measures that best fit the data.

Of course, we have all the problems derived from the cardinal case. For example, we can obtain several solutions.

3.2 Method 2

A problem can be argued for Method 1: A decision maker can easily compare objects over a concrete criterion; then, the scales obtained are rather exact and also the normalization step gives no problem. The situation changes when dealing with the overall score; in that case, a decision maker is likely to give wrong information when he compares the "overall neutral" and the "overall satisfactory" with other actions. The reason is that in many situations the decision maker is not able to imagine an object that is, say, satisfactory for all criteria.

For example, suppose that we are comparing cars and that we have as criteria the color, the price, the speed and the size. Then, a car that is big enough to be "satisfactory" for the criterum size, is also rather expensive; hence, the decision maker can find trouble to imagine a car "satisfactory" for size and also cheap (and thus "satisfactory" for price), as no car satisfies these two conditions simultaneously.

Thus, the overall scale obtained could be wrong and the fuzzy measure obtained would not be correct. In other words, some problems can arise in the normalization step for the overall scale. Then, we can wonder what happens when the overall scale is not fixed and we allow some degrees of freedom.

We suppose here that we have used the MACBETH approach in order to obtain scales over the criteria, and also that we have found an overall scale, but now we do not fix values for $\nu((\mathbf{0}_N, \mathbf{1}_\emptyset))$ and $\nu((\mathbf{1}_N, \mathbf{0}_\emptyset))$. Hence, the overall scale is not uniquely determined, and since ν is an interval scale, we have a family of possible scales, namely

$$\nu^* = \theta_1 \nu + \theta_2, \ \forall \theta_1, \theta_2 \in \mathbb{R}.$$

Then, we look not only for a fuzzy measure μ but also for the best scale ν^* fitting the data or, taking account that ν^* is completely determined by θ_1 and θ_2, we look for μ, θ_1 and θ_2 that best describe the ordinal information. This leads us to solve the following problem:

$$\min_{\mu \in \mathcal{FM}, \theta_1, \theta_2} \sum_{i=1}^{m} (\mathcal{C}_\mu(x^i) - \theta_1 \nu(x^i) - \theta_2)^2.$$

Of course, as we have more degrees of freedom, this will translate into a reduction of the quadratic error.

Let us now see two different ways to solve this problem.

First, suppose that μ is fixed. Then, $\mathcal{C}_\mu(x^i)$ is fixed, too (as we know the evaluations of each object over each criterion). Thus, we have the pairs $\{(\nu(x^i), \mathcal{C}_\mu(x^i))\}$ and we want to find θ_1, θ_2 minimizing the quadratic error. This problem is equivalent to find the coefficients of θ_1, θ_2 in the regression problem $y = \theta_1 x + \theta_2$.

Then, for a fixed μ, we can obtain the best values for θ_1 and θ_2, i.e. we can find the best overall scale ν^*. It must be noted that both θ_1 and θ_2 depend linearly on μ.

Hence, we can find an expression for the quadratic error in terms of μ. This quadratic error can be found from the regression model. Finally, it suffices to minimize this quadratic error. Let us introduce some notations:

If $x^i_{(0)} = 0 \leq x^i_{(1)} \leq \dots \leq x^i_{(n)}$ and $A^i_{(j)} = \{(1), \dots, (j)\}$, let us define a vector $\mathbf{f}_i$ by

$$\mathbf{f}_i(A) = \begin{cases} x^i_{(j)} - x^i_{(j-1)} & \text{if } A = A^i_{(j)} \\ 0 & \text{if } A \neq A^i_{(1)}, \dots, A^i_{(n)} \end{cases}$$

From $\mathbf{f}_i$, we define other vectors $\mathbf{h}_i$, $i = 1, \dots, l$ and a matrix $\mathbf{H}$ by

$$\mathbf{h}_i(A) = \mathbf{f}_i(A) - \frac{\sum_{j=1}^{l} \mathbf{f}_j(A)}{l}, \quad \mathbf{H} = \sum_{i=1}^{l} \mathbf{h}_i \mathbf{h}_i^T.$$

On the other hand, let us denote

$$\bar{\nu} = \frac{\sum_{i=1}^{l} \nu(x^i)}{l}, \quad b_i = \frac{(\nu(x^i) - \bar{\nu})}{\sum_{i=1}^{l}(\nu(x^i) - \bar{\nu})^2}.$$

Then, we define new vectors $\mathbf{g}_i = b_i \mathbf{h}_i$ and $\mathbf{g} = \sum_{i=1}^{l} \mathbf{g}_i$. With these notations, we obtain:

Proposition 1. *If we consider the quadratic error criterion, the problem of identifying μ reduces to solve the quadratic problem:*

$$\begin{aligned} &minimize & &\mathbf{u}^T \mathbf{H} \mathbf{u} - \mathbf{u} \mathbf{g} \mathbf{g}^T \mathbf{u} \\ &under\ the\ constraints & &\mathbf{A}_u \mathbf{u} + \mathbf{b}_u \geq 0 \end{aligned} \tag{2}$$

where $\mathbf{u}$ is the vector containing the values of $\mu(A)$, $\forall A \subset N$, and $\mathbf{A}_u$, $\mathbf{b}_u$ are the same as in (1).

We have seen that Method 2 provides a more general method than Method 1. Moreover, as θ_1 and θ_2 are variables, we obtain a reduction in the error. However, an important problem may arise: In Method 2 we can obtain a solution for which the value of θ_1 is not positive. A negative value of θ_1 means an inversion in the preferences of the decision maker and thus the result obtained cannot be taken as a good representation of the data. In this sense, the following can be proved:

Proposition 2. *Consider $x^i, x^j \in X$. If $C_\mu(x^i) > C_\mu(x^j)$ whenever $x^i \succ x^j$, then, necessarily, $\theta_1 > 0$.*

This means that if the fuzzy measure μ obtained by Method 2 does not produce any inversion in the preferences given by the decision maker, then θ_1 must attain a positive value. However, the reciprocal is not true as the following example shows:

Example 1. Suppose $|X| = 2$ and that we have 4 data, namely

x_1^i	x_2^i	$\nu(x^i)$
0.9	0	0.225
1	0.5	0.25
0.1	0.8	0.025
0	0.5	0.2

Tedious calculation leads to the solution $\theta_2 = 1.5, \theta_1 = 0.2, \mu(1) = 0.994, \mu(2) = 0.989$. But in this case, we have

$$C_\mu(0.9, 0) = 0.891, C_\mu(1, 0.5) = 1, C_\mu(0.1, 0.8) = 0.8, C_\mu(0, 0.5) = 0.5,$$

and thus we have changed the preferences of the decision maker as $x^3 \succ x^4$, while $\nu(x^3) < \nu(x^4)$.

This simple example shows an inversion in the preferences obtained by the solution and the preferences given by the decision maker, even if $\theta_1 > 0$. If more data are given, then the probability of inversion grows.

Another problem that arises is the problem of uniqueness. It is possible that several measures lead to the same error. Let us study the structure of the set of solutions. Let us denote

$$S = \{(\mu, \theta_1(\mu), \theta_2(\mu)) | \mu \text{ optimal solution}\}.$$

Now, an interesting result can be proved about S.

Proposition 3. *S is a convex set, i.e. if $(\mu_1, \theta_1^1, \theta_2^1), (\mu_2, \theta_1^2, \theta_2^2) \in S$, then for any $\alpha \in [0, 1]$, $\alpha(\mu_1, \theta_1^1, \theta_2^1) + (1 - \alpha)(\mu_2, \theta_1^2, \theta_2^2)$ is a solution, too.*

In particular, the set of possible values of θ_1 is an interval. It can be proved that the extreme points of this interval can be found through the following linear program with variable $\mathbf{u}$:

$$\max / \min_\mu (l \sum_{i=1}^{l} \mathbf{f}_i^T \nu(x^i) - \sum_{i=1}^{l} \mathbf{f}_i^T \sum_{i=1}^{l} \nu(x^i)) \mathbf{u}$$

s.t. μ is a fuzzy measure.

Of course, this program provides the extreme points for **possible** θ_1. The real optimal values for θ_1 are given when the fuzzy measures considered are also optimal. However, this result will be used in next example to see that it is possible to find situations in which all the possible values of θ_1 are non-positive:

Example 2. Let us suppose that we have two objects and two criteria. Let us suppose that the available information is given by

x_1^i	x_2^i	$\nu(x^i)$
0.6	0.5	0.4
0.6	0.3	0.6

Then, the corresponding values of $\mathbf{f}_1, \mathbf{f}_2$ considering the binary order are

$$\mathbf{f}_1(1) = 0.1, \ \mathbf{f}_1(2) = 0, \ \mathbf{f}_1(1,2) = 0.5.$$

$$\mathbf{f}_2(1) = 0.3, \ \mathbf{f}_2(2) = 0, \ \mathbf{f}_2(1,2) = 0.3.$$

Then, the value that multiplies $\mu(1)$ is 0.04, the value multiplying $\mu(2)$ is 0 and the value multiplying $\mu(1,2)$ is -0.04. Thus, our objective function is

$$\max_{\mu} \left[0.04\mu(1) - 0.04\mu(1,2) \right].$$

As μ is a fuzzy measure, we can never obtain a positive value for θ_1.

This example shows that we can find situations in which θ_1 takes only non-positive values, and consequently, we cannot model properly the ordinal information. However, it must be remarked that in this example there is an incoherence in the preferences of the decision maker, as x^2 is preferred to x^1 while regarding the scores over the criteria, x^1 should be considered better than x^2.

4   Conclusions and open problems

In this paper we have studied the problem of identifying a fuzzy measure when only ordinal information is provided. The idea of the proposed method is to translate the information given in natural language to cardinal information, and apply then the method that we have developed in [7] to identify the fuzzy measure. The translation of ordinal information to cardinal information is given through the MACBETH approach.

Our method consists in two steps:

1. In the first step we use the MACBETH approach to translate our ordinal information in a cardinal information.
2. In the second step we look for fuzzy measures that best fit this cardinal information given in the first step.

Of course, all expressions obtained could be written in terms of the Möbius inverse [8], Shapley interaction [4] and so on.

We feel that our approach mixes both the richness of the MACBETH approach and fuzzy measures. However, much research must be done:

- An implementation of these algorithms. This should allow us to compare more concretely our algorithms with other methods in terms of computational cost, number of iterations, ... On the other hand, a comparison between our algorithms could be performed; for example, if the solution obtained in the four algorithms is more or less the same, it does not worth to work with Method 2 because Method 1 is much more simpler.

- It could be interesting to find conditions for a unique solution and expressions for the set of solutions in the general case.
- Of course, we have developed all the algorithms for general fuzzy measures, but the same process can be done for special families of fuzzy measures, just adding more constrains. In this sense, the special case of 2-additive measures [4] could be of interest as the measures obtained are very useful for interpretations.

Acknowledgements

The research in this paper has been supported in part by FEDER-MCYT grant number BFM2001-3515.

References

1. A. Chateauneuf. Comonotonic axioms and RDEU theory for arbitrary consequences. *Journal of Mathematical Economics*, to appear.
2. G. Choquet. Theory of capacities. *Annales de l'Institut Fourier*, (5):131–295, 1953.
3. C. A. Bana e Costa and J.-C. Vansnick. A theoretical framework for Measuring Attractiveness by a Categorical Based Evaluation TecHnique (MACBETH). In Jo ao Clímaco, editor, *Multicriteria Analysis*, pages 15–24. Springer, 1997.
4. M. Grabisch. *k*-order additive discrete fuzzy measures. In *Proceedings of 6th Int. Conf. on Information Processing and Management of Uncertainty in Knowledge-Based Systems (IPMU)*, pages 1345–1350, Granada (Spain), 1996.
5. M. Grabisch. Alternative representations of discrete fuzzy measures for decision making. *Int. J. of Uncertainty, Fuzziness and Knowledge-Based Systems*, 5:587–607, 1997.
6. M. Grabisch and J.-M. Nicolas. Classification by fuzzy integral-performance and tests. *Fuzzy Sets and Systems, Special Issue on Pattern Recognition*, (65):255–271, 1994.
7. P. Miranda and M. Grabisch. Optimization issues for fuzzy measures. *International Journal of uncertainty, Fuzziness and Knowledge-Based Systems*, 7(6):545–560, 1999.
8. G. C. Rota. On the foundations of combinatorial theory I. Theory of Möbius functions. *Zeitschrift für Wahrscheinlichkeitstheorie und Verwandte Gebeite*, (2):340–368, 1964.
9. D. Schmeidler. Integral representation without additivity. *Proc. of the Amer. Math. Soc.*, (97(2)):255–261, 1986.
10. M. Sugeno. *Theory of fuzzy integrals and its applications*. PhD thesis, Tokyo Institute of Technology, 1974.

Exploring an Alternative Method for Handling Inconsistency in the Fusion of Expert and Learnt Information

Jonathan Rossiter

AI Group, Dept. of Engineering Mathematics
University of Bristol, Bristol, BS8 1TR, UK,
email: Jonathan.Rossiter@bris.ac.uk

Abstract. This paper presents an approach to reasoning with learnt and expert information where inconsistencies are present. Information is represented as an uncertain taxonomical hierarchy where each class is a concept specification either defined by an expert or learnt from data. We show through simple examples how learnt information and uncertain expert knowledge can be represented and how conclusions can be reasoned from the fused hierarchy. This reasoning mechanism relies on a default assumption to rank conclusions based on the position of contributing information in the class hierarchy. We examine the aggregation function of the default reasoning process and propose improvements that result in more natural fusions of expert and learnt hierarchical information.

1 Introduction

In this paper we consider the problem of fusing expert knowledge with knowledge that has been learnt from data. We represent both expert and learnt knowledge within a hierarchy of uncertain classes. Information fusion is performed in this environment in order to reason with structured information from different sources. Crucially we exploit the structure of the hierarchy itself to resolve inconsistencies that occur across the fused hierarchy.

The first section discusses our choice of uncertain taxonomies for the representation of both expert and learnt information. We then briefly explain how inconsistencies can occur in uncertain hierarchies and suggest the use of Cao's default reasoning algorithm [3] for reasoning with partially inconsistent hierarchies. Following this section we look at two illustrative problems that show how these mechanisms can be applied in the fusion of expert and learnt knowledge. Finally we discuss alternative methods for the aggregation of subsets of consistent information that give improved performance over Cao's original algorithm.

2 Learnt and expert information

In this paper we make the assumption that expert knowledge can be represented as a hierarchy of classes within which uncertainty is integrated.

Uncertainty in this framework takes the form of fuzzy sets and probability intervals where fuzzy sets represent property values and probability intervals express both the memberships of objects in classes and the applicability of properties to an object or a class. More information about these theories of uncertain class hierarchies can be found in [2], [4], [1] and [8]. Of course there may be practical reasons why we cannot represent some specific knowledge in this framework. For example, we may have some knowledge but we may not know how this knowledge fits into the class hierarchy. Alternatively we may be certain that some knowledge is completely inconsistent with the class hierarchy. In this paper we sidestep this issue and we only consider the cases where knowledge defines some conceptual features known to be (to some degree) consistent with some known part of the uncertain hierarchy.

We represent expert knowledge by constructing simple rules based on matching discrete fuzzy sets on labels where each element is a *(label : membership)* pair defining the membership of a point in a continuous piece-wise fuzzy set. For example, we first construct a vocabulary of label fuzzy sets $L = \{very\_small,\ small,\ medium,\ large,\ very\_large\}$ where each member is a label describing a fuzzy set. Hence $[-1.5 : 0\ -0.75 : 1\ 0 : 0]$ would define a simple triangular fuzzy set corresponding to the label *small*. We then specify expert knowledge as discrete fuzzy sets on these labels. Hence $legal = \{small : 1\ medium : 0.5\}$ defines the property *legal* in terms of labels *small* and *medium*. In this way expert knowledge can be encoded in a transparent form that corresponds directly to real world linguistic statements such as 'the point is *legal* if it is *small* in the x axis and *large* in the y axis, or it is *large* in the x axis and *small* in the y axis'. These uncertain linguistic rules can then be inserted into the hierarchy of knowledge at the appropriate point.

Learnt knowledge, on the other hand, can be represented in any number of ways depending on the learning mechanism used. Common representations are graphical, numeric or connectionist. One focus of machine learning that has relevance to transparent information fusion is machine learning within a framework of computing with words. We have presented such a machine learner in [6]. This method learns simple rules from data where the rules contain fuzzy sets on labels. Since this representation of learnt knowledge is the same as our proposed representation of expert knowledge we will use this method for learning from data in all examples in this paper. Unfortunately this learning mechanism does result in some decomposition error. For the purposes of this information fusion work this (normally undesirable) characteristic is not a problem. We are simply aiming to develop new approaches to information fusion rather than develop competitive learning algorithms.

One question we have yet to answer is how to fuse two structured hierarchies. The learnt models we are dealing with here are simple fuzzy rulesets which model singular concepts. In this respect the ruleset defines, or contributes to a definition of, a single uncertain class. This makes the fusion operation much simpler since we need only consider the general problem of inserting a single learnt class into an expert hierarchy. Even so, this problem still involves substantial computation to determine where to insert this single class. In this paper we make the further assumption that we know the concept that the learnt class defines and that this concept has also been defined by the expert independently. That is, if a learnt class c_L defines the concept $conc(c_L)$ and the expert hierarchy contains classes $C_E = \{c_{E1} \ldots c_{En}\}$, there must exist $c_{Em} \in C_E$ such that $conc(c_{Em}) = conc(c_L)$. This being the case we fuse the learnt class into the expert hierarchy by inserting c_L as a sibling of c_{Em}, i.e. c_L and c_{Em} share the same immediate superclass.

3 Inconsistency in uncertain class hierarchies

A major problem with fusion of this kind is that multiple independent sources of knowledge, be it from experts or from some learning mechanism, are unlikely to be entirely consistent. In constructing and fusing uncertain class hierarchies these inconsistencies may only become apparent when the hierarchy is queried. Since properties are inherited with some degree of applicability expressed as a support interval any inconsistency becomes evident when intersecting these support intervals.

Let us take the the classes C_1 and C_2 which each contain a definition for the property ψ, and an object O which has some probability of membership in each class, i.e. $Pr(C_1|O)$ and $Pr(C_2|O)$. We first determine the probability that O will inherit ψ from C_1 and C_2 separately using an interval form of Jeffrey's rule [5][4], i.e. we calculate $p_1 = Pr(\psi_{c_1}|O)$ and $p_2 = Pr(\psi_{c_2}|O)$. An inconsistency between the two inherited version of P is defined by the intersection of intervals p_1 and p_2 yielding the empty interval, i.e. $p_1 \cap p_2 = [\,]$. Clearly we require a mechanism to deal with such inconsistencies. One approach is to calculate the maximally consistent subset of inherited properties. Unfortunately this process is exponential in time complexity for uncertain hierarchies where it is possible for an object to inherit from *all* classes. There are a number of other approaches, such as Dempster's rule, which can deal with independent and conflicting information with more practical time complexity but these are often rather blunt in application. Typically these approaches do not take into account the structural information defined in the hierarchy itself. A far better approach is to examine the hierarchy and use this to constrain property inheritance with respect to consistency. We would say that classes higher up the hierarchy define general concepts while classes lower down the hierarchy define more specific concepts. In this regard it seems natural where conflicts occur to draw our most concrete conclusions using inherited properties from classes lower down the hierarchy and discount properties from classes higher up. In this paper we implement such as approach in the form of Cao's default reasoning algorithm based on hierarchical ranking, as first presented in [3].

3.1 Default reasoning in fused uncertain hierarchies

In this paper we consider how this default reasoning algorithm impacts upon the fusion of expert and learnt information in uncertain class hierarchies. The default reasoning algorithm derives a set of preferred consistent subsets $P = \{p_1, \ldots, p_n\}$ and combines these to form the single support interval $Pr(\psi|O)$, which denotes the support for property ψ being applicable to object O, using Equation 1. Broadly the effect of Equation 1 is to intersect (i.e. restrict) the supports *up* branches of the hierarchy and then to union (i.e. widen) the resulting intervals *across* the hierarchy. A more detailed discussion of this algorithm with respect to information fusion is found in [7].

$$Pr(\psi|O) = \bigcup_{p_i \in P} (\bigcap x \mid x \in p_i) \qquad (1)$$

3.2 Two simple examples

We study the behaviour of this algorithm using the two simple example class hierarchies of expert and learnt information shown in Figures 3 and 4 which define the conceptual shapes shown in Figures 1 and 2 respectively.

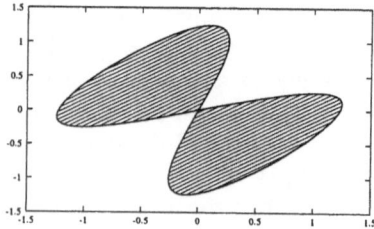

Fig. 1. Figure eight shape

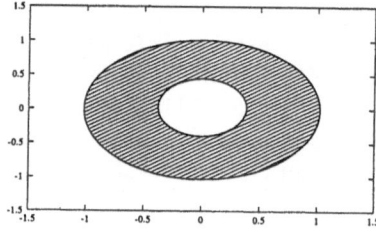

Fig. 2. Doughnut shape

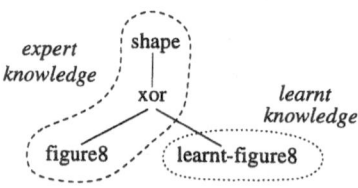

Fig. 3. Figure eight hierarchy

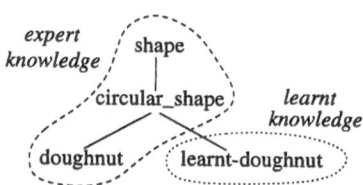

Fig. 4. Doughnut hierarchy

Figures 5 and 7 show the classification results using just the expert hierarchies and Figures 6 and 8 show the results using just the learnt classes. The

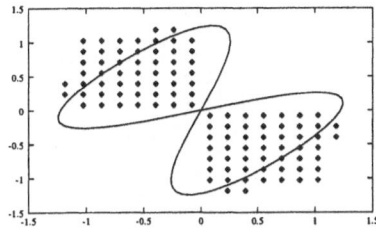

Fig. 5. Expert's figure eight: 81.5%

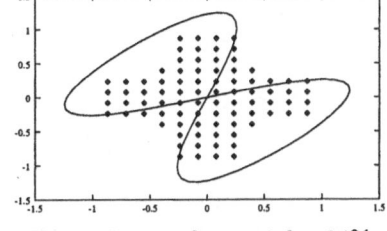

Fig. 6. Learnt figure eight: 84%

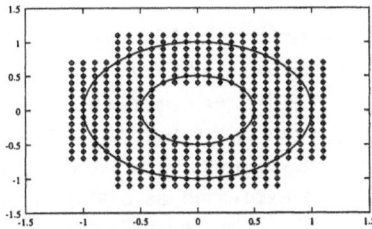

Fig. 7. Expert's doughnut: 81.3%

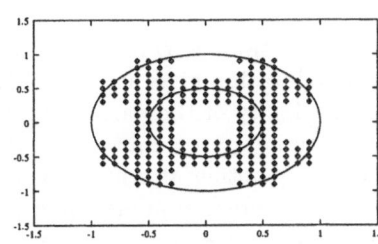

Fig. 8. Learnt doughnut: 85.5%

accuracy values show the percentage of test points classified correctly. Notice how the expert knowledge, represented using simple linguistic fuzzy rules, captures the general shapes but, especially in the doughnut case, tends to over-generalise the structure of the concepts. The learnt models, on the other hand, exhibit decomposition error which results in characteristic symmetrical patterns.

When we fuse the expert and learnt classes and apply default reasoning to resolve inconsistencies we generate the classification results shown in Figures 9 and 10. It is interesting to observe the effects of the fusion and reasoning processed on the structure of the classification results. Fusion results in slightly worse results for the figure eight problem and slightly better results for the doughnut problem. The important point here is that we have fused learnt and expert knowledge in a common hierarchical representation and have been able to draw measured conclusions about the shape of the concepts being represented through the stratified selection of consistent knowledge. Further to this, we can examine the linguistic rules that have been combined and this may be of great use to experts who wish to learn more about a problem domain from the learnt model.

4   Alternative default fusion strategies

While the aggregation function shown in Equation 1 is formally reasonable it is questionable in practice. The tendency in the union operation is to widen the support region without taking into account the intervals of uncertainty *within* the supports. Take for example the support intervals $[l_1, u_1]$ and $[l_2, u_2]$

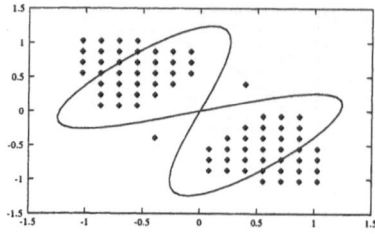

Fig. 9. Fused figure eight: 79.5%

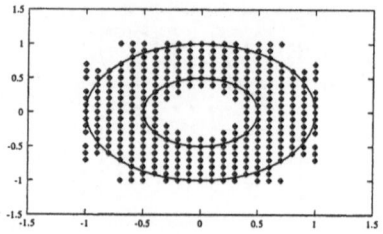

Fig. 10. Fused doughnut: 88.8%

which we combine by interval union to yield $[\mathrm{MIN}(l_1, l_2), \mathrm{MAX}(u_1, u_2)]$. If the union operation gives us $[l_1, u_1]$ then this clearly does not take into account the specificity of the narrower interval $[l_2, u_2]$. A better approach may be to join these supports in a more considerate way, such as through an interval disjunction. Another alternative is to defuzzify the supports since we commonly require the final support for a property to be expressed as a singleton. In the following alternatives the support logic conjunction is implemented as in Equation 2 and the support logic disjunction is implemented as in Equation 3. The defuzzification aggregation $\vee^*$ is shown in Equation 4.

$$[l, u] \wedge [n, p] = [l \times n, \quad u \times p] \tag{2}$$

$$[l, u] \vee [n, p] = [l + n - l \times n, \quad u + p - u \times p] \tag{3}$$

$$[l, u] \vee^* [n, p] = \frac{l + u + n + p}{4} \tag{4}$$

Motivation for the use of logical rather than set operations is based on the selective nature of the default reasoning algorithm. The algorithm selects a set of consistent subsets which form a theory for the applicability of property ψ. Determining the applicability of ψ can therefore be thought of as the resolution of this theory into a single conclusion. Logical operations are one alternative for resolving this theory. Motivation for the use of the defuzzifying mechanism in Equation 4 is based on the assumption of a uniform prior distribution over the probability interval $[l, u]$. Given such a uniform distribution a reasonable choice for a singleton representation of the applicability of ψ is the mid point between the upper and lower bounds.

Table 1 show the classification results for alternative aggregation operations to those in Equation 1. Method 1 equates to Equation 1. These results are shown graphically in Figures 11 to 18. The results show a marked improvement over Equation 1 when using methods 3 and 4. This indicates most strongly that aggregation of supports from consistent subsets through Cao's default reasoning algorithm can be improved by applying alternative operations to set union and intersection. It is also the case that in the two examples the structure of the fused shapes was much improved using these alternative approaches. This is shown most strongly in Figures 13, 14 16, 17 and 18.

Method	Operations	Figure Eight (%)	Doughnut (%)
1	$(\cap, \cup)$	79.5	88.8
2	$(\cap, \vee)$	79.5	88.4
3	$(\cap, \vee^*)$	86	90.4
4	$(\wedge, \vee)$	85	92.1
5	$(\wedge, \vee^*)$	84.5	89.2

Table 1. Classification results

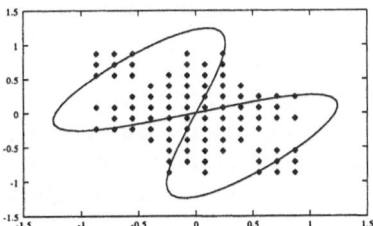

Fig. 11. Figure 8, method 2: 79.5%

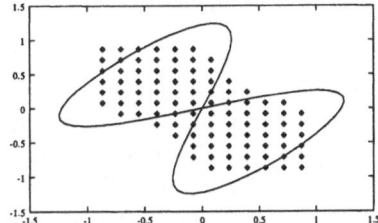

Fig. 12. Figure 8, method 3: 86%

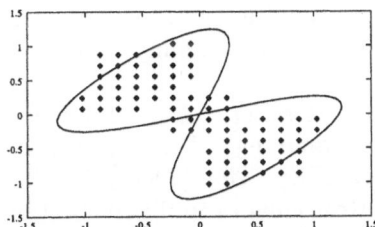

Fig. 13. Figure 8, method 4: 85%

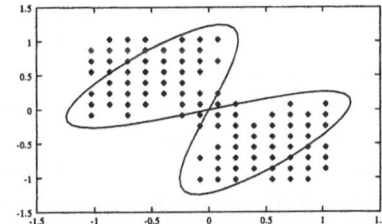

Fig. 14. Figure 8, method 5: 84.5%

5   Conclusions

We have shown how expert and learnt knowledge can be represented and fused in simple uncertain class hierarchies. Inconsistencies that commonly come to

336

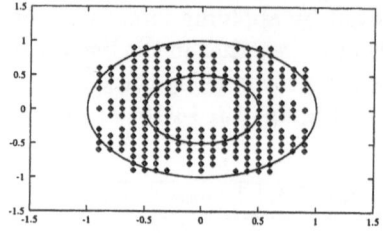

Fig. 15. Doughnut, method 2: 88.4%

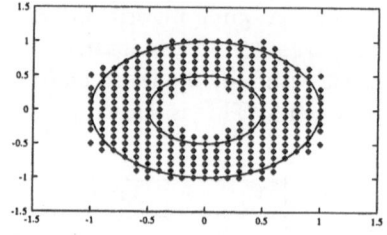

Fig. 16. Doughnut, method 3: 90.4%

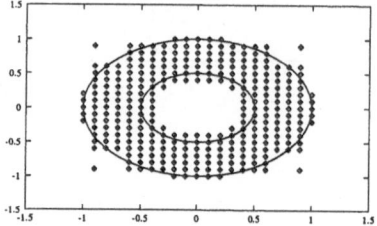

Fig. 17. Doughnut, method 4: 92.1%

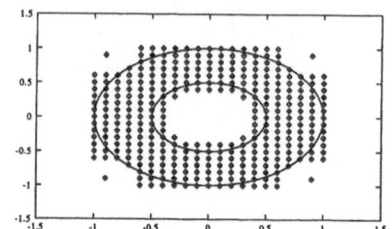

Fig. 18. Doughnut method 5: 89.2%

light only after the information fusion process has finished can be handled in polynomial time using Cao's default reasoning algorithm. We present some alternatives methods for the aggregation of consistent subsets of applicable properties which generally perform better in the domain of fused expert and learnt uncertain hierarchical information.

References

1. Baldwin. J.F., Cao, T.H., Martin, T.P., Rossiter, J.M.: "Implementing Fril++ for Uncertain Object-Oriented Logic Programming", Proceedings of IPMU2000
2. Baldwin. J.F., Cao, T.H., Martin, T.P., Rossiter, J.M.: "Towards Soft Computing Object-Oriented Logic Programming". Proc. of the Ninth IEEE Int. Conf. on Fuzzy System, FUZZ-IEEE 2000, (2000), 768–773
3. Cao, T.H.: "Uncertain Inheritance and Reasoning as Probabilistic Default Reasoning". Int. J. of Intelligent Systems, **16** (2001) 781–803
4. Cao, T.H., Rossiter, J.M., Martin, T.P., Baldwin. J.F.: "Inheritance and Recognition in Uncertain and Fuzzy Object-Oriented Models". Proc. of the 1st Int. Joint Conf. of the Int. Fuzzy Systems Ass. and the North American Fuzzy Information Processing Soc., IFSA/NAFIPS, (2001)
5. Jeffrey, R.C. "The logic of decision" McGraw-Hill, 1965
6. Rossiter, J.M., Cao, T.H., Martin, T.P., Baldwin. J.F.: "Object-oriented modelling with words". Proc. of the Tenth IEEE Int. Conf. on Fuzzy System, FUZZ-IEEE 2001, (2001)

7. Rossiter, J.M.: "Fusing Partially Inconsistent Expert and Learnt Knowledge in Uncertain Hierarchies". Proc. of the Third Int. Conf. on Intelligent Data Engineering and Automated Learning, IDEAL2002, (2002)
8. Rossiter, J.M., Cao, T.H., Martin, T.P., Baldwin. J.F.: "A Fril++ Compiler for Soft Computing Object-Oriented Logic Programming", Proceedings of Sixth International Conference on Soft Computing, IIZUKA2000, pp. 340-345

Approximate Reasoning Schemes: Classifiers for Computing with Words

Andrzej Skowron, Marcin S. Szczuka

Institute of Mathematics
Warsaw University
Banacha 2, 02-097 Warsaw, Poland

Abstract. In the paper we discuss classifiers relevant to approximate reasoning. The approach is based on rough-fuzzy hybridization. We discuss its possible applications to computing with words.

1   Introduction

We propose to use classifiers for rough-fuzzy concepts (see, [8]) as a tool in searching for approximate reasoning rules, called productions. From such productions approximate reasoning schemes can be derived. They are the basic constructions in rough-neuro computing [4] based on rough mereological approach [6]. The approach can be treated as a way to Computing with Words (see, e.g., [11], [12]). The proposed approach splits approximate reasoning into the following stages. In the first stage classifiers for relevant concepts should be induced using the existing statistical and other methods (see, e.g., [2], [3], [7]). Next, productions are extracted from data. It is important to note that it is possible to develop productions in such a way that they are using only linguistic names. Using the productions approximate reasoning schemes can be derived. On the top level solely linguistic names appear in the reasoning scheme.

2   Information Granule Systems and Parameterized Approximation Spaces

In this section, we present a basic notion for our approach, i.e., information granule system. An information granule system is a tuple

$$S = (G, R, Sem) \tag{1}$$

where

1. G is a finite set of parameterized constructs (e.g., formulas) called information granules;
2. R is a finite (parameterized) relational structure;
3. Sem is a semantics of G in R.

We assume that with any information granule system there are associated:

1. H a finite set of granule inclusion degrees with a partial order relation $<$ which defines on H a structure used to compare the inclusion degrees; we assume that H consists of the lowest degree 0 and the largest degree 1;

2. $\nu_p \subseteq G \times G$ a binary relation *to be a part to a degree at least p* between information granules from G, called *rough inclusion*. (Instead of $\nu_p(g, g')$ we also write $\nu(g, g') \geq p$.)

Components of an information granules system are parameterized. It means that we deal with parameterized formulas and a parameterized relational system. The parameters are tuned to make it possible to construct finally relevant information granules, i.e., granules satisfying specification or/and some optimization criteria. Parameterized formulas can consist of parameterized sub-formulas. The value set of parameters labelling a sub-formula is defining a set of formulas. By tuning parameters in optimization process or/and information granule construction a relevant subset of parameters is extracted and used for construction of the target information granule.

There are two kinds of computations on information granules. These are computations on information granule systems and computations on information granules in such systems, respectively. The first ones are aiming at construction of a relevant information granule systems defining parameterized approximation spaces for concept approximations used on different levels of target information granule constructions and the goal of the second ones is to construct information granules over such information granule systems to obtain target information granules, e.g., satisfying a given specification (at least to a satisfactory degree).

Examples of complex granules are tolerance granules created by means of similarity (tolerance) relation between elementary granules, decision rules, sets of decision rules, sets of decision rules with guards, information systems or decision tables (see, e.g., [8]). The most interesting class of information granules are information granules approximating concepts specified in natural language by means of experimental data tables and background knowledge.

One can consider as an example of the set H of granule inclusion degrees the set of binary sequences of a fixed length with the relation ν to be a part defined by the lexicographical order. This degree structure can be used to measure the inclusion degree between granule sequences or to measure the matching degree between granules representing classified objects and granules describing the left hand sides of decision rules in simple classifiers. However, one can consider more complex degree granules by taking as degree of inclusion of granule g_1 in granule g_2 the granule being a collection of common parts of these two granules g_1 and g_2.

New information granules can be defined by means of operations performed on already constructed information granules. Examples of such operations are set theoretical operations (defined by propositional connectives). However, there are other operations widely used in machine learning or pattern recognition [3] for construction of classifiers. These are the *Match* and *Conflict_res* operations. We will discuss such operations in the following section. It is worthwhile mentioning yet another important class of operations, namely, operations defined by data tables called decision tables [8]. From these decision tables, decision rules specifying operations can be induced. More complex operations on information granules are so called transducers [1]. They have been introduced to use background knowledge (not necessarily in the form of data tables) in construction of new granules. One can consider theories or their clusters as information granules. Reasoning schemes in natural language define the most important class of operations on information granules to be investigated. One of the basic problems for such operations and schemes of reasoning is how to approximate them by available information granules, e.g., constructed from sensor measurements.

In an information granule system, the relation ν_p to be a part to a degree at least p has a special role. It satisfies some additional natural axioms and additionally some axioms of mereology [6]. It can be shown that the rough mereological approach built on the basis of the relation to be a part to a degree generalizes the rough set and fuzzy set approaches. Moreover, such relations can be used to define other basic concepts like closeness of information granules, their semantics, indiscernibility and discernibility of objects, information granule approximation and approximation spaces, perception structure of information granules as well as the notion of ontology approximation. One can observe that the relation to be a part to a degree can be used to define operations on information granules corresponding to generalization of already defined information granules. For details the reader is referred to [4].

Let us finally note that new information granule systems can be defined using already constructed information granule systems. This leads to a hierarchy of information granule systems.

3 Classifiers as Information Granules

An important class of information granules create classifiers. One can observe that sets of decision rules generated from a given decision table $DT = (U, A, d)$ (see, e.g., [8] can be interpreted as information granules. The classifier construction from DT can be described as follows:

1. First, one can construct granules G_j corresponding to each particular decision $j = 1, \ldots, r$ by taking a collection $\{g_{ij} : i = 1, \ldots, k_j\}$ of left hand sides of decision rules for a given decision.
2. Let E be a set of elementary granules (e.g., defined by conjunction of descriptors) over $IS = (U, A)$. We can now consider a granule denoted by

$$Match(e, G_1, \ldots, G_r)$$

for any $e \in E$ being a collection of coefficients ε_{ij} where $\varepsilon_{ij} = 1$ if the set of objects defined by e in IS is included in the meaning of g_{ij} in IS, i.e., $Sem_{IS}(e) \subseteq Sem_{IS}(g_{ij})$; and 0, otherwise. Hence, the coefficient ε_{ij} is equal to 1 if and only if the granule e matches in IS the granule g_{ij}.
3. Let us now denote by $Conflict\_res$ an operation (resolving conflict between decision rules recognizing elementary granules) defined on granules of the form $Match(e, G_1, \ldots, G_r)$ with values in the set of possible decisions $1, \ldots, r$. Hence,

$$Conflict\_res(Match(e, G_1, \ldots, G_r))$$

is equal to the decision predicted by the classifier

$$Conflict\_res(Match(\bullet, G_1, \ldots, G_r))$$

on the input granule e.

Hence, classifiers are special cases of information granules. Parameters to be tuned are voting strategies, matching strategies of objects against rules as well as other parameters like closeness of granules in the target granule.

The classifier construction is illustrated in Fig. 1 where three sets of decision rules are presented for the decision values $1, 2, 3$, respectively. Hence, we have $r = 3$. In figure to omit too many indices we write α_i instead of g_{i1}, β_i instead of g_{i2}, and γ_i instead of g_{i3}, respectively. Moreover, $\varepsilon_1, \varepsilon_2, \varepsilon_3$, denote $\varepsilon_{1,1}, \varepsilon_{2,1}, \varepsilon_{3,1}$; $\varepsilon_4, \varepsilon_5, \varepsilon_6, \varepsilon_7$ denote $\varepsilon_{1,2}, \varepsilon_{2,2}, \varepsilon_{3,2}, \varepsilon_{4,2}$; and $\varepsilon_8, \varepsilon_9$ denote $\varepsilon_{1,3}, \varepsilon_{2,3}$, respectively.

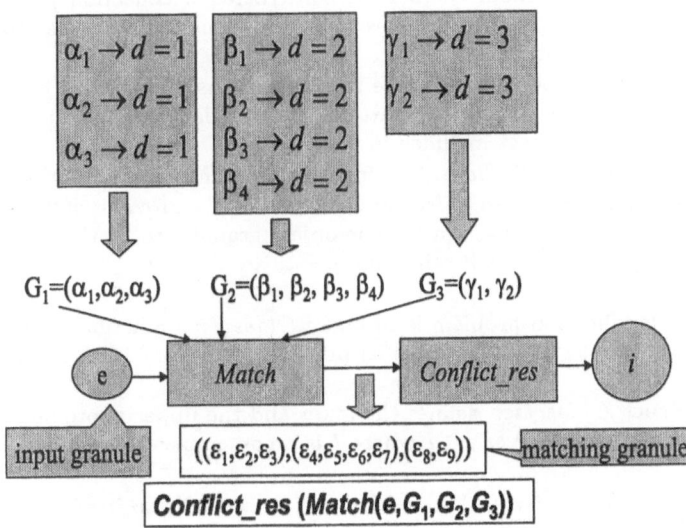

Fig. 1. Classifiers as Information Granules

The reader can now easily describe more complex classifiers by means of information granules. For example, one can consider soft instead of crisp inclusion between elementary information granules representing classified objects and the left hand sides of decision rules or soft matching between recognized objects and left hand sides of decision rules.

4 Approximation Spaces in Rough-Neuro Computing

In this section we would like to look more deeply on the structure of approximation spaces in the framework of information granule systems.

Such information granule systems are satisfying some conditions related to their information granules, relational structure as well as semantics. These conditions are the following ones:

1. Semantics consists of two parts, namely relational structure R and its extension R^*.
2. Different types of information granules can be identified: (i) object granules (denoted by x), (ii) neighborhood granules (denoted by n with subscripts), (iii) pattern granules (denoted by pat), and (iv) decision class granules (denoted by c).
3. There are decision class granules $c_1, \cdots, c_r$ with semantics in R^* defined by a partition of object granules into r decision classes. However, only the restrictions of these collections to the object granules from R are given.
4. For any object granule x there is a uniquely defined neighborhood granule n_x.
5. For any class granule c there is constructed a collection granule $\{(pat, p) : \nu_p^R(pat, c)\}$ of pattern granules labelled by maximal degrees to which pat is included in c (in R).
6. For any neighborhood granule n_x there is distinguished a collection granule $\{(pat, p) : \nu_p^R(n_x, pat)\}$ of pattern granules labelled by maximal degrees to which n_x is at least included in pat (in R).
7. There is a class of $Classifier$ functions transforming collection granules (corresponding to a given object x) described in two previous steps into the power-set of $\{1, \cdots, r\}$. One can assume object granules to be the only arguments of $Classifier$ functions if other arguments are fixed.

The classification problem is to find a $Classifier$ function defining a partition of object granules in R^* as close as possible to the partition defined by decision classes.

Any such $Classifier$ defines the lower and the upper approximations of union of decision classes c_i over $i \in I$ where I is a non-empty subset of $\{1, \cdots, r\}$ by

$$\underline{Classifier}(\{c_i\}_{i \in I}) = \{x \in \bigcup_{i \in I} c_i : \emptyset \neq Classifier(x) \subseteq I\}$$

$$\overline{Classifier}(\{c_i\}_{i \in I}) = \{x \in U^* : Classifier(x) \cap I \neq \emptyset\}.$$

The positive region of $Classifier$ is defined by

$$POS\,(Classifier) = \underline{Classifier}(\{c_1\}) \cup \cdots \cup \underline{Classifier}(\{c_r\}).$$

The closeness of the partition defined by the constructed $Classifier$ and the partition in R^* defined by decision classes can be measured, e.g., using ratio of the positive region size of $Classifier$ to the size of the object universe. The quality of $Classifier$ can be defined taking, as usual, only into account objects from $U^* - U$:

$$quality(Classifier) = \frac{card(POS\,(Classifier) \cap (U^* - U))}{card((U^* - U))}.$$

One can see that approximation spaces have many parameters to be tuned in order to construct the approximation of high quality class granules.

One more interesting issue is the direct connection between descriptions using classifier-based granules and the characterization in terms of the Dempster-Shafer theory of evidence. This inter-connection derives from the relationships that exist between rough set theory and evidence theory as described in e.g. [9]. We may introduce belief and plausibility functions that characterize granules defined by classifiers in the following way (with previous notation):

$$Bel_{Classifier}(I) = \frac{|\{x \in U^* : Classifier(x) \subseteq I\}|}{|U^*|}$$

$$= \frac{|\{\underline{Classifier}(\{c_i\}_{i \in I})|}{|U^*|}$$

$$Pl_{Classifier}(I) = \frac{|\{x \in U^* : Classifier(x) \cap I \neq \emptyset\}|}{|U^*|}$$

$$= \frac{|\{\overline{Classifier}(\{c_i\}_{i \in I})|}{|U^*|}$$

5 Standards, Productions, and AR-schemes

AR-schemes have been proposed as schemes of approximate reasoning in rough neurocomputing (see, e.g., [4], [8]). The main idea is that the deviation of objects from some distinguished information granules, called standards or prototypes, can be controlled in appropriately tuned approximate reasoning. Several possible standard types can be chosen. Some of them are discussed in the literature (see, e.g., [4]). We propose to use standards defined by classifiers. Such standards correspond to lower approximations of decision classes or (definable parts of) boundary regions between them.

Rules for approximate reasoning, called productions, are extracted from data (for details see [8]). Any production has some premisses and conclusion. In the considered case each premiss and each conclusion consists of a triple *(classifier, standard, deviation)*. This idea in hybridization with rough-fuzzy information granules (see, e.g., [8]) seems to be especially interesting. The main reasons are:

- standards are values of classifiers defining approximations of cut differences and boundary regions between cuts [8],
- there is a natural linear order on such standards defined by classifiers.

To explain the meaning of productions let us consider the following example of a production with two premisses:

if $(C_1, stand_1, \varepsilon_1)$ **and** $(C_2, stand_2, \varepsilon_2)$ **then** $(C, stand, \varepsilon)$

In the production classifiers C_1, C_2, C are labelled by standards $stand_1$, $stand_2$, $stand$ and deviations ε_1, ε_2, ε. The deviation ε is showing the range in which (in the considered linear order) can the deviation move the standard $stand$. The intended meaning of such production is that if the deviation of input from standards $stand_1$, $stand_2$ are respectively at most ε_1, ε_2 then the conclusion deviates from $stand$ to degree at most ε.

From production (extracted from data) AR-schemes can be derived (see, e.g., [8]).

One more important step that can be performed in order to bring this framework closer to the idea of pure computing with words is by substituting the degrees of closeness (deviations $\varepsilon, \varepsilon_1, \varepsilon_2$ in our case) by linguistic variables. What we want to make possible is the formulation of granule production in a purely linguistic way, for example:

> **if** similarity between C_1 output and standard $stand_1$ is *high*
> **and** similarity between C_2 output and standard $stand_2$ is *low*
> **then** similarity between C output and standard $stand$ is *medium*

To achieve this task we have to define partitions for the ranges of deviation as the deviation is used to measure similarity between classifier and corresponding standards. Let us consider the deviation ε for the classifier C output and standard $stand$. It is quite natural to assume that the subsets of ε range are ordered linearly. Also, their layout should be fuzzy-like. We may e.g. take three such sets stating represented as $\{low, medium, high\}$. As these sets may (and in fact should) overlap, in turn we get more possible linguistic values e.g. $\{low, low\ or\ medium, medium, medium\ or\ high, high\}$.

The retrieval of proper sets for deviation ranges should be devised as an interactive data-driven process. By analysis of standards and classifiers and matching them against the training data we attempt to establish an initial layout for deviations. This layout (the choice and setting of subsets) is then verified and possibly modified in order to achieve high compliance with the underlying data sets. The choice of proper parameters for the sets of deviation ranges may be based on various known techniques in data analysis such as clustering, statistical analysis, density analysis etc.

6   Conclusion

We have proposed to use standards defined by classifiers. Such standards can next be used in the process of extracting of productions from data and for deriving AR-schemes. This is also a step towards implementation of the general idea of computing with words.

Acknowledgements

The research has been supported by the State Committee for Scientific Research of the Republic of Poland (KBN) research grant 8 T11C 025 19 and by the Wallenberg Foundation.

References

1. Doherty, P., Łukaszewicz, W., Skowron, A., Szałas, A.: *Combining rough and crisp knowledge in deductive databases*, (to appear in [4]).
2. Hastie, T., Tibshirani, R., Friedman, J. H.: *The Elements of Statistical Learning: Data Mining, Inference, and Prediction*, Springer Series in Statistics, Springer Verlag, Berlin 2001.
3. Mitchell, T.M.: *Machine Learning*. Mc Graw-Hill, Portland, 1997.
4. Pal, S.K., Polkowski, L., Skowron, A. (Eds.): *Rough-Neuro Computing: Techniques for Computing with Words*. Springer-Verlag, Berlin, 2002. (to appear).

5. Pawlak, Z.: *Rough Sets. Theoretical Aspects of Reasoning about Data.* Kluwer Academic Publishers, Dordrecht, 1991.
   Towards an
6. Polkowski, L., Skowron, A.: Rough mereology: a new paradigm for approximate reasoning. *International J. Approximate Reasoning* **15**(4), 1996, 333–365.
7. Skowron, A.: Rough sets in KDD. in: Z. Shi, B. Faltings, and M. Musem (Eds.), *16-th World Computer Congress (IFIP'2000): Proceedings of Conference on Intelligent Information Processing (IIP'2000)*, Publishing House of Electronic Industry, Beijing, 2000, 1–17.
8. Skowron, A.: Toward intelligent systems: Calculi of information granules. *Bulletin of the International Rough Set Society* **5**(1-2), 2001, 9–30.
9. Skowron, A., Grzymała–Busse, J.W.: From rough set theory to evidence theory. In: R.R. Yaeger, M. Fedrizzi, and J. Kacprzyk (eds.), *Advances in the Dempster Shafer Theory of Evidence*, John Wiley & Sons, Inc., New York (1994) 193–236 2001, 57–86.
10. L.A. Zadeh. Fuzzy sets. *Information and Control* **8**, 1965, pp. 333–353.
11. Zadeh, L.A.: Fuzzy logic = computing with words. *IEEE Trans. on Fuzzy Systems* **4**, 1996, 103–111.
12. Zadeh, L.A.: A new direction in AI: Toward a computational theory of perceptions. *AI Magazine* **22**(1), 2001, 73–84.

Using Consensus Methods to User Classification in Interactive Systems

Janusz Sobecki, Ngoc Thanh Nguyen

Wrocław University of Technology, Department of Information Systems
Wybrzeże S. Wyspiańskiego 27, 50-370 Wrocław, Poland.
E-mail : sobecki@pwr.wroc.pl, thanh@ pwr.wroc.pl

Abstract. In this paper we describe the consensus problem solution applied to the web-based system interface construction for each class of users. We assume that we have a multiagent system that collects knowledge about user interactions with different systems in the form of user profiles which serve to user classification. This multiagent system also collects the interface profiles of all users of these systems, which are used for the adaptive interface construction. Because of the differences among the users and their experiences, the knowledge stored in the form of interface profiles is inconsistent, even for the users belonging to the same class, so the most efficient interface could be found when we reconcile the knowledge by means of consensus methods.

Key words: Consensus method, user interface, user profile classification

1   Introduction

In adaptive hypermedia the user classification is applied in almost all approaches [3]. Most of the methods try to assign users appropriate predefined user model [5,9], sometimes called the usage model [3]. The classification could be based on so called user data, i.e. user description delivered by the user himself, or the usage data gathered during the whole interaction process. The usage date could be based on very different elements of the user interaction with the web systems, i.e. URL's of the visited pages, chosen hyperlinks, filled-in forms, words from visited pages, advertisement clicked, etc. [12].

There are many methods used for the user model acquisition: user supplied, acquisition rules, plan recognition and stereotype recognition, as well as usage model acquisition: machine learning, Hidden Markov Models, graph-based induction. One of the quite popular method used in the adaptive hypermedia is so called clique method, i.e. used for filtering problems [3]. The clique filtering is made in three phases: find similar neighbors, select comparison group of neighbors and finally compute prediction based on (weighted) representations of selected neighbors. In this paper however we propose adaptive   web-based interface construction that is based on the consensus method used for appropriate prediction of the class representations using also the interface usability function for the 'interface knowledge' reconciliation.

Generally, consensus is understood as a general agreement in matters of opinion or testimony and a powerful tool for solving many problems for which the solvers are in conflicts [6]. The conflict (or a conflict situation) takes place, if at least some participants (agents) from a certain group generate different opinions or different views on the same subject or issue. What is inherently hidden in this definition is that a conflict can be visible and considered, when the opinions of participants (or conflict agents) are communicated with the same class of carriers and referred to the same set of criteria. In other words, a common formal language is needed to represent conflicting opinions. In a more precise approach, it needs to be possible to represent the agents' opinions by means of the same mathematical structures, e.g. relations over the same sets, or functions with the same signatures. A conflict situation can be defined by means of an information system ($\mathcal{U}, \mathcal{A}$), in which $\mathcal{U}$ is a set of agents, and $\mathcal{A}$ is a set of attributes representing issues of conflict.

In this paper we present the consensus problem solution applied to the web-based system adaptive interface construction for each class of users. First we present the consensus problem and then its application to interface construction using consensus structure. Then we describe the user classification based on different methods. Finally we address the problem of the interface utility values determination for short.

2 Determination of Interface Profiles' Consensus

2.1 Basic Notions

We assume that a real world is described by means of a finite set A of attributes and a set V of *elementary values*, where $V = \bigcup_{a \in A} V_a$ (V_a is the domain of attribute a). Let $\Pi(V_a)$ denote the set of subsets of set V_a and $\Pi(V_B) = \bigcup_{b \in B} \Pi(V_b)$ for any $B \subseteq A$. We accept the following assumption: For each attribute a its value is a set of elementary values from V_a, thus it is an element of set $\Pi(V_a)$. By an elementary value we mean a value which is not divisible in the system. Thus it is a relative notion, for example, one can assume the following values to be elementary: time units, numbers, partitions etc.

We define the following notions: Let $B \subseteq A$, a tuple of type B is a function $r: B \rightarrow \Pi(V_B)$ where $(\forall b \in B)(r(b) \subseteq V_b)$. Instead of $r(b)$ we will write r_b and a tuple of type B will be written as r_B. The set of all tuples of type B is denoted by $TYPE(B)$. A tuple is elementary if all attribute values are empty sets or 1-element sets. The set of elementary tuples of type B is denoted by $E\text{-}TYPE(B)$. Empty tuple, whose all values are empty sets, is denoted by symbol ϕ. Partly empty tuple, whose at least one value is empty, is denoted by symbol θ. A non-empty set R of tuples of type B is called a relation of type B, thus $R \subseteq TYPE(B)$. The sum of 2 tuples r and r' of type B is a tuple r'' of type B $(r''=r \cup r')$ such that $(\forall b \in B)(r''_b = r_b \cup r'_b)$. A product of 2 tuples r and r' of type B is also a tuple r'' of type B $(r''=r \cap r')$ such that $(\forall b \in B)(r''_b = r_b \cap r'_b)$. Let $r, r' \in TYPE(B)$, we say that tuple r is included in tuple r' (that is $r \prec r'$), iff $(\forall b \in B)(r_b \subseteq r'_b)$.

2.2 Definition of Conflict System

We assume that some real world is commonly considered by agents which are placed in sites of a distributed system. The subjects of agents' interest consist of events which occur (or have to occur) in the world. The task of these agents is based on determining the values of event attributes (an event is described by an elementary tuple of some type). The elements of the system defined below should describe this situation [7].

Definition 1. *By a conflict system we call the following quadruple:*

$$Conflict\_Sys = (A, X, P, Z)$$

where:
- A – *a finite set of attributes, which includes a special attribute Agent; each attribute $a \in A$ has a domain V_a (a non-empty and finite set of elementary values) such that values of attribute a is a subset of V_a; values of attribute Agent are* 1*element sets, which identify the agents.*
- X – *a finite set of conflict carriers, $X = \{\Pi(V_a) : a \in A\}$.*
- P – *a finite set of relations on carriers from X, each relation is of some type A (for $A \subseteq A$ and Agent$\in A$).*
- Z – *a finite set of logic formulas for which the model is relation system (X,P).*

The purpose of Definition 1 relies on representing two kinds of information: the first consists of information about conflicts in the distributed system, which require solving, and the second includes the information needed for consensus determining.

In the conflict system an event is described by an elementary tuple of type $B \subseteq A \setminus \{Agent\}$. The values of attributes represent the parameters of the event. For example the following tuple describes the event "*For the class c_1 the best interface should consist of the following parameters: the window size is 240x320, the sound volume – 0, the number of columns – 1 and the template - classical*".

Class	Window_size	Sound_volume	Number_of_col.	Template
c_1	240x320	0	1	classical

Relations belonging to set P are classified in such a way that each of them includes relations representing similar events. For identifying relations belonging to a given group the symbols "$+$" and "$-$" should be used as the upper index. If P is the name of a group, then relation P^+ is called a positive relation (contains positive knowledge) and P^- negative relation (contains negative knowledge). If $r \in P^+$ then we have the following interpretation: In the opinion of agent r_{Agent} one or more events included in r_A should take place. If $r \in P^-$ then we say that in the opinion of agent r_{Agent} none of the events included in r_A should take place. The same agent cannot simultaneously state that the same event should take place and should not take place. It means that the same event cannot be classified by the same agent into positive and negative relations simultaneously.

2.3 Conflict Profiles

We define a conflict situation which contains information about a concrete conflict as follows:

Definition 2. *A conflict situation is a pair* $<\{P^+,P^-\}, A\rightarrow B>$ *where* $A,B\subseteq A$, $A\cap B=\varnothing$ *and* $r_A\neq\theta$ *and* $r'_A\neq\theta$ *for every tuples* $r\in P^+$ *and* $r'\in P^-$.

According to the above definition a conflict situation consists of agents (conflict body) which appear in relations P^+ and P^- (conflict content) representing the positive and negative knowledge of agents referring to subjects represented by set A of attributes. These relations are the basis of consensus. Expression $A\rightarrow B$ means that the agents are not agreed referring to combinations of values of attributes from A with values of attributes from B, and the purpose of the consensus choice is that for a tuple type A there should be assigned at most one tuple of type B.

For a given situation s we determine the set of these agents which take part in the conflict as follows:

$$Agent(s) = \{a\in V_{Agent}: (\exists r\in P^+)(r_{Agent}=\{a\}) \vee (\exists r\in P^-)(r_{Agent}=\{a\})\},$$

and the set of *subject elements* (or *subjects* for short) as follows:

$$Subject(s) = \{e\in E\text{-}TYPE(A): (e\neq\theta) \wedge [(\exists r\in P^+)(e\prec r) \vee (\exists r\in P^-)(e\prec r)]\}.$$

Set *Subject(s)* then includes such subject elements, which have been occupied by agents. For example for situation $<\{Wind^+, Wind^-\}, \{Region\}\rightarrow\{Time, Wind\_Speed\}>$ the subjects are these regions for which the agents present their forecast for time and speed of the wind. Now for each subject $e\in Subject(s)$ let us determine sets with repetitions $Profile(e)^+$ and $Profile(e)^-$ which include the positive and negative knowledge of agents on subject e, as follows:

$$Profile(e)^+ = \{r_{B\cup\{Agent\}}: (r\in P^+) \wedge (e\prec r_A)\},$$
$$Profile(e)^- = \{r_{B\cup\{Agent\}}: (r\in P^-) \wedge (e\prec r_A)\}.$$

These sets are called positive and negative profiles of given conflict subject e.

2.4 Consensus Definition and Determination

Below we present the definition of consensus [8].

Definition 3. *Consensus on subject* $e\in Subject(s)$ *of situation* $s=<\{P^+,P^-\},A\rightarrow B>$ *is a pair of two tuples* $(C(s,e)^+, C(s,e)^-)$ *where* $C(s,e)^+,C(s,e)^-\in TYPE(A\cup B)$ *and the following conditions are fulfilled:*
a) $C(s,e)^+_A = C(s,e)^-_A = e$,

b) $C(s,e)^+_B \cap C(s,e)^-_B = \phi$,

c) Tuples $C(s,e)^+$ and $C(s,e)^-$ *fulfil logic formulas from set* **Z**,

d) *One or more of the postulates* **P1-P6**, *described in detail in* [8], *are satisfied:*

Let ϕ be a distance function between tuples from relations belonging to set P, the following theorem should enable to determine a consensus satisfying all postulates **P1-P6** [7].

Theorem 1. *If there is defined a distance function* ϕ *between tuples of TYPE(B), then for given subject e of situation* $s=<\{P^+,P^-\},A{\rightarrow}B>$ *tuples* $C(s,e)^+$ *and* $C(s,e)^-$ *which satisfy conditions a)-c) of Definition 3 and minimize the expressions*

$$\sum_{r \in profile(e)^+} \phi(r_B, C(s,e)^+_B) \quad and \quad \sum_{r \in profile(e)^-} \phi(r_B, C(s,e)^-_B) \quad should \ create \ a \ consensus$$

satisfying all of postulates **P1-P6**.

2.5 Consensus Determining for User Profiles

For our web-based information system we specify the parameters of the conflict system as follows:

- $A = \{Agent, System, Class, A_1, A_2,..., A_n\}$,

where:

-   *Agent* represents interface agents,
-   *Class* represents classes of users, and
-   $A=\{A_1,A_2,...,A_n\}$ is a set of attributes describing the interface profiles for user service.

- $P = \{Profile^+, Profile^-\}$ where
$Profile^+, Profile^- \subseteq \Pi(V_{Agent}) \times \Pi(V_{Interface}) \times \Pi(V_{Class}) \times \Pi(V_{A1}) \times \Pi(V_{A2}) \times...\times \Pi(V_{An})$.

We interpret a tuple of relation *Profile*$^+$, for example, $<Agent:a_1, Class:c_1, A_1:a_1, A_2:a_2,..., A_n:a_n>$ as follows: in opinion of agent a_1 the appropriate interface profile for serving users from class c_1 on the web-based information system should be the tuple $<A_1:a_1, A_2:a_2,..., A_n:a_n>$. A tuple $<Agent:a_2, Class:c_1, A_1:a'_1, A_2:a'_2,..., A_n:a'_n>$ belonging to relation *Profile*$^-$ means that according to agent a_2 on the information system for users from class c_1 the profile $<A_1:a'_1, A_2:a'_2,..., A_n:a'_n>$ is appropriate interface for their service. The interface quality evaluation is made by the utility function described in the following section.

- **Z**: Logical formulas representing conditions which have to be satisfied by the tuples belonging to the relations from **P**.

A conflict situation is then defined as follows:

$$s=(\{Profile^+, Profile^-\}, \{Class\} \rightarrow \{A_1,A_2,...,A_n\}).$$

The set $\{Profile^+, Profile^-\}$ is then the basis of consensus, the set of attribute $\{Class\}$ represents and the consensus subject, and set $\{A_1, A_2,..., A_n\}$ describes the content of consensus. The problem of the user classification will be considered in the following section.

3   User Classification

The acquisition of the user or usage models is usually connected with the user classification problem. This problem has been addressed by the specialists from the area of information retrieval (IR) since early 60-ties of the last century [2], but also today, in the era of the e-economy and the Information Society this research is continued by many specialists from: HCI, user modeling as well as marketing [4].

The classification used in the user modeling could have different properties. First, it could be made manually, i.e. by the users themselves or by some experts, or automatically. Second, it could use predefined classes or create them automatically, also the number of classes could be specified or dynamically determined according to the user population characteristics. Third it could be based on the user data, usually delivered by the users themselves or on the usage data, collected during the user interactions with web-based systems. Fourth, we can also consider other set partitions than classification, i.e. partitions with repetitions. Finally, we can use very different classification methods, from standard methods developed in the IR area such as Dattola or Rocchio methods [2] or applied by the adaptive hypermedia such as mentioned above machine learning tools or Hidden Markov Models [3].

The selection of the particular classification method could depend on many different circumstances. The manual classification that is made by the users themselves is the most easy to implement. It could be made by the user's simple selection of one of the predefined classes, i.e. student, lecturer or absolvent at an university site, or by submitting so called user data by filling out the questionnaire containing several questions on the users personal data as well as different preferences and interests. That could serve as an input data for further classification with any of the above mentioned methods. Jonathan Robbin the marketing specialists from the U.S. noticed over thirty years ago that the address zip code may serve as a very well indicator of the peoples social status, interests and various marketing behavior [10]. However, these works are now criticized [13], the web-based system users are often asked to enter their zip code, along more or less personal questions. Users usually are very reluctant to giving any information about themselves because they treat them as too privet or simply don't want to loose time doing this. A good solution to this problem is the automatic user information gathering and classification.

In contrary to the manual gathering the automatic methods could be made by the client and/or system sides using very different technologies from CGI, PHP, ASP, Flash to Cookies and DoubleClick [13]. Having the method to record users' web behavior we are able to classify them using different data concerning their interactions with different web-based systems, i.e. gathering data for usage model construction. Users working with any web-based system are making different actions; the most popular is of course browsing. There are however others, we can say more qualified actions: filling-in different forms that are part of the pages, joining mailing list, purchase of any goods that are offered on the Web, etc.

In our classification problem we can assume that each time the user visits the particular page its agent can store its URL in the local files. These documents are usually in form of HTML document's, with quite deal of natural language words. These words placed in the vector could serve as the user profile.

The process of user classification results in grouping heterogeneous set of users U={u₁,...,uₙ} into disjoint subsets of the set U by some measure of similarity or distance. Finding an appropriate measure, as well as a user representation, are key problems of users classification, the function P4 used in the Dattola classification algorithm is following:

$$s(u_j, u_k) = \sum_{i=1}^{m} \min(u_{k,i}, u_{j,i}) .$$

The process of classification could be computational difficult, with even exponential with respect to the number of elements to be classified [1,2], so practically other sub-optimal solutions have to be used. They usually are based on the selection of some initial partition as for example in the Dattola algorithm [2]. Then for each vector (user profile) the distance function between the vector and each of the class centers (centroides) is determined. The vector is joined to the class with the center that has the smallest value of the function. There are many possible distance functions that could be considered, their list can be found in the work [2]. In our system architecture for the centroides could serve, for example, the profiles of randomly selected users who personalized the system interface by their own.

4   Utility Function

In order to measure the quality of the system interface the domain of the HCI [5,9] has worked out many methods most of them, however, are not straightforward and could not be measured by a simple function. In the social adaptive interfaces construction it is necessary to evaluate interfaces in the automatic manner.

Web-based systems could be built and maintained for different reasons but most of them have straightforward commercial aspects. For example: the web portal that earns money on advertisements, Internet shop that sells different goods or Internet auction servers that enables exchange of different goods between their users, and many other ways. Each of these applications has different goals and so different things express their achievements.

For web portals, usually, the more time the user spends on browsing the portal pages the better, because in the meantime he or she have at least visual contacts with advertisements. So for such systems the utility function should consider the following parameters: number of pages visited, number of advertisements downloaded by the user's browser, number of followed links presented on the advertisements, frequency of visits on the site, etc. The companies such as DoubleClick could offer sophisticated measures in this case.

For e-shops the most important is the amount of money spent by the user. There are however other possible factors that should be considered in the utility function construction: number of visits in the shop, number of visited pages, number of goods that were put into the basket, speed of shopping, utilization of lists and retrieval mechanisms of goods, purchase of goods being the special offer, etc.

For e-auction sites usually, like in eBay, there's no charge to browse, bid on or buy items, but clients have to pay fees to list and sell items. Some start-ups and those auctions that cooperate with the web portals could not charge any fees. Despite of this, the utility function of any online trading center should consider: value and number of sold or bought items by the user, number of bids made by the user, number of auctions the user took part, frequency of bids made by the user, etc.

The above-mentioned examples are quite typical for e-commerce but because of their different business models also different utility function of their interface systems should be considered. So the utility function should be specified particularly for every individual application.

The problems with the utility function construction increase when we consider multiplatform access to these systems. GUI for PC is today world-known standard but some other platforms, i.e. handhelds and navigation systems mounted in cars are quite new, so that standards are still being developed. In consequence the utility function construction should also consider these specific interface gestures.

5 Summary

In the paper we presented consensus methods applied in the adaptive interface construction using the classification of the population of users. We also considered the problem of the utility function construction in the web-based systems to be able to evaluate the interfaces in the automatic manner.

References

1. Brown SM et. al. (1998) Using explicit requirements and metrics for interface agent user model for interface agent user model correction. In: Proc. of the second international conference on Autonomous Agents, Minneapolis, Minesota, United States pp. 1-7.
2. Dabrowski M, Laus-Maczynska K (1978) Information retrieval and classification. Survey of methods (in polish). WNT Warszawa.
3. Kobsa, A., Koenemann, J., Pohl, W. (2001) Personalized Hypermedia Presentation Techniques for Improving Online Customer Relationships. The Knowledge Engineering Review 16(2) pp. 111-155.
4. Maglio, P.P., Barrett, R., Campbell, C.S., Selker, T. (2000). SUITOR: an attentive information system. In: Proceedings of the 2000 international conference on Intelligent user interfaces. pp. 169 – 176.
5. Newman WM, Lamming MG (1996) Interactive system design. Addison-Wesley Harlow.
6. Nguyen N.T. (2002). Methods for Consensus Choice and their Applications in Conflict Resolving in Distributed Systems. Wroclaw: Wroclaw University of Technology Press.

7. Nguyen NT (2001). Conflict profiles' susceptibility to consensus in consensus systems. Bulletin of International Rough Sets Society **5**(1/2) pp. 217-224.
8. Nguyen, N.T. (2000). Using Consensus Methods for Solving Conflicts of Data in Distributed Systems. Lecture Notes on Computer Science **1963** pp. 409-417
9. Peerce J et. al. (1996) Human-computer interaction. Addison-Wesley Harlow.
10. Quinn L.M., Pawasarat J. (2001). Confronting Anti-Urban Marketing Stereotypes: A Milwaukee Economic Development Challenge. June 2001. http://www.uwm.edu/Dept/ETI/purchasing/markets.htm.
11. Sobecki J (1999). Interactive multimedia information planning. In: Valiharju T. (ed.) Digital Media in Networks 1999, Univ. of Tampere. pp. 38-44.
12. Sobecki J, Nguyen NT (2001). Consensus-based adaptive user interface for universal access systems. In: Stephanidis,C (ed.) Proc. of 9th International Conference on HCI and 1st International Conference on Universal Access in HCI. LEA London. vol.3 pp. 112-116.
13. Whalen D.(2002). The Unofficial Cookie FAQ, Version 2.54 Contributed to Cookie Central by David Whalen, http://www.cookiecentral.com/faq/#2.7.

Aggregation Methods to Evaluate Multiple Protected Versions of the Same Confidential Data Set

Aïda Valls[1], Vicenç Torra[2], and Josep Domingo-Ferrer[1]

[1] Dept. Comput. Eng. and Maths - ETSE, Universitat Rovira i Virgili
   Av Paisos Catalans 26, 43007 Tarragona (Catalonia, Spain)
   e-mail: {jdomingo, avalls}@etse.urv.es
[2] Institut d'Investigació en Intel·ligència Artificial - CSIC
   Campus UAB s/n, 08193 Bellaterra (Catalonia, Spain)
   e-mail: vtorra@iiia.csic.es

Abstract. This work is about disclosure risk for national statistical offices and, more particularly, for the case of releasing multiple protected versions of the same micro-data files. This is, several copies of a single original data file are released to several data users. Each user receives a protected copy, and the masking method for each copy is selected according to the research interests of the user: the selected masking method is such that it minimizes the information loss for his/her particular research.

Nevertheless, multiple releases of the same data increase the disclosure risk. This is so, because coalitions of data users can reconstruct original data and, thus, find the original (non-masked) information. In this work we propose a tool for evaluating this reconstruction.

1 Introduction

Data dissemination is a mandatory requirement for statistical offices: they collect data to be published. However, the release of this data has to be done in such a way that there is no disclosure. In other words, no sensitive data is linked to the original respondent.

For example, the publication of incomes, professions and ZIP codes for the inhabitants of a town should not allow the inference of the exact income of a particular inhabitant. Moreover, publication is forbidden if there is a single person in the town for a given pair (ZIP, profession). This is so because the release of such data implies disclosure (knowing the profession and the ZIP code of a person implies knowing his/her income).

To avoid disclosure, masking methods are applied (see [3], [13] for reviews of masking methods and [4] for a comparative study on masking methods performance). Masking methods introduce distortion to the data prior to its publication so that the information is not disclosured. Distortion should be kept small so that published data is valid for researchers and users (they can infer the same conclusions that would be inferred from the original data) but on the other hand should be protected *enough* so that disclosure is not possible. Statistical Disclosure Control (SDC) studies methods that attempt to perform such a nontrivial distortion.

1.1 Artificial Intelligence and Soft Computing for Statistical Disclosure Control

The fields of Artificial Intelligence and Soft Computing provide several tools that are useful for Statistical Disclosure Control. These tools can be broadly classified in three categories (a more detailed review is given in [6]):

Methods to overcome distortion: Tools and methods for data mining and machine learning have been developed to be resilient to errors in datafiles (either due to accidental or to intentional distortion). Among other uses, data mining and modeling techniques can be used to correct errors (if data do not follow data models) and to fill missing values. Also, information fusion in general and aggregation operators in particular can be used to increase the accuracy of the data. This is particularly appropriate when there are multiple releases of the same data or there is data from multiple sources.

Methods to evaluate disclosure risk: In general, all methods that can be used to overcome distortion are appropriate to evaluate disclosure risk. The better results of a method to overcome distortion, the worse the protection and the larger the disclosure risk. This is so, because if a method can recontruct the original data it means that data can be disclosured.

Methods for re-identification (see e.g. [17]) also fall in this category. They are used to link records that correspond to the same individual but belong to different files. These techniques are used to compare the masked data with the original one. The more records are re-identified, the worse the protection achieved. Some of the existing re-identification methods are the probabilistic based ones, distance based ones and clustering based ones. Recently, techniques based on soft computing have been used [12].

Methods to cause distortion: Although that most artificial intelligence and soft computing techniques are used in SDC for restoring the original data and establishing the disclosure risk, they can also be used for causing intentional distortion to data. This is the case of using aggregation operators [5] for masking numerical data, and machine learning techniques for increasing the performance of a particular masking method [9].

In this work we focus on the first two categories. In particular, we consider the evaluation of disclosure risk in the case of multiple releases of the same data. This is, several copies of a single original data file are released to several data users. Each user receives a protected copy, and the masking method for each copy is selected according to the research interests of the user: the selected masking method is such that it minimizes the information loss for his/her particular research.

Nevertheless, in this situation the following property has to be taken into account:

Property 1. [6] If there is a knowledge integration technique that can reconstruct an original data set out of n different distorted versions of the data set, then statistical confidentiality is compromised if more than n different SDC-protected versions of the same confidential data set are released.

This is so, because coalitions of data users can reconstruct original data and, thus, find the original (non-masked) information. Therefore, the following property also holds:

Property 2. [6] Information loss in SDC is inversely proportional to the reconstruction capabilities of Knowledge Integration and Re-identification techniques. Disclosure risk is proportional to these reconstruction capabilities.

In this work, we describe our information fusion system ClusDM and show its application to reconstruct the original data from multiple releases of the same data. This works extends our previous results presented in [7].

To do so, Section 2 describes our information fusion tool and Section 3 describes the application to a set of multiple protected data. The work finishes in Section 4 with some conclusions.

2   Our information fusion approach

Although our system, ClusDM, has been developed for its application in decision making environments, some of its components can be used for other information fusion applications. In this work we describe its application for fusing multiple releases of the same data. In this section, we give an overview of the general capabilities of the system.

From an abstract point of view, the data fusion component is applied to data matrices V that contain values for each pair (object, attribute). These matrices can be modeled as a function:

$$V : \mathbf{O} \rightarrow D(A_1) \times D(A_2) \times \cdots \times D(A_m)$$

where $\mathbf{O} = \{O_1, \cdots, O_n\}$ denotes the set of objects, $A_1, A_2, \cdots A_m$ are the attributes and $D(A_i)$ denotes the domain of attribute A_i.

Given a matrix V of this form, the data fusion component builds a new attribute A_C that corresponds to the aggregation of $A_1, \cdots, A_m$. When data is numerical, the attribute A_C is numerical. Instead, if the data is categorical (or some attributes are numerical and other are categorical) the new attribute is categorical.

358

In the case of numerical information, the system applies the principle of irrelevant alternatives. This is, the aggregated value for each object $(A_C(O_i))$ only depends on the values for that object $(A_1(O_i), \cdots, A_m(O_i))$. This is, there exists a function $\mathbb{C}$ such that:

$$A_C(O_i) = \mathbb{C}(A_1(O_i), \cdots, A_m(O_i))$$

Our system implements several aggregation operators $\mathbb{C}$. Among others, it includes the weighted mean and the OWA operator [14].

In the case of considering categorical and mixed information, the system does not satisfy the condition of irrelevant alternatives. This condition is usually applied (e.g. in [8]) for technical reasons because it simplifies the computations as each object is operated without considering the values of the others. However, while this condition is acceptable for numerical data (in particular, when values correspond to measurements), this is not so true for categorical values. In the categorical case, the values establish equivalences between objects (they are indistinguishable according to a given criteria) and, in the case of ordinal scales, preferences between objects. Therefore it seems natural to keep these relationships in the aggregated attribute.

In order to keep these similarities, the condition of irrelevant alternatives is dropped and we apply clustering for obtaining an aggregated attribute. Once the set of clusters is obtained, the system assigns linguistic labels to each cluster.

According to this, our approach to information fusion considers two steps. We give some additional details of these steps:

Clustering: To obtain the clusters for categorical data, the system assumes an underlying semantics for linguistic labels based on negation functions following [10]. This is, negation functions that are not a one-to-one mapping as in multivalued logic but a one-to-many functions. From the point of view of the user, these negation functions can be interpreted as antonyms following [2].
The clustering of the data is directed by the attributes and their number of categories.

Assignment of linguistic labels: For the assignment, the system is able to select the most appropriate vocabulary considering the ones used in each variable. In this application, we select the vocabulary of the original criterion, because we are trying to see if we can re-identify the original categories. The next step consists of selecting a category to describe each cluster and, if it is needed, to adapt the vocabulary by splitting the category [16]. This assignment process is explained in detail in [15].

3 Experimentation

We have considered data from the *American Housing Survey 1993* [1] and applied our approach to multiple realeases of a single variable (the variable

DEGREE). In the rest of this section we review the application of the approach to a set of 20 records.

Table 1 includes the original variable DEGREE (in column *o.v.*). The set of linguistic terms used by this variable is $L = \{coldest, cold, cool, mild, mixed, hot\}$. In Table 1, the original values are replaced by the position of the category in the set L. Thus, value 1 stands for coldest, 2 for *cold*, 3 for *cool* and so on. This variable has been masked using 4 different masking methods. In particular, we have applied Top and Bottom coding, Global recoding and Post-Randomization Method. Several parameterizations where considered. Additionally, the original values of record f have been updated so the value for column $P4$ is now equal to 3. Table 1 includes the masked variables for 7 different pairs (techniques, parameterizations). The interested reader is referred to [7] for details.

Now, we have applied the ClusDM method to obtain an aggregated value for each record, in order to check if we can reconstruct the original values from these 7 different released variables.

name	o.v.	B4	T4	G4	R10	P8	P9	P4	a.v.	name	o.v.	B4	T4	G4	R10	P8	P9	P4	a.v.
a	3	&	&	3	3	3	3	3	3	k	3	&	&	3	4	3	3	3	3
b	3	&	&	3	2	3	3	3	3	l	3	&	&	3	2	3	3	3	3
c	3	&	&	3	3	3	3	3	3	m	3	&	&	3	3	3	3	3	3
d	3	&	&	3	3	3	3	3	3	n	2	&	2	2	2	2	2	2	2
e	4	4	&	4	4	4	4	4	4	o	3	&	&	3	3	3	3	3	3
f	4	4	&	4	4	4	4	3	4	p	2	&	2	2	2	2	2	2	2
g	4	4	&	4	3	4	4	4	4	q	3	&	&	3	3	3	3	3	3
h	4	4	&	4	4	4	4	4	4	r	5	&	&	n	5	3	3	4	3
i	4	4	&	4	4	4	4	4	4	s	2	&	2	2	2	2	2	2	2
j	1	&	1	n	1	1	1	1	1	t	2	&	2	2	2	3	4	2	2

Table 1. Records used: First column corresponds to a name for the record; second column is the original value (o.v.); columns 3-9 are masked variables; column 10 is the aggregated value (a.v.)

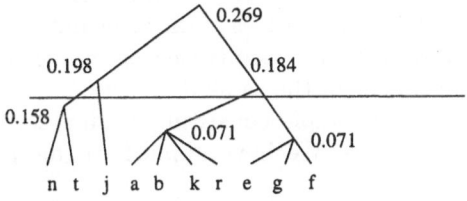

Fig. 1. Dendrogram for the clustering of the records in Table 1

Using the Taxonomic distance and the centroid method, we obtain the dendrogram in Figure 1. Then, an α-cut has to be selected in the tree to obtain a partition of the elements. The α-cut is selected so that the number of clusters is equal to 4 because this is the average number of linguistic labels used in columns $B4$-$P4$. The number of categories used in each column is displayed in Table 2. The selected α-cut is also displayed in Figure 1. The obtained partition is defined by four sets (named A, B, C and D) as follows: $A = \{n, t\}$, $B = \{a, b, k, r\}$, $C = \{j\}$, $D = \{e, f, g\}$. This partition satisfies the conditions required in [7] for a correct partition selection: (i) records with all the variables with the same value should correspond to different clusters (e.g. record a and e) and (ii) clusters should be defined according to the dendrogram.

Note that for the sake of simplicity, we only include in the dendrogram and in the partition one of those elements that are indistinguishable (i.e., it appears the element a but does not appear c because it has the same values for all columns).

Column	B4	T4	G4	R10	P8	P9	P4
Number of used labels	2	3	3	5	4	4	4

Table 2. Number of categories used in each columns

Once the clusters have been obtained, a category has to be assigned to each one. This is done considering the distance between each cluster and the *ideal* element (the one that has larger value for all categories). Then, taking into account this distance and the location of the clusters in relation to the semantics of linguistic labels (see [15]) the following assignment is given: Class C is assigned to category 1, Class A is assigned to 2, Class B is assigned to 3 and Class D is assigned to 4. These assignments are shown in Table 1.

4 Conclusions

In this work we have reviewed the applicability of artificial intelligence and soft computing techniques for statistical disclosure control. We have shown that information fusion techniques can be used by coallitions of users to overcome distortion in multiple releases of the same data. This is, information fusion techniques can be used for reconstructing an original data set out of n different distorted versions of the same data set.

We have described an example consisting of 20 records with 7 different releases of the same variable. We have applied to this records the system ClusDM.

The results obtained support the Property 2 stated in the introduction: disclosure risk is proportional to the reconstruction capabilities of information fusion systems. In this case, we have shown that the original data was reconstructed except for record r that is assigned to Category 3 instead of 5.

Acknowledgements

The authors are partially supported by the EU project CASC: Contract: IST-2000-25069 and CICYT project STREAMOBILE (TIC2001-0633-C03-01/02)

References

1. Census Bureau, (1993), American Housing Survey 1993, Data publicly available from the U. S. Bureau of the Census through the Data Extraction System, http://www.census.gov/DES/www/welcome.html
2. de Soto, A.R., Trillas, E., (1999), On antonym and negate in fuzzy logic, Int. J. of Int. Systems, 14:3, 295-303
3. Domingo-Ferrer, J., Torra, V., (2001), A Quantitative Comparison of Disclosure Control Methods for Microdata, 111-133, in Confidentiality, Disclosure, and Data Access: Theory and Practical Applications for Statistical Agencies, P. Doyle, J. I. Lane, J. J. M. Theeuwes, L. M. Zayatz (Eds.), Elsevier.
4. Domingo-Ferrer, J., Torra, V., (2001), Disclosure Control Methods and Information Loss for Microdata, 91-110, in Confidentiality, Disclosure, and Data Access: Theory and Practical Applications for Statistical Agencies, P. Doyle, J. I. Lane, J. J. M. Theeuwes, L. M. Zayatz (Eds.), Elsevier.
5. Domingo-Ferrer, J., Torra, V., (2002), Aggregation techniques for statistical confidentiality, in "Aggregation operators: New trends and applications", (Ed.), R. Mesiar, T. Calvo, G. Mayor, Physica-Verlag, Springer.
6. Domingo-Ferrer, J., Torra, V., (2002), On the Connections between Statistical Disclosure Control for Microdata and Some Artificial Intelligence Tools, submitted.
7. Domingo-Ferrer, J., Torra, V., Valls, A., (2002), Semantic based aggregation for statistical disclosure control, submitted.
8. Dubois, D., Koning, J-L., (1991), Social choice axioms for fuzzy set aggregation, Fuzzy Sets and Systems, vol.43, pp.257-274.
9. F. Sebe, J. Domingo-Ferrer, J. M. Mateo-Sanz, V. Torra, Post-Masking optimization of the tradeoff between information loss and disclosure risk in masked microdata sets, Lecture Notes in Computer Science 2316, 163-171.
10. Torra, V., (1996), Negation functions based semantics for ordered linguistic labels, Int. J. of Intelligent Systems, 11 975-988.
11. Torra, Towards the re-identification of individuals in data files with common variables, Proc. of the 14th European Conference on Artificial Intelligence (ECAI2000), Berlin, Germany, 2000.
12. Torra, V., (2000), Re-identifying Individuals using OWA Operators, Proc. of the 6th Int. Conference on Soft Computing, Iizuka, Fukuoka, Japan, 2000.
13. Willenborg, L., De Waal, T., (1996), Statistical Disclosure Control in Practice, Springer LNS 111.

14. Yager, R. R., (1988), On ordered weighted averaging aggregation operators in multi-criteria decision making, IEEE Trans. on SMC, 18 183-190.
15. Valls, A., Moreno, A., Sanchez, D., A multi-criteria decision aid agent applied to the selection of the best receiver in a transplant, Proc. of the 4th Int. Conference on Enterprise Information Systems, ICEIS, 431-438, Ciudad Real, Spain, 2002.
16. Valls, A., Torra, V., (2000), Explaining the consensus of opinions with the vocabulary of the experts, Proc. IPMU 2000, Madrid, Spain, 2000.
17. Winkler, W. E., (1995), Advanced methods for record linkage, American Statistical Association, Proceedings of the Section on Survey Research Methods, pp. 467-472.

On Multivariate Fuzzy Time Series Analysis and Forecasting

Berlin Wu[1] - Yu-Yun Hsu[2]

[1] Department of Mathematical Sciences, National Chengchi University, Taiwan
[2] Department of Mathematical Sciences, National Dong Hwa University, Taiwan

Abstract. In this paper, we propose an integrated procedure for multivariate fuzzy time series modeling and its theory structure through fuzzy relation equations in this research. Combining the data of closing price and trading volume, we apply this method to construct multivariate fuzzy time series model for Taiwan Weighted Stock Index and forecast future trend while comparing the forecasting performance by average forecasting accuracy. We strongly believe that this model will be profound of meaning in forecasting future trend of financial market.

Keywords: *Fuzzy relation, fuzzy Markov relation matrix, multivariate fuzzy time series, fuzzy rule base, average forecasting accuracy.*

1 Introduction

In time series analysis, the trend of data can be the basis of detecting events' occurrence such as increasing, decreasing, seasonal cycles or outliers. Hence, by observing certain characteristics, an optimal fitting model can be selected from a prior model family, such as ARIMA models, ARCH models, Threshold models, and so forth. While the error in data collection, time lag or the correlation among variables can show estimated numbers as precise numbers, but actually they are a set of possible numbers in some intervals. Under such a situation, an attempt to construct a mathematical model via the traditional models and analytical methods to interpret the data and trends of a time series may result in the risk of producing over-fitting models.

The concept of fuzzy sets (logic), first proposed by Zadeh (1965), provides a more realistic and moderate approach, by referring to fuzzy measure and classification concept human brain utilizes in dynamic surroundings, to handle the phenomenon of multi-complexities and uncertainties. Because fuzzy theory has intrinsic features of linguistic variables, it can minimize trouble on dealing with uncertain problems. Therefore, fuzzy theory has been widely applied in many fields such as aerospace, mechanical engineering, medical science, power generation, and geology, etc. Among these fields, the application of fuzzy control systems is even more popular; see Nguyen and Sugeno (1998).

In the humanities and social sciences, fuzzy statistics and fuzzy correlation has gradually got attention. This is a natural result because the complicated phenomenon of humanities and society is hard to be fully explained by traditional models. Regarding stock market as an example, the essence of closing price is uncertain and indistinct. Moreover, there are many factors influence closing price, such as trading volume and exchange rate, etc. Therefore, if we merely consider closing price of yesterday to construct our forecasting model, not only will we misestimate the future trend, but also we will suffer unnecessary loss. While literatures in the past have been focusing on univariate fuzzy time series but lesser on multivariate dynamic data. In view of this, we propose an integrated procedure for multivariate fuzzy time series modeling and its theory structure through fuzzy relation equations in this research. Furthermore, combining the data of closing price and trading volume, we apply this method to construct multivariate fuzzy time series model for Taiwan Weighted Stock Index and forecast future trend while comparing the forecasting performance by average forecasting accuracy. We strongly believe that this model will be profound of meaning in forecasting future trend of financial market.

2 Fuzzy Time Series Analysis

2.1 Fuzzy Time Series

Before developing multivariate fuzzy time series model and forecasting, we must give some related definitions for fuzzy time series,

Definition 2.1 Fuzzy time series

Let $\{X_t \in R, t = 1,2,...,n\}$ be a time series, Ω be the range of $\{X_t \in R, t = 1,2,...,n\}$ and { $P_i; i = 1,2,...,r, \bigcup_{i=1}^{r} P_i = \Omega$ } be an ordered partition on Ω. Let $\{L_i, i = 1,2,...,r\}$ denote linguistic variables with respect to the ordered partition set. For $t = 1,2,...,n$, if $\mu_i(X_t)$, the grade of membership of $\{X_t\}$ belongs to L_i, satisfies $\mu_i : R \to [0,1]$ and $\sum_{i=1}^{r} \mu_i(X_t) = 1$, then $\{FX_t\}$ is said to be a fuzzy time series of $\{X_t\}$ and written as

$$FX_t = \mu_1(X_t)/L_1 + \mu_2(X_t)/L_2 + \cdots + \mu_r(X_t)/L_r,$$

where / is employed to link the linguistic variables with their memberships in FX_t, and the + indicates, rather than any sort of algebraic addition, that the listed pairs of linguistic variables and memberships collectively.

For convenience, let us denote FX_t as $FX_t = (\mu_1, \mu_2,...,\mu_r)$.

Example 2.1
Consider the time series $\{X_t\} = \{0.7, 1.9, 2.7, 4.2, 3.5, 3.1, 4.4, 3.7\}$. Let $\Omega = [0,5]$ and choose an ordered partition set $\{[0,1), [1,2), [2,3), [3,4), [4,5]\}$ on Ω. Let $\{L_1, L_2, L_3, L_4, L_5\}$ denote linguistic variables:

Very low= $L_1 \propto [0,1)$;

Low= $L_2 \propto [1,2)$; Medium= $L_3 \propto [2,3)$

High= $L_4 \propto [3,4)$; Very high= $L_5 \propto [4,5]$.

We evaluate the mean $\{m_1 = 0.5, m_2 = 1.5, m_3 = 2.5, m_4 = 3.5, m_5 = 4.5\}$ of the ordered partition set. Since X_1 is between 0.5 and 1.5, and

$$\frac{1.5-0.7}{1.5-0.5} = 0.8 \in L_1 \qquad \frac{0.7-0.5}{1.5-0.5} = 0.2 \in L_2,$$

we get the fuzzy set FX_1 with respect to X_1 is (0.8,0.2,0,0,0). Similarly, we can get the following Table 2.1.

Table 2.1 Fuzzy time series $\{FX_t\}$ of $\{X_t\}$

		Very low	Low	Medium	High	Very high	
$FX_{1=}$	(	0.8	0.2	0	0	0	)
$FX_{2=}$	(	0	0.6	0.4	0	0	)
$FX_{3=}$	(	0	0	0.8	0.2	0	)
$FX_{4=}$	(	0	0	0	0.3	0.7	)
$FX_{5=}$	(	0	0	0	1	0	)
$FX_{6=}$	(	0	0	0.4	0.6	0	)
$FX_{7=}$	(	0	0	0	0.1	0.9	)
$FX_{8=}$	(	0	0	0	0.8	0.2	)

Definition 2.2 Fuzzy relation

Consider an ordered partition set $\{P_i, i = 1,2,...,r\}$ on Ω. Let $G = (\mu_1,...,\mu_r)$ and $H = (v_1,...,v_r)$ be fuzzy sets, then a fuzzy relation R between G and H is

$$R = G^t \circ H = [R_{ij}]_{r \times r},$$

where μ_i, v_j denote memberships, t denotes transpose, and $R_{ij} = \min(\mu_i, v_j)$.

2.2 Calculation of Fuzzy Markov Relation Matrix $\Re$

From Section 2.1, we can find that fuzzy relation is the key for constructing good fuzzy time series models. If we can precisely handle fuzzy relation matrix through fuzzy relation, then fuzzy time series models will provide a better fitting result. Besides, there are many different ways for calculating a fuzzy relation matrix. Dubois and Prade (1991), Wu (1986) had proposed some methods to calculate fuzzy relation matrix but none of them is based on the same premises.

Definition 2.3 Fuzzy Markov relation matrix

The fuzzy time series $\{FX_t, t = 1,2,...,n\}$ is an autoregressive process of order one, AR(1), that is, FX_t depends only on FX_{t-1}, for all t. Let FX_t has finite memberships $\mu_i(X_t)$, $i = 1,2,...,r$, than the fuzzy Markov relation matrix can be written as

$$\Re = [\Re_{ij}]_{r \times r} = \max_{2 \leq t \leq n}[\min(\mu_i(X_{t-1}), \mu_j(X_t))]_{r \times r}.$$

3 Multivariate Fuzzy Time Series Modeling and Forecasting

3.1 Multivariate Fuzzy Time Series Modeling

Definition 3.1 The FVAR(1) time series

$\{(FX_{1,t}, FX_{2,t}, \cdots, FX_{k,t})\}$ is a multivariate fuzzy autoregressive process of order one (FVAR(1)) if

$$\left(FX_{1,t}, FX_{2,t},..., FX_{k,t}\right) = \left(FX_{1,t-1}, FX_{2,t-1},..., FX_{k,t-1}\right)\begin{pmatrix} \Re_{11} & \cdots & \cdots & \Re_{1k} \\ \vdots & \ddots & & \vdots \\ \vdots & & \ddots & \vdots \\ \Re_{k1} & \cdots & \cdots & \Re_{kk} \end{pmatrix},$$

for all t, where $\Re_{ij}$ denote fuzzy Markov relation matrix of $\{FX_{i,t}\}$ and $\{FX_{j,t}\}$, $i, j = 1,2,\cdots,k$. Since $\left(FX_{1,t}, FX_{2,t}, \cdots, FX_{k,t}\right)$ depends only on $\left(FX_{1,t-1}, FX_{2,t-1},\cdots, FX_{k,t-1}\right)$ for all t, it is also said to be a multivariate fuzzy Markov process.

3.2 Principle of Qualitative Identification by Fuzzy Rule Base

In the multivariate fuzzy time series, one of the most important points is how to transform fuzzy numbers (membership functions) into corresponding linguistic variables (attributions). In general, it is determined by the position of the greatest membership function. However, if there is more than one greatest membership function, how to make choice to determine the attribution? So far, there is no certain rule to follow. To which, this research defines a linguistic vector indicator function to handle the situation when there is more than one greatest membership function.

Definition 3.2 Linguistic vector indicator function

Let $L_i = \{(L_{i1}, \cdots, L_{ir}); L_{ij} : \text{linguistic variable}; j = 1, ..., r\}$ be a linguistic vector of $\{FX_{i,t}\}$ and $F\hat{X}_{i,t}$ be a vector of memberships in L_i, $i = 1, ..., k$. Then $F\tilde{X}_{i,t} = \{(I_{i1_t}, \cdots, I_{ir_t}); I_{ij_t} = 1 \text{ or } 0; j = 1, ..., r\}$ is said to be linguistic vector indictor function, $i = 1, ..., k$, and

$$I_{ij_t} = \begin{cases} 1, & \text{if } \mu_{L_{ij}}(F\hat{X}_{i,t}) \geq k \\ 0, & \text{if } \mu_{L_{ij}}(F\hat{X}_{i,t}) < k \end{cases},$$

where $\mu_{L_{ij}}(F\hat{X}_{i,t})$ denote membership of $F\hat{X}_{i,t}$ in L_{ij}.

According to Definition 3.2, we can transfer fuzzy numbers predicted by multivariate fuzzy time series models to linguistic vector indicator functions. Yet, how to determine corresponding linguistic variables through linguistic vector indicator functions? Unfortunately, there is no rule we can follow so far. To solve this, we use Definition 3.2 and establish threshold function by fuzzy reasoning to obtain a fuzzy rule base and further analyze its outputting linguistic variables. Finally, we can use this threshold function H_t to establish the following fuzzy rule base.

Fuzzy rule base

For $i = 1, ..., k$,

(1) If $F\tilde{X}_{i,t} \in \{(1,0,0,0,0), (1,1,0,0,0), (1,0,1,0,0), (1,1,1,0,0)\}$, then the outputting linguistic variable is" plunge (very low) "

(2) If $F\tilde{X}_{i,t} \in \{(0,1,0,0,0), (1,1,0,1,0), (1,1,1,0,1), (1,1,0,0,1), (1,0,0,1,0), (1,1,1,1,0), (0,1,1,0,0), (1,0,1,1,0)\}$, then the outputting linguistic variable is" drop (low) "

(3) If $F\tilde{X}_{i,t} \in \{(0,0,1,0,0), (1,0,1,0,1), (1,0,0,0,1), (1,1,1,1,1), (0,1,0,1,0), (1,1,0,1,1), (0,1,1,1,0)\}$, then the outputting linguistic variable is" draw

(medium) "

(4) If $F\widetilde{X}_{i,t} \in \{(0,0,0,1,0),\ (0,1,0,1,1),\ (1,0,1,1,1),\ (1,0,0,1,1),\ (0,1,0,0,1),$ $(0,1,1,1,1),\ (0,0,1,1,0),\ (0,1,1,0,1)\}$, then the outputting linguistic variable is" soar (high) "

(5) If $F\widetilde{X}_{i,t} \in \{(0,0,0,0,1),\ (0,0,0,1,1),\ (0,0,1,0,1),\ (0,0,1,1,1)\}$, then the outputting linguistic variable is " surge (very high) "

3.3 Multivariate Fuzzy Time Series Forecasting

Forecasting provides indispensable information in decision-making process. Especially, a precise forecasting result can provide decision makers precious information to make correct decision and appropriate reaction. To which, we use multivariate fuzzy times series model for forecasting to realize its predictive effects. In this research, the multivariate fuzzy time series model of forecasting is defined as follows,

Definition 3.3 Forecasting FVAR(1) time series
For the multivariate fuzzy autoregressive process of order one model

$$\left(FX_{1,t}, FX_{2,t}, \ldots, FX_{k,t}\right) = \left(FX_{1,t-1}, FX_{2,t-1}, \ldots, FX_{k,t-1}\right) \begin{pmatrix} \Re_{11} & \cdots & \cdots & \Re_{1k} \\ \vdots & \ddots & & \vdots \\ \vdots & & \ddots & \vdots \\ \Re_{k1} & \cdots & \cdots & \Re_{kk} \end{pmatrix}.$$

and observations $\left(FX_{1,t}, FX_{2,t}, \ldots, FX_{k,t}\right)$, $t = 1,2,\ldots,n$, then

(1) One-step prediction is

$$\left(FX_{1,n}(1), FX_{2,n}(1), \ldots, FX_{k,n}(1)\right) = \left(FX_{1,n}, FX_{2,n}, \ldots, FX_{k,n}\right) \begin{pmatrix} \Re_{11} & \cdots & \cdots & \Re_{1k} \\ \vdots & \ddots & & \vdots \\ \vdots & & \ddots & \vdots \\ \Re_{k1} & \cdots & \cdots & \Re_{kk} \end{pmatrix}.$$

(2) Two-step prediction is

$$\left(FX_{1,n}(2), FX_{2,n}(2), \ldots, FX_{k,n}(2)\right) = \left(FX_{1,n}, FX_{2,n}, \ldots, FX_{k,n}\right) \begin{pmatrix} \Re_{11} & \cdots & \cdots & \Re_{1k} \\ \vdots & \ddots & & \vdots \\ \vdots & & \ddots & \vdots \\ \Re_{k1} & \cdots & \cdots & \Re_{kk} \end{pmatrix}^2$$

(3) l -step prediction is

$$\left(FX_{1,n}(l), FX_{2,n}(l),..., FX_{k,n}(l)\right) = \left(FX_{1,n}, FX_{2,n},..., FX_{k,n}\right) \begin{pmatrix} \mathfrak{R}_{11} & \cdots & \cdots & \mathfrak{R}_{1k} \\ \vdots & \ddots & & \vdots \\ \vdots & & \ddots & \vdots \\ \mathfrak{R}_{k1} & \cdots & \cdots & \mathfrak{R}_{kk} \end{pmatrix}^{l}.$$

An integrated process for multivariate fuzzy time series modeling

Step 1: Observe time series $\{X_{1,t}\} \cdots, \{X_{k,t}\}$. Decide the range Ω_i and the linguistic variables $\{L_{i1}, L_{i2}, \cdots, L_{ir}\}$ of $\{X_{i,t}\}$, $i = 1, 2, \cdots, k$.

Step 2: Calculate the fuzzy time series $\{FX_{i,t}\}$ of $\{X_{i,t}\}$, $i = 1, 2, \cdots, k$ and detect the linguistic variable according to the position of the greatest membership in $FX_{i,t}, t = 1, 2,..., n$.

Step 3: Calculate the fuzzy relations between $\{FX_{i,t}\}$ and $\{FX_{j,t}\}$, $i, j = 1, 2, \cdots, k$.

Step 4: By Step 3, according to all fuzzy relations between $\{FX_{i,t}\}$ and $\{FX_{j,t}\}$, $i, j = 1, 2, \cdots, k$, we can get the fuzzy Markov relation matrix, then constructing a multivariate fuzzy time series model.

Step 5: Examining $F\tilde{X}_{i,t}$, $i = 1, 2, \cdots, k$. If the number of "1" is only one, we can detect the corresponding linguistic variable immediately, otherwise detect the corresponding linguistic variables by fuzzy rule base.

Step 6: Forecasting by multivariate fuzzy time series model.

Step 7: Stop.

4 An Empirical Application for Taiwan Weighted Stock Index

4.1 Data Analysis

These data source comes from Taiwan Stock Exchange Corporation, including daily price limit and trading volume difference of weighted index from January 15 2000 to February 21 2000 with size 24.

4.2 Fuzzy Model Construction

After fuzzifying these data of daily price limit and trading volume difference of weighted index, we can apply the method mentioned in Section 2.1 to each fuzzy set L_{ij} ($i = 1, 2$; $j = 1, 2,..., 5$) and calculate data's corresponding membership functions in L_{ij} ($i = 1, 2$; $j = 1, 2,..., 5$).

4.3 Forecasting Performance

Because this research is to explore the qualitative trend of time series, we put transformed membership functions through fuzzy rule base in fuzzy systems for getting their corresponding linguistic variables to facilitate analysis. We already comprehensively defined and introduced fuzzy rule base in Section 3.2 and compared it with Autoregressive Integrated Moving Average (ARIMA) usually used in analyzing time series data. The results derived from above principles is shown at Table 4.1.

Table 4.1 The comparison of fitted value for price limit of Taiwan Weighted Stock Index

Date	Real value	Best ARIMA MA(1)	FVAR(1)
2000-1-17	Soar	Surge	Soar
2000-1-18	Drop	Surge	Draw
2000-1-19	Drop	Surge	Soar
2000-1-20	Drop	Plunge	Draw
2000-1-21	Soar	Draw	Soar
2000-1-24	Soar	Draw	Soar
2000-1-25	Drop	Drop	Draw
2000-1-26	Surge	Draw	Soar
2000-1-27	Draw	Draw	Soar
2000-1-28	Draw	Soar	Drop
2000-1-29	Drop	Draw	Drop
2000-1-31	Soar	Draw	Draw
2000-2-1	Soar	Draw	Draw
2000-2-9	Surge	Draw	Soar
2000-2-10	Draw	Draw	Soar
2000-2-11	Draw	Plunge	Drop
2000-2-14	Plunge	Plunge	Plunge
2000-2-15	Drop	Surge	Drop
2000-2-16	Soar	Draw	Draw
2000-2-17	Soar	Draw	Draw
2000-2-18	Drop	Drop	Soar
2000-2-19	Draw	Draw	Draw
2000-2-21	Plunge	Surge	Plunge
Right		0.26	0.35
Accuracy		0.68	0.82

The prediction for price limit and trading volume difference of weighted index in future four periods are shown at Table 4.2 and 4.3.

Table 4.2 The comparison of real and predictive values for price limit of Taiwan Weighted Stock Index

Date	Real value	Best ARIMA MA(1)	FVAR(1)
2000-2-22	Plunge	Surge	Drop
2000-2-23	Drop	Surge	Draw
2000-2-24	Drop	Draw	Drop
2000-2-25	Plunge	Draw	Draw

Table 4.3 The comparison of real and predictive values for trading volume difference of Taiwan Weighted Stock Index

Date	Real value	Best ARIMA MA(1)	FVAR(1)
2000-2-22	Very low	Low	Very low
2000-2-23	Low	Very high	Low
2000-2-24	Medium	Medium	High
2000-2-25	Medium	High	Low

From Table 4.2 and Table 4.3, we can get the average forecasting accuracy for price limit and trading volume difference of weighted index are 0.75 and 0.875, respectively. This result illustrates that the multivariate fuzzy time series model in this research can offer certainly accurate forecasting. The major reason why the prediction cannot hit real value is that we only consider the greatest memberships and omit other memberships. Therefore, only with reasonable forecasting model can we decide investment strategy from forecasting results. Otherwise, without the direction of clear outlines, investors will face a plight about which information they should take.

5 Conclusion

In the scientific research and analysis, the uncertainty and fuzziness contained in statistical data is often the obstacle for traditional model construction. If we use quasi-accurate value for cause and effect analysis or quantitative measurement, it will result in bias of cause and effect, misleading of decision model, and enlarging difference between prediction and real situation. Manski (1990) has pointed out that as numerical data contains the risk of over-demand and over-interpretation, the adoption of fuzzy numbers can help us avoid such a risk. Thus, it is very important to carefully examine fuzziness and robustness of numerical data in investigating quantitative method in social sciences. However, for those hard to explain cognitive questions, we can more clearly express them through membership functions and fuzzy statistical analysis. Hence, qualitative measurement and fuzzy statistics should be an advanced way to better describe human thinking and feeling.

In fact, using fuzzy and simple linguistic data for constructing forecasting model will often increase the fuzziness of result in each period. Generally speaking, this kind of fuzziness seems a normal phenomenon. In the contrary, if the concept of handling numerical data is unchanged and forecasting method is not innovated, they will deter the objectivism of quantitative method and the possibility for long-term forecasting. In this research, we illustrate the establishment of fuzzy time series, fuzzy relations, and fuzzy Markov relation matrix and further construct a multivariate fuzzy time series model. It is worth to mention that we base on experience rule and derivative method to consider a threshold function for building fuzzy rule base and transform fuzzy numbers to corresponding linguistic variables through fuzzy rule base. Lastly, we build a good integrated process of modeling and use this process to construct a forecasting model for price limit and trading volume difference of Taiwan Weighted Stock Index. Using daily price limit and trading volume difference of weighted index from January 15 2000 to February 21 2000 as historical data, we also establish a appropriate multivariate fuzzy times series model. Furthermore, we use average forecasting accuracy to compare the performances between multivariate fuzzy time series models and traditional ARIMA models and clearly prove that multivariate fuzzy time series model has better forecasting performance than traditional ARIMA models. Not only does this research provide a new forecasting method for investors, but also offer better forecasting results for investors to make decision with correct information.

References

Klir, G. F. and T. A. Folger (1988) Fuzzy Sets, Uncertainly and Information. Englewood Cliffs, NJ:Prentice Hall.

Nguyen, H. and M. Sugeno (1998) Fuzzy Modeling and Control. CRC Press.

Song, Q. and B. S. Chissom (1994) Forecasting enrollments with fuzzy time series—Part II. Fuzzy Sets and Systems, 62, 1-8.

Tseng, T. and C. Klein (1992) A new algorithm for fuzzy multicriteria decision making. International Journal of Approximate Reasoning, 6, 45-66.

Tseng, F., G. Tzeng, H.Yu, and B. Yuan (2001) Fuzzy ARIMA model for forecasting the foreign exchange market. Fuzzy Sets and Systems, 118, 9-19.

Wang, Z. and G. J. Klir (1992) Fuzzy Measure Theory. New York: Plenum Press.

Wu, B. and M. Chen (1999) Use of fuzzy statistical technique in change periods detection of nonlinear time series. Applied Mathematics and Computation, 99, 241-254.

Wu, B. and S. Hung (1999) A fuzzy identification procedure for nonlinear time series: with example on ARCH and bilinear models. Fuzzy Sets and Systems, 108, 275-287.

Wu, B. and C. Sun (1996) Fuzzy statistics and computation on the lexical semantics. Language, Information and Computation (PACLIC 11), 337-346. Seoul, Korea.

Wu, W. (1986) Fuzzy reasoning and fuzzy relational equations, Fuzzy Sets and Systems, 20, 67-78.